Lecture Notes in Networks and Systems

Volume 254

The series "Lecture Notes in Networks and Systems" publishes the latest developments in Networks and Systems—quickly, informally and with high quality. Original research reported in proceedings and post-proceedings represents the core of LNNS.

Volumes published in LNNS embrace all aspects and subfields of, as well as new challenges in, Networks and Systems.

The series contains proceedings and edited volumes in systems and networks, spanning the areas of Cyber-Physical Systems, Autonomous Systems, Sensor Networks, Control Systems, Energy Systems, Automotive Systems, Biological Systems, Vehicular Networking and Connected Vehicles, Aerospace Systems, Automation, Manufacturing, Smart Grids, Nonlinear Systems, Power Systems, Robotics, Social Systems, Economic Systems and other. Of particular value to both the contributors and the readership are the short publication timeframe and the world-wide distribution and exposure which enable both a wide and rapid dissemination of research output.

The series covers the theory, applications, and perspectives on the state of the art and future developments relevant to systems and networks, decision making, control, complex processes and related areas, as embedded in the fields of interdisciplinary and applied sciences, engineering, computer science, physics, economics, social, and life sciences, as well as the paradigms and methodologies behind them.

Indexed by SCOPUS, INSPEC, WTI Frankfurt eG, zbMATH, SCImago.

All books published in the series are submitted for consideration in Web of Science.

More information about this series at http://www.springer.com/series/15179

Jemal H. Abawajy · Kim-Kwang Raymond Choo ·
Haruna Chiroma
Editors

International Conference on Emerging Applications and Technologies for Industry 4.0 (EATI'2020)

Emerging Applications and Technologies for Industry 4.0

Springer

Editors
Jemal H. Abawajy
Faculty of Science, Engineering and Built Environment
Deakin University
Geelong, VIC, Australia

Kim-Kwang Raymond Choo
Department of Information Systems and Cyber Security
The University of Texas at San Antonio
San Antonio, TX, USA

Haruna Chiroma
Mathematical Sciences Department
Abubakar Tafawa Balewa University
Bauchi, Nigeria

ISSN 2367-3370 ISSN 2367-3389 (electronic)
Lecture Notes in Networks and Systems
ISBN 978-3-030-80215-8 ISBN 978-3-030-80216-5 (eBook)
https://doi.org/10.1007/978-3-030-80216-5

This Springer imprint is published by the registered company Springer Nature Switzerland AG
The registered company address is: Gewerbestrasse 11, 6330 Cham, Switzerland

Preface

Nigeria Computer Society is the first to sponsor Lecture Notes in Networks and Systems conference in West Africa. The conference is initially scheduled to hold physically in Uyo, Akwa Ibom state, Nigeria, but held virtually with Lagos, Nigeria, as the control centre as a result of COVID-19 significant disruption in global activities. The conference held between 11 to 13 of August 2020 with the themed: ***Emerging Applications and Technologies for Industry 4.0 (EATI 2020).*** The EATI 2020 conference is aimed at providing international platform for both public and private sectors to discuss the advancement of digital technologies in Industry 4.0. The conference encourages virtual interaction among students, novice and expert researchers, industry experts and IT professionals for the exchange of knowledge.

Research works that discussed emerging applications and technologies for Industry 4.0 with high quality were submitted to the Nigeria Computer Society EATI 2020 by experts from government, research institutes, academia and private sector across the world from countries such as Slovenia, Norway, Iran, Indonesia, South Africa, Qatar, Malaysia, UK, Saudi Arabia, Sweden, Australia, Pakistan and Nigeria in the following thematic areas: applications and techniques in cyber-intelligence; applications and techniques in Internet of Things; applications and techniques in Industry 4.0; applications and techniques in information system; applications and techniques in high-performance computing and networks; applications and techniques in computational science. The Nigeria Computer Society EATI 2020 conference attracted 96 papers. All submitted papers were subjected to initial plagiarism screening before passing through rigorous peer review process by outstanding scholars scattered across the world. A total of 21 high-quality papers with significant scientific contributions (21.8%) were accepted for inclusion after revision in Springer—Advances in Intelligent Systems and Computing; 76 papers (79.2%) were rejected.

The organizers of this conference will like to thank all the national and international participants as well as guest speakers that accepted to present at the EATI 2020 international conference. We also appreciate our partners Arit of Africa, Computer Professionals Registration Council of Nigeria, Computer Warehouse

Group, Mainone, Management Edge Limited, National Information Technology Development Agency, Oracle Corporation and Sidmach Technologies Limited.

We wish to really appreciate our editors, technical programme committee members and reviewers for ensuring the quality of the papers accepted for presentations as the peer review process underpins the scientific quality and reputation of the EATI 2020 conferences. The national president of Nigeria Computer Society, Professor Adesina Simon Sodiya, chairman research and development committee, professor Awodele Oludele, chairman, conferences committee, Ayodeji Rex Abitogun and all members of his committee who have taken the organization of this conference as task that must be accomplished, thank you all. The National Executive Council of the Nigeria Computer Society is immensely indebted to the dynamic conference planning committee which worked assiduously, even in the midst of the challenges posed by COVID-19 to make the conference a reality online.

Thank you and God bless.

Jemal H. Abawajy
Kim-Kwang Raymond Choo
Haruna Chiroma

Organization

Steering Committee

Jemal H. Abbawajy	Deakin University, Australia
Kim-Kwang Raymond Choo	University of Texas at San Antonio, USA
Laurence Tiang	St. F. X. University, Canada
Mohammed Atiquzzaman	University of Oklahoma, USA
Haruna Chiroma	Abubakar Tafawa Balewa University, Nigeria

Advisory

Tokhi	Mohammad Osman London South Bank University, UK
Tutut Herawan	Universitas Krisnadwipayana, Indonesia

Honorary Chairs

Adesina Simon Sodiya	Federal University of Agriculture Abeokuta, Nigeria
Tarik A. Rashid	University Kurdistan Hewler, Kurdistan

General Chairs

Jemal H. Abbawajy	Deakin University, Australia
Awodele Oludele	Babcock University, Nigeria
Ibrahim A. T. Hashem	Taylors University, Malaysia
Kim-Kwang Raymond Choo	University of Texas at San Antonio, USA

Workshop and Special Session Chairs

Julius Olatunji Okesola	First Technical University, Nigeria
Olumide Longe	American University of Nigeria

Technical Programme Committee Chairs

Philippe Fournier-Viger	Harbin Institute of Technology Shenzhen, China
Kim-Kwang Raymond Choo	University of Texas at San Antonio, USA

Technical Programme Committee

Xin-She Yang	Middlesex University, UK
Ameer Al-Nemrat	University of East London, UK
Mohammed Ali Al-Garadi	University of California San Diego, USA
William Liu	Auckland University of Technology, New Zealand
Alberto Sánchez Campos	Universidad Rey Juan Carlos, Spain
Mohammad Shojafar	Ryerson University, Canada
Massimo Panella	University of Rome, Italy
Nuno M. Garcia	University of Beira Inteiror, Portugal
Nurul Sarkar	Auckland University of Technology, New Zealand
Lukas Pichl	International Christian University, Tokyo, Japan
Abdullah Khan	University of Agriculture Peshawar, Pakistan
Mohamed Elhoseny	Mansoura University, Egypt
Tufan Kumbasar	Technical University, Turkey
Absalom E. Ezugwu	University of KwaZulu-Natal, South Africa
Laurance Tiang	St. F. X. University, Canada
Wenjia Niu	Beijing Jiaotong University, China
Omprakash Kaiwartya	Nottingham Trend University, UK
Min Yu	Chinese Academy of Sciences, China
Andrew E. Fluck	University of Tasmania, Australia
Yu-Da Lin	National Kaohsiung University of Science and Technology, Taiwan
Shu Li	Chinese Academy of Sciences, China
Shafi'i M. Abdulhamid	Community College Qatar, Qatar
Mohamed M. Mostafa	Gulf University for Science and Technology, Kuwait
Hassan Chizari	University of Gloucestershire, UK
Xin Han	Xi'an Shiyou University, China
Tarik A. Rashid	University Kurdistan Hewler, Kurdistan
Gang Li	Deakin University, Australia
Ibrar Yaqoob	Kyung Hee University, South Korea

Mueen Uddin	Effat University Jeddah, Saudi Arabia
Jemal Abawajy	Deakin University, Australia
Ibrahim A. T. Hashem	Taylors University, Malaysia
Philippe Fournier-Viger	Harbin Institute of Technology Shenzhen, China
Muhammad Asif Zahoor Raja	COMSATS University, Pakistan
Tutut Herawan	Universitas Krisnadwipayana, Indonesia
Gai-Ge Wang	University of Alberta, Canada
Adem Acır	Gazi University, Turkey
Ahmad Taher Azar	Prince Sultan University, Saudi Arabia
Mohamed Elhoseny	Mansoura University, Egypt
Roberto A. Vázquez	Universidad La Salle, Mexico
Liyana Shuib	University of Malaya, Malaysia
Serestina Viriri	University of KwaZulu-Natal, South Africa
Thierry Oscar Edoh	RFW-University of Bonn, Germany
Junaid Ahsenali Chaudhry	ChonBuk National University, Jeonju, South Korea
Zheng Xu	Tsinghua University, China
Chaowei Phil Yang	George Mason University, USA
Hengshu Zhu	Baidu Inc., China
Morshed Chowdhury	Deakin University, Australia
Kim-Kwang Raymond Choo	The University of Texas at San Antonio, USA
Mohammed Atiquzzaman	University of Oklahoma, USA
Rafiqul Islam	Charles Sturt University, Australia
Osvaldo Gervais	University of Perugia, Italy
Prabhat Mahanti	University of New Brunswick, Canada
Eneko Osaba Icedo	Foundation of Tecnalia Research and Innovation, Derio, Spain

Proceeding Editors

Jemal H. Abbawajy	Deakin University, Australia
Haruna Chiroma	Abubakar Tafawa Balewa University, Nigeria
Kim-Kwang Raymond Choo	University of Texas at San Antonio, USA

Publicity Chairs

Gand Li	Deakin University, Australia
Segun I. Popoola	Manchester Metropolitan University, Manchester, UK
Adamu Ibrahim Abubakar	International Islamic University, Malaysia
Tola Ajagbe	iPass Ltd, Nigeria

National Organizing Committee Chairs

Ayodeji Rex Abitogun	Management Edge Ltd, Abuja, *Nigeria*
Jide Awe	Jidaw Systems Ltd, Lagos, Nigeria
Femi Williams	Chams Plc, Lagos, Nigeria

Community Chairs

Sunday Agholor	Federal College of Education Abeokuta, Nigeria
Stanley Adiele Okolie	Federal University of Owerri, Nigeria
Ademola O. Adesina	Olabisi Onabanjo University, Ago Iwoye, Nigeria

Contents

The Relevance of Nature-Inspired Metaheuristic Algorithms in Smart Sport Training

Iztok Fister Jr.(✉)

Faculty of Electrical Engineering and Computer Science, University of Maribor, Koroška cesta 46, 2000 Maribor, Slovenia
iztok.fister1@um.si

Abstract. The main contribution of this paper is to show the linkage between the domains of Smart Sport Training and Nature-Inspired Metaheuristic Algorithms. Every year, the Smart Sport Training domain is becoming more and more crowded by different intelligent solutions that help, support and encourage people in maintaining their healthy lifestyle, as well as their sporting activities. On the other hand, nature-inspired algorithms are powerful methods for solving different kinds of optimization problems. In this paper, we show the applicability of nature-inspired algorithms in solving different intelligent tasks in the domain of Smart Sport Training. Recent progress and selected applications are outlined systematically, and the current implications of these developments are substantiated by their real usage.

Keywords: Artificial Sport Trainer · Nature-inspired algorithms · Health and fitness · Smart Sport Training · Optimization

1 Introduction

This paper corresponds with the talk that I gave at the International Conference On Emerging Applications & Technologies for Industry 4.0 (EATI) 2020. The paper is split into two parts. The first part of this paper is devoted to the foundations of nature-inspired metaheuristic algorithms and their challenges, while the second part is devoted to the description and examples of practical use of nature-inspired metaheuristic algorithms in the domain of Smart Sport Training (SST) [10,23].

In recent years, we have made effort to propose a digital twin that would have similar abilities as a human sport trainer. We named this solution Artificial Sport Trainer, or, simply, AST [10]. AST is based on computational intelligence methods [7], where nature-inspired metaheuristics play the most crucial role among the other computational intelligence methods [7], e.g. fuzzy systems. After a few years of AST design and development, AST is now a collection of smaller units (building blocks), where each one covers an aspect of sport training, i.e.

J. H. Abawajy et al. (Eds.): EATI 2020, LNNS 254, pp. 1–8, 2021.
https://doi.org/10.1007/978-3-030-80216-5_1

planning the sport training sessions, meal planning or injury prevention. These building blocks are linked and integrated with each other in order to simulate the abilities of a human sport trainer. Loosely speaking, the strongest confident argument for the adoption of AST is that AST makes decisions solely based on data. Therefore, AST tends to be fully autonomous and not tailored to a specific group of athletes.

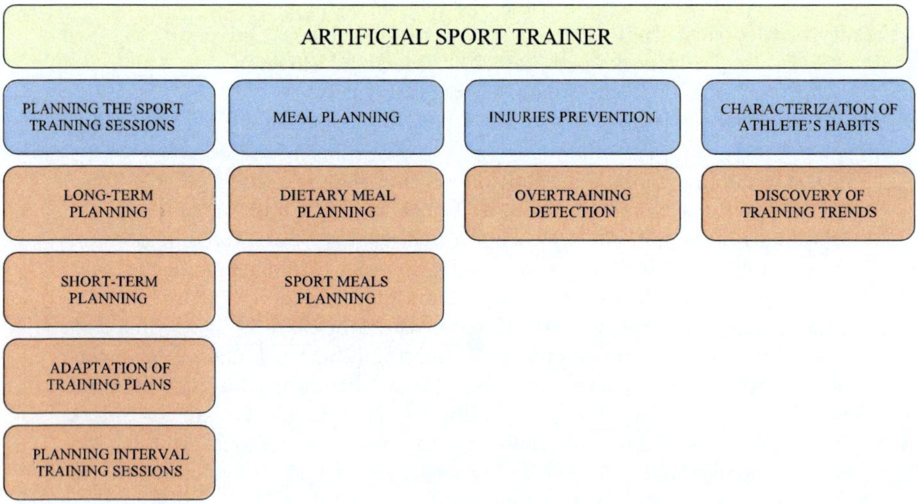

Fig. 1. Artificial Sport Trainer in a nutshell.

Figure 1 depicts the current functionalities that are supported in AST. In the current state of the AST, the majority of the emphasis is given to the planning of sport training sessions, as well as meal planning. Nevertheless, new functionalities are still under heavy development and, therefore, more aspects of training are planned for future integration with existing AST functionalities.

AST is just one example of a more complete intelligent system in sport that is mostly based on nature-inspired metaheuristics. There are many different applications using nature-inspired algorithms in sport, but they do not offer a comprehensive view of sports training such as AST. In other words, most of the applications cover only one aspect of sport training.

In the continuation of this article, in Sect. 2, a short overview of nature-inspired algorithms is provided, while Sect. 3 outlines the features of smart sport training as well as, some of the selected applications of nature-inspired metaheuristics in the SST domain. In Sect. 4 the transition of knowledge from research papers to the real-world is discussed using a real example. The paper is concluded with a summary in Sect. 5.

2 Nature-Inspired Metaheuristic Algorithms

Nature frequently serves as an inspiration to researchers in the development of new metaheuristic algorithms. These computer algorithms mimic the particular behavior of some fascinating behaviors of different animal species, natural evolution, physics and chemistry-based phenomena, sport and also sociological phenomena respectively [2]. The main members of this family of algorithms are: Evolutionary Algorithms (EA) [6] and Swarm Intelligence Algorithms [3]. Although many sub-members exist in each group, what all algorithms have in common is that they consist of the initial population which undergoes to the variation operators that are specific for the particular algorithm. After the modification of a population using the variation operators, the fittest members of the population are then selected for the next generation. A very approximate and unified algorithmic presentation of a nature-inspired metaheuristic algorithm is presented in Algorithm 1.

Algorithm 1. Simple nature-inspired metaheuristic algorithm.

INITIALIZE_population_with_random_candidates;
EVALUATE_each_candidate;
while TERMINATION_CONDITION_not_met **do**
    MODIFY_candidates_using_specific_variation_operators;
    EVALUATE_modified_candidates;
    SELECT_candidates_for_the_next_generation;
end while

Despite the age of this research area, there is still a lack of universal taxonomy which would group all nature-inspired metaheuristic algorithms in exact groups, either by the inspiration or principle of their architecture. There are some efforts [16,21] that propose some classification/taxonomies but the whole research community has not yet accepted any final taxonomy.

Finally, it should be noted that the entire research area has also had some dark moments. The infinitely large pool of possibilities for using algorithms in various applications, has consequently, encouraged researchers to develop new algorithms that can also be presented as nature-inspired metaheuristic algorithms. After the year 2000, the number of new algorithms increased drastically (see Fig. 2)[1]. Usually, these "new/novel" algorithms mimic the selected inspiration or metaphor from nature (a particular biological or physical system), whereby the authors have hidden their internal operation under the description of the inspired system's behavior. Numerous research papers (some of them [14,22,24]) have raised questions about the uniqueness and true scientific value of these newly created algorithms.

[1] References taken from https://github.com/fcampelo/EC-Bestiary. Only references with a valid DOI were considered in this study.

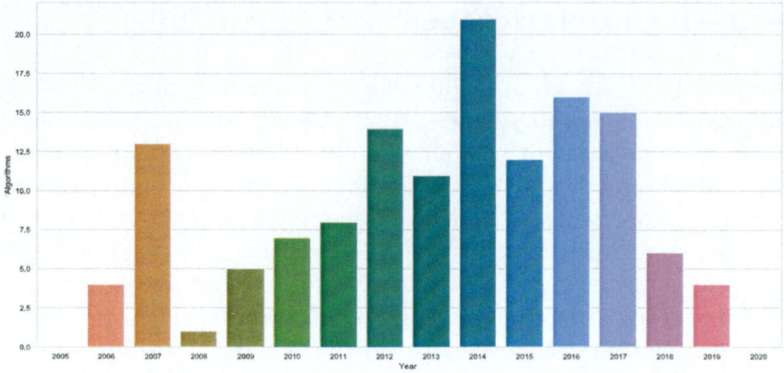

Fig. 2. The emergence of new nature-inspired algorithms.

3 Smart Sport Training

Smart Sport Training is a type of sports training, which utilizes the use of wearables, sensors, and Internet of Things (IoT) devices, and/or intelligent data analysis methods and tools to improve training performance and/or reduce a workload, while maintaining the same or better training performance [23]. SST is not a very old field of research, but, according to Fig. 3[2], it has been gaining in popularity in recent years. One of the main implications of this is the rapid development of mobile and ubiquitous computing, which makes it easy to track and monitor data during sports activities on the one hand, and on the other hand, the development of different intelligent data analysis methods. Intelligent data analysis methods allow us to make different predictions in individual and team sports, search for hidden knowledge in large databases of sport performances, make sport more sociable, and recommend appropriate training for an athlete that is based on either his/her existing trainings, or is created from scratch.

3.1 Applications of Nature-Inspired Metaheuristics in Smart Sport Training

One of the advances of nature-inspired metaheuristic algorithms is that they can easily be used in areas where there is a lack of domain-specific knowledge about the problem to be solved. Many problems in the domain of sport are of that kind. The added value of these algorithms is also their easy integration in support systems for decision-making, while these methods are also very scalable, yet offer the possibility of easy parallelization.

[2] * on Fig. 3 denotes the current year that is still in progress. Therefore, the number of research works is not final.

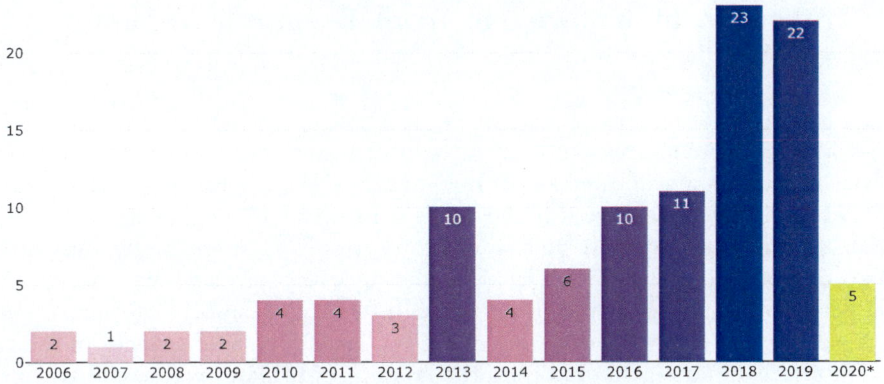

Fig. 3. AI in sport according to the years

The domain of SST is very wide. On the one hand it represents different sports, while, on the other hand, it represents many hard problems that can be solved by the use of nature-inspired metaheuristic algorithms. Table 1 presents some selected examples of SST applications that are based on nature-inspired metaheuristics besides the AST that was presented in the previous sections.

Table 1. Some selected applications of nature-inspired algorithms in SST.

Applications of nature-inspired algorithms in different sports		
Cricket	Cricket team selection	[1]
Cycling	Planning the training sessions	[10, 19]
	Diet planning	[8]
	Characteristics mining	[11]
Football	Planning	[5]
Running	Performance analysis	[12]
	Planning the optimum running speed	[4]
Soccer	Simulation of soccer kicks	[18]
Triathlon	Planning training sessions	[9]

According to Table 1, nature-inspired metaheuristics can be used for SST applications in individual sports (cycling, running, triathlon as an example) as well as in team sports (cricket, football, soccer as an example). However, current research trends [23] show more research directed towards the individual sports, presumably due to the easier control of individual athletes in experiments as well as easier data collection in individual sports (see the example of a cycling dataset [15]).

4 Transition of Knowledge from Research to the Real-World

There is no doubt that researchers in the research area that link metaheuristics and the SST domains have produced a lot of valuable work in recent years. However, there is a real question of how many of the applications presented in research papers are deployed in the real-world. Most of the research that was presented until recently were ideas and prototypes [23]. Interestingly, only a few works reported some validation-level results in a real environment. We suspect that the flow of knowledge produced through research is slowly entering the real world [23] due to many barriers, including the insensitivity of coaches in the use of computer support or in obtaining the consent of athletes who could test new methods in their sport training.

(a) (b) (c)

Fig. 4. An example of a real-world application (a) GUI application (b) Smart watch (c) Bike settled on a Tacx trainer

Figure 4 depicts an example of knowledge transfer from research ideas/ prototypes to practical use. The ideas of automatic interval planning [13] and adaptation of sport training plans [17] were applied for the realization of training of cyclists. Firstly, the proposed training plan was generated and later an initial training plan was adapted according to the performances of cyclists. In order to effectively validate the proposed ideas, a GUI application (Fig. 4a) was developed. The GUI application interacts with the cyclist on the one hand, while also showing the cyclist's heart rate in real-time, which is transmitted to a computer from a smart watch (Fig. 4b) via ANT+ protocol [20]. The cyclist's bike is settled on a Tacx trainer[3]. Heart rate data is stored each second, and after each interval is finished, the algorithm adapts the training for all subsequent intervals.

[3] https://tacx.com/.

A practical evaluation of the proposed solution revealed three important findings:

- The sport trainer was able to gather more insights and characteristics of his/her athletes when training intervals.
- The overall training was more dynamic, because each next training load of an interval was determined according to the previous intervals.
- Cyclists performed the whole training session of the same or even better quality as if they were training without this solution.

5 Conclusion

This paper briefly presented the interplay between nature-inspired metaheuristic algorithms and their use in SST. The SST domain is now becoming very popular within the research community. Additionally, the whole research area also has considerable backing in the real-world, since sport is one of the biggest businesses around, and therefore open to technological ideas. On the other hand, the connection of SST in the support of a healthy lifestyle is also an important mission for this research. For the future, there are still a plenty of challenges. Firstly, there are a lot of sports with no research in the domain of SST, which means there are many potential possibilities for future research. Secondly, the most important thing would be to ensure the quick flow of ideas in research papers into the real world. Last, but not least, obtaining test datasets and their dissemination in order to allow the easier replication of published results also remains an important cornerstone for the future of this area.

Acknowledgments. The author wishes to express his thanks for the financial support from the Slovenian Research Agency (Research Core Funding No. P2-0057).

References

1. Ahmed, F., Jindal, A., Deb, K.: Cricket team selection using evolutionary multi-objective optimization. In: Panigrahi, B.K., Suganthan, P.N., Das, S., Satapathy, S.C. (eds.) SEMCCO 2011. LNCS, vol. 7077, pp. 71–78. Springer, Heidelberg (2011). https://doi.org/10.1007/978-3-642-27242-4_9
2. Alexandros, T., Georgios, D.: Nature inspired optimization algorithms related to physical phenomena and laws of science: a survey. Int. J. Artif. Intell. Tools **26**(06), 1750022 (2017)
3. Blum, C., Merkle, D.: Swarm intelligence. In: Blum, C., Merkle, D., (eds.) Swarm Intelligence in Optimization, pp. 43–85 (2008)
4. Brzostowski, K., Drapała, J., Grzech, A., Świątek, P.: Adaptive decision support system for automatic physical effort plan generation—data-driven approach. Cybern. Syst. **44**(2–3), 204–221 (2013)
5. Connor, M., Fagan, D., O'Neill, M.: Optimising team sport training plans with grammatical evolution. In: 2019 IEEE Congress on Evolutionary Computation (CEC), pp. 2474–2481. IEEE (2019)

6. Eiben, A.E., Smith, J.E., et al.: Introduction to Evolutionary Computing, vol. 53. Springer, Heidelberg (2003). https://doi.org/10.1007/978-3-662-05094-1
7. Engelbrecht, A.P.: Computational Intelligence: An Introduction. John Wiley & Sons, Hoboken (2007)
8. Fister, D., Rauter, S., Fister, I., Fister Jr., I.: Generating eating plans for athletes using the particle swarm optimization. In: 17th International Symposium on Computational Intelligence and Informatics (CINTI), pp. 193–198 (2016)
9. Fister, I., Brest, J., Iglesias, A., Fister Jr., I.: Framework for planning the training sessions in triathlon. In: Proceedings of the Genetic and Evolutionary Computation Conference Companion, pp. 1829–1834 (2018)
10. Fister, I., Fister Jr., I., Fister, D.: Computational Intelligence in Sports. Springer, Heidelberg (2019). https://doi.org/10.1007/978-3-030-03490-0
11. Fister, I., Fister Jr., I., Fister, D.: BatMiner for identifying the characteristics of athletes in training. In: Computational Intelligence in Sports. ALO, vol. 22, pp. 201–221. Springer, Cham (2019). https://doi.org/10.1007/978-3-030-03490-0_9
12. Fister Jr., I., Fister, D., Deb, S., Mlakar, U., Brest, J., Fister, I.: Making up for the deficit in a marathon run. In: Proceedings of the 2017 International Conference on Intelligent Systems, Metaheuristics and Swarm Intelligence, pp. 11–15 (2017)
13. Fister Jr., I., Fister, D., Iglesias, A., Galvez, A., Rauter, S., Fister, I.: Population-based metaheuristics for planning interval training sessions in mountain biking. In: International Conference on Swarm Intelligence, pp. 70–79 (2019)
14. Fister Jr., I., Mlakar, U., Brest, J., Fister, I.: A new population-based nature-inspired algorithm every month: Is the current era coming to the end? In: StuCoSReC: Proceedings of the 2016 3rd Student Computer Science Research Conference. University of Primorska, Koper, pp. 33–37 (2016)
15. Fister Jr., I., Rauter, S., Fister, D., Fister, I.: A collection of sport activity datasets with an emphasis on powermeter data. Technical report, University of Maribor (2017). http://www.iztok-jr-fister.eu/static/publications/Sport5.zip
16. Fister Jr., I., Yang, X.-S., Fister, I., Brest, J., Fister, D.: A brief review of nature-inspired algorithms for optimization. Elektrotehniški vestnik **80**(3), 116–122 (2013)
17. Fister Jr., I., Iglesias, A., Osaba, E., Mlakar, U., Brest, J., Fister, I.: Adaptation of sport training plans by swarm intelligence. In: Mendel 2017 (2017)
18. Khemka, N., Jacob, C., Cole, G.: Making soccer kicks better: a study in particle swarm optimization. In: Proceedings of the 7th Annual Workshop on Genetic and Evolutionary Computation, pp. 382–385 (2005)
19. Kumyaito, N., Yupapin, P., Tamee, K.: Planning a sports training program using adaptive particle swarm optimization with emphasis on physiological constraints. BMC Res. Notes **11**(1), 9 (2018)
20. Mehmood, N.Q., Culmone, R.: An ant+ protocol based health care system. In: 2015 IEEE 29th International Conference on Advanced Information Networking and Applications Workshops, pp. 193–198. IEEE (2015)
21. Molina, D., Poyatos, J., Del Ser, J., García, S., Hussain, A., Herrera, F.: Comprehensive taxonomies of nature-and bio-inspired optimization: Inspiration versus algorithmic behavior, critical analysis and recommendations. arXiv preprint arXiv:2002.08136 (2020)
22. Piotrowski, A.P., Napiorkowski, J.J., Rowinski, P.M.: How novel is the "novel" black hole optimization approach? Inf. Sci. **267**, 191–200 (2014)
23. Rajšp, A., Fister, I.: A systematic literature review of intelligent data analysis methods for smart sport training. Appl. Sci. **10**(9), 3013 (2020)
24. Sörensen, K.: Metaheuristics—the metaphor exposed. Int. Trans. Oper. Res. **22**(1), 3–18 (2015)

Analysis of Variable Learning Rate Back Propagation with Cuckoo Search Algorithm for Data Classification

Maria Ali[1], Abdullah Khan[1(✉)], Asfandyar Khan[1], and Saima Anwar Lashari[2]

[1] Institute of Computer Sciences and Information Technology, Faculty of Management and Computer Sciences, Agriculture University Peshawar Pakistan, Peshawar, Pakistan
mariaali@codeforpakistan.org, {Abdullah_khan, asfandyar}@aup.edu.pk

[2] College of Computing and Informatics, Saudi Electronic University, Riyadh, Kingdom of Saudi Arabia
S.LASHARI@seu.edu.sa

Abstract. For the data classification task back propagation (BP) is the most common used model to trained artificial neural network (ANN). Various parameters were used to enhance the learning process of this network. However, the conventional algorithms have some weakness, during training. The error function of this algorithm is not explicit to locate the global minimum, while gradient descent may cause slow learning rate and get stuck in local minima. As a solution, nature inspired cuckoo search algorithms provide derived free solution to optimize composite problems. This paper proposed a novel meta-heuristic search algorithm, called cuckoo search (CS), with variable learning rate to train the network. The proposed variable learning rate with cuckoo search algorithm speed up the slow convergence and solve the local minima problem of the backpropagation algorithm. The proposed CS variable learning rate BP algorithms are compared with traditional algorithms. Particularly, diabetes and cancer benchmark classification problems datasets are used. From the analyses results it show that proposed algorithm shows high efficiency and enhanced performance of the BP algorithm.

Keywords: Classification · Optimization · Artificial Neural Network · Learning rate · Cuckoo search · Variable learning Rate

1 Introduction

Artificial Neural Networks (ANNs) is one of the sub field of artificial intelligence [1–4]. ANN inspired by human nerves system which is consists of parallel processing unit or nodes called artificial neurons. ANN tries to mimic how the brain works [5]. ANN have been attracted the attention of various researchers, and many researchers use this technique in different applications [6]. All these field include such as prediction of landslides, volcanos, and traffic [7, 8], medicine [9], bioinformatics [10, 11], engineering [12], biology, ecology, physics, chemistry, agronomy, economy, medicine, mathematics

J. H. Abawajy et al. (Eds.): EATI 2020, LNNS 254, pp. 9–21, 2021.
https://doi.org/10.1007/978-3-030-80216-5_2

and computers science [10, 13]. ANN consists of parallel input/output processing units called neuron having ability of dynamic experimental process [14]. It is consisting of interconnected nodes that communicate with one and more than one hidden layer, to transform data to output layer through hidden-layer from the input layer [15]. During the learning, the network learns from weights by shifting or changing weights. MLP is class of feed forward neural networks (FFNN), which consist of more than one layer of nodes used to perform difficult calculation. Multilayer Perceptron (MLP) is a supervised network to train how to transfer input data into desired output [4, 16]. However, MLP consists of three layers such as: one input one or more hidden layers and one output layers [17–19]. The most popular training algorithm, for MLP, is Back Propagation (BP). It's commonly used because for minimizing error and the output of neural network to desired pattern, if not than the operations are repeated again and again, and weight are change continually until the error is minimized [20, 21]. As BP algorithm uses Gradient Descent (GD) technique that causes some limitations include a slow learning and convergence speed also distinguished that stuck on local minima problems, which are directly linked with neuron diffusion in the hidden layer [20, 22, 23]. Different parameters were used to enhance the training process of the BP such as activation function, learning rate and momentum [24, 25]. Learning rate is known as highly vital hyper parameter used in the neural networks training [15, 26].

ANN is trained using BP algorithm, by adjusting learning rate (η) parameter in order to obtain better information and providing more efficient learning procedures to solve different problem [26, 27]. Yu *et al.* in [28] proposed dynamic optimization of the learning rate, to enhance the existing BP. Ye in [29] stated that the BP fails to improve the search of optimal weight combination by a constant learning rate (η). Yu and Liu in [30] introduced an efficient learning technique, which is concerned with the effect of learning rate and momentum on network training time. In [31] described that with adaptive learning rate and momentum term (BPALM) decrease the training time of the conventional BPNN. Similarly, Yemeni and Hong in [32] proposed an auto adapted learning rate for the BP, the author mention that during the training the adjustment of the network weights is associated with error gradient. Furthermore, Thota *et al.* in [33] used an optimal time varying learning rate. This study used different value to show the consistency of the BP model. But from the simulation analyses it show that mostly model depends on the selection of best learning rate η value. Abbas *et al.* in [20] used variable learning rate to enhance the learning process of BP algorithm, but still this models have some limitation.

Similarly, rather than this parameter various researcher has suggested many other techniques that are used to increase the training process of the BP algorithm [34]. Among all of them the most commonly used algorithms are an optimization algorithm. Different optimization algorithms have been developed and integrated with BP algorithm. But the most common of them is Cuckoo Search (CS) due to its performance and efficiency which developed by X. S. Yang [35, 36]. CS algorithm includes novel iterative solution and using levy flight to support a range of solutions and avoid stagnation. CS is based on nature inspired bird's behavior used broadly to fix optimization problems. It is an extremely useful in fixing global optimization because it capable to keep balance among global and local arbitrary walks using exchanging parameter [37, 38]. Therefore, this paper proposed a CS algorithm for variable learning rate BP [20] to solve the local minima problem and enhance the training process of the variable learning rate BP algorithm. The remaining paper as given. Section 2 described the learning algorithm used in this paper, while Sect. 3 explains proposed algorithm. Similarly, Sect. 4 summarizes result and discussion of the used algorithm. Finally, Sect. 5 gives the conclusion of this paper.

2 Learning Algorithms

2.1 Back Propagation Algorithm

The most common and effective technique used for training ANN is back-propagation (BP). BP is applied to wide range of problem. This model is the mostly used, fit, and competent for multilayered networks [39]. ANN using BP-algorithm is training model for the effectiveness of Multi-Layered Perceptron (MLP) network [40]. A Back-Propagation (BP) which used gradient descent (GD) method to reduce the error of the model [15, 24]. As back propagation algorithm based on GD technique which causes some limitations include a slow learning and convergence speed. Therefore, convergence of BP are depending on the connectivity of network, weights, biases, learning rate parameters and activation function [41, 42].

Different parameters are applied to enhance the training process of the BP. But learning rate is one of them which is proposed in this study. The next section will explain the learning rate and the various types of the learning rate [20] are discuss below.

2.2 Learning Rate

One of the most efficient procedure to speed up the convergence of BP-learning is the learning rate. However, the weight size should be selected properly throughout the training procedure. If the learning rate η value is selected very large the error accelerates the training procedure and comparatively slower the convergence. On the other hand, if very small learning rate value is used then the algorithm will take longer time to convergence or may not at all convergence [15].

Learning rate η is known as highly vital hyper-parameter used in the neural networks training. Smith in [26] proposed to illustrates a novel technique for set the learning rate, given name cyclical learning rates, which find the most excellent values for the global learning rates. Cyclical learning rates of fixed values gets better classification instead to tune networks often in fewer iterations. Abbas *et al.* [20] proposed various type of the variable learning rate to improve the learning process of the BP. All variable learning rate is discussing as below.

2.3 Variable Learning Rate

i. Linear Increasing Learning Rate

The Linear Increasing Learning Rate is calculated as;

$$\eta_{LILR} = alpa1 + (alpa2 - alpa1) * \frac{maxepoch - epoch}{epoch} \tag{1}$$

In this technique the learning rate value initiate from a small value 0.1 and then gradually increase that value to a large value 0.9. Where alpa1 and alpa2 are constant parameters.

ii. Linear Decreasing Learning Rate

So, the Linear Decreasing learning Rate is calculated as; In this technique large value 0.9 linearly decreased to small value 0.1.

$$\eta_{LDLR} = alpa2 - ((alpa2 - alpa1) * \frac{maxepoch - epoch}{epoch} \tag{2}$$

iii. Chaotic Learning Rate

The Chaotic learning Rate is calculated as; This approach was initiated by "chaotic optimization mechanism" the equation of chaotic learning rate is randomly selected.

$$\eta_{CLR} = \left((alpa2 - alpa1) * maxepoch - \frac{epoch}{epoch} + alpa1 * B \right) \tag{3}$$

iv. Random Chaotic Learning Rate

The Random Chaotic Learning Rate is calculated as; In this approach a random number of Rand () is also generated.

$$\eta_{RCLR} = alpa2 - alpa1 * (maxepoch - epoch)/epoch * 0.01 \tag{4}$$

v. Oscillating Learning Rate

The Oscillating Learning Rate are calculated as;

$$\eta_{OLR} = (alpa2 + alpa1)/2 + ((alpa2 - alpa1)/2) * cos(2 * pi * epoch)/T \tag{5}$$

2.4 Cuckoo-Search (CS)

Cuckoo search is an optimized algorithm developed by X. S. Yang (2009) [35, 36], which is based on nature inspired birds behavior used broadly to fix optimization problems in disparate areas of engineering [37]. It is an extremely useful in fixing global optimization because it capable to keep balance among global and local arbitrary walks using exchanging parameter [43–45]. The CS algorithm which mimic the natural ways of birds. The Cuckoo Search is based on three basic rules that are given below:

1. Each cuckoo lay one egg at a time and randomly puts in a chosen nest.
2. The best nests with high quality of eggs will carry over to the next generations;
3. The number of available host nests is fixed, and the egg laid by a cuckoo is discovered by the host bird with a probability $pa \in [0, 1]$. The host bird can either throw the egg away or abandon or can completely build a new nest in a new location [36, 46, 47].

3 The Proposed CSVLRBP Algorithm

In this research a new hybrid model is proposed which is based on CS variable learning rate BP algorithm (CSVLRBP). In the proposed CSVLRBP algorithm, the ANN network structure is design, weights, biases value and the CS algorithm populations are initializing. In this model each nest represented a probable solution (i.e., the preliminary weight and biases for VLRBPNN). First the best primary weights, and biases are initialized with cuckoo search and these values are assign to the VLRBPNN. The proposed CSVLRBP calculated the error and compare the error function with target error. If the target error is achieved the network will stop the training. If not than in the next cycle, CS will update the weights with initializing new weights values to the network, and CS will continue searching the best weights until the final cycle/epoch of the system is reached or target low MSE is achieved. Figure 1 shows the flow of the proposed CSVLRBP algorithm. The numerical calculation of the proposed model is given below. The calculation of weights and bias based on BP technique. The network output in the output layer is estimated for each unit t as;

$$err_t = (T_t - Y_t) \tag{6}$$

$$Err_t = Y_t(1 - Y_t)(T_t - Y_t) \tag{7}$$

The hidden layer for each unit j as:

$$Er_j = h_j\left(1 - h_j\right)\sum\nolimits_k Err_t W_{xj} \tag{8}$$

Hence the weights and 'biases' are changed simply for the output weights.

$$\Delta w_{kj} = \eta_{LILR} * w_{xj} + Er_j h_j \tag{9}$$

$$\Delta b_{kj} = \eta_{LILR} * b_{xj} + Er_j h_j \tag{10}$$

The input weights for the CSVLBP are calculated as;

$$\Delta W_k = \eta_{LILR} * W_{xi} + Er_j net_i \tag{11}$$

$$\Delta \mathrm{b}_k = \eta_{LILR} * b_{x(ji)} + Er_j net_i \tag{12}$$

So, the other variables learning Rate (η) can be calculated in proposed algorithm using Eq. (1) up to Eq. (5). The network performances index is calculated as;

$$V(x) = \frac{1}{2}\sum\nolimits_{t=1}^{R} (T_t - X_t)^T (T_t - Y_t) \tag{13}$$

$$V_F(x) = \frac{1}{2}\sum\nolimits_{t=1}^{R} E^T.E \tag{14}$$

The MSE performance-index in the proposed method is calculated as follow

$$V_\mu(x) = \frac{\sum_{j=1}^{N} V_F(x)''}{P_i} \tag{15}$$

Where,y_r is the network 'output' when the tth'input'net_i is presented'. the performance index and average-performance index are represented as $V_\mu(x)$, $V_F(x)$, and the number of cuckoo populations in i^{th} iteration is P_i. The MSE list of ith iteration can be calculated as;

$$MSErr_i = \left\{V_\mu^1(x), V_\mu^2(x), V_\mu^3(x) \ldots\ldots V_\mu^n(x)\right\} \tag{16}$$

Thus, the Cuckoo Search nest x_j is calculated as;

$$x_j = Min\left\{V_\mu^1(x), V_\mu^2(x), V_\mu^3(x) \ldots\ldots V_\mu^n(x)\right\} \tag{17}$$

And the rest of the MSE is supposed as other cuckoo nest. A novel solution x_i^{t+1} for Cuckoo i is produced by means of a levy flight according to the following Equation;

$$x_i^{t+1} = x_i^t + \alpha \otimes levy(\lambda) \tag{18}$$

So, the other Cuckoo movement x_i toward sx_j can be drawn from Eq. (19).

$$X = \begin{cases} x_i + rand \cdot (x_j - x_i) rand_i > p\alpha \\ x_i \; else \end{cases} \tag{19}$$

Through levy flight the CS can move from x_i towards x_j it can be written as;

$$\nabla X_i = \begin{cases} x_i + \alpha \otimes levy(\lambda) \sim 0.01.\left(\frac{U_j}{|V_j|^{\frac{1}{\mu}}}\right).(X - X_{best}) rand_i > p_\alpha \\ x_i \quad else \end{cases} \tag{20}$$

Where,∇V_i is a minute movement of x_i towards x_j. The bias and weights for each layer is then tuned as;

$$W_x^{n+1} = {}'' W_x^n - \nabla X_i \tag{21}$$

$$b_x^{n+1} = {}' b_x^n - \nabla X_i \tag{22}$$

The pseudo-code for the CSVLRBP is given as:

1. ***Start***
2. *Initialize CS population size and VLRBPNN structure*
3. *Input the training-data*
4. *While MSE<stop criteria*
5. *Get ahead of the cuckoo best-nests as best weights-to network*
6. *Feed-Forward Propagation network-runs using-the best weights-initialized with-Cuckoo Search*
7. *The output is calculated of one-layer from--its earlier one--and the output calculation is starts from the last layer of the network and move 'backwards'.*
8. *Update bias and weights using **Equation (9) to (12)***
9. *Calculation error Using **Equation (13)***
10. *Minimize the error by adjusting network parameter using CS.|*
11. *Generate Cuckoo egg (x_j) by taking levy flight from random nest.* $X_i = x_j$
12. *Abandon a fraction pα ϵ* [0 1] *of the worst nest. Build new nest at new location via levy flight to replace the old one.*
13. *Evaluate the fitness of the nest, Chose a random nest* i
14. ***If***
15. $X_j > X_i$ *Then*
16. $x_i \leftarrow x_j$
17. $X_i \leftarrow V_j$
18. ***End if***
19. *Cuckoo Search keeps on calculating the greatest probable weight at every epoch until the network is meet best solution.*
20. ***End while.***
21. ***End***

4 Result and Discussions

Experiments were carried out several time with different architecture of different variants of BP algorithms. Accuracy, and MSE performance parameters are used to evaluate the performance of the proposed models. The dataset is divided in portioning is 70% for training and 30% for testing.

4.1 Preliminaries

This Research focuses on the performance of the used models in term of MSE and accuracy. The benchmarked datasets taken from UCI machine learning repository. To carried out the simulation result The Workstation used equipped with a 2-GHz processor, 2-GB of RAM while the operating system used was Microsoft Windows 10. For simulations purposes, MATLAB R2018a software was used to run the proposed algorithms, Breast Cancer [48], and Pima Indian Diabetes [49], classification datasets were used;

4.2 Wisconsin Breast Cancer Classification Problem

This data set was designed by "William H. Wolberg for gathering of information from Microscopic Research of breast tissue sample pick" for the detection of breast cancer

[48]. This Data set was taken from UCIMLR database, the problem seeks to analyze breast cancer by try to classify a cancer as either benign or malevolent based on the incessant clinical variable. The data set consists of nine inputs node having two outputs node with 699 cases. The input characteristics are the cluster thickness or thinness, consistency of cell magnitude and shape of cell, amounts of marginal adhesion, single size epithelial cell, and frequency of naked nuclei, bland chromatin, mitoses, and normal nucleoli. The particular network structural design used for the classification problem of breast cancer contain nine inputs node, five hidden nodes, two outputs node and the target error are set to be 0.00001.

Table 1. Performance of the proposed algorithms for Breast Cancer dataset

Training dataset			Testing dataset	
Algorithms	Accuracy	MSE	Accuracy	MSE
CSLILRBP	99.14	0.05	99.16	0.05
CSLDLRBP	99.79	0.029	99.70	0.009
CSCLRBP	97.96	0.001	99.16	0.002
CSRCLRBP	99.53	0.13	98.34	0.003
CSOLRBP	99.16	0.05	99.68	0.003
LILRBP	95.15	0.23	95.14	0.25
LDLRBP	94.92	0.26	94.92	0.26
CLRBP	94.81	0.28	95.06	0.24
RCLRBP	95.14	0.26	95.01	0.26
OLRBP	94.91	0.28	95.13	0.26

Table 1 demonstrates the performance of the proposed algorithms such as (CSLILRPB, CSLDLRBP, CSCLRBP, CSRCLRBP, CSOLRBP) both for training and testing dataset which illustrate better performance than simple VRLBP algorithms. The proposed algorithms have achieved 99.14, 99.79, 97.96, 99.53 and 99.16% of accuracy with MSE of (0.05, .029, 0.001. 0.013, 0.05). While the other algorithms fall behind in performance with MSE's of 0.23, 0.26, 0.28, 0.26 and 0.28, and accuracy of 95.15, 94.92, 94.81, 95.14, and 94.91% correspondingly for training dataset. Similarly, for the testing data the proposed algorithms have achieved 99.16, 99.70, 99.16, 98.34 and 99.68% of accuracy with MSE of (0.05, 0.009, 0.002, 0.003, 0.003). While the other algorithms fall behind in performance with MSE's of 0.25, 0.26, 0.24, 0.26 and 0.26, and accuracy of 95.14, 94.92, 95.06, 95.01, and 95.13%. The accuracy and MSE performances for all algorithms are shown in Fig. 1.

4.3 Diabetes Classification Problem

This dataset contains the entire information of the chemical change in a female body whose inconsistency can roots diabetes [49]. This dataset consists of 768 cases, 8 input nodes and 2 output nodes, therefore the network for this system is set to-8 input neurons -5 hidden neurons and-2 output neurons, the error for diabetes dataset is set to 0.00001.

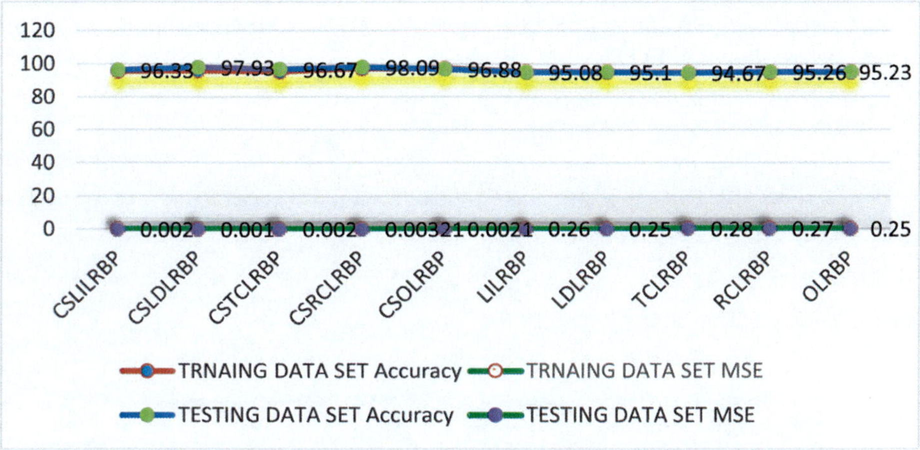

Fig. 1. "Comparison" of average MSE and accuracy for Cancer dataset

Table 2. Performance the proposed algorithms for Diabetes dataset

Training dataset			Testing dataset	
Algorithms	Accuracy	MSE	Accuracy	MSE
CSLILRBP	97.49	0.002	96.33	0.001
CSLDLRBP	95.86	0.001	97.93	0.002
CSTCLRBP	95.34	0.002	96.67	0.001
CSRCLRBP	97.27	0.003	98.09	0.0032
CSOLRBP	97.27	0.004	96.88	0.001
LILRBP	95.10	0.26	95.08	0.26
LDLRBP	95.22	0.25	95.10	0.24
CLRBP	94.88	0.28	94.67	0.30
RCLRBP	94.91	0.27	95.26	0.24
OLRBP	95.10	0.25	95.23	0.25

Table 2 show the performance of the proposed algorithms such as (CSLILRPB, CSLDLRBP, CSCLRBP, CSRCLRBP, CSOLRBP) both for training and testing dataset which illustrate better performance than simple VRLBP algorithms. The proposed algorithms have achieved 97.49, 95.86, 95.34, 97.27 and 97.27% of accuracy with MSE of (0.002, 0.001, 0.002, 0.003, 0.004). While the other algorithms fall behind in performance with MSE's of 0.26, 0.25, 0.28, 0.27 and 0.25, and accuracy of 95.10, 95.22,

94.88, 94.91, and 95.10% respectively for training dataset. Similarly, for the testing data the proposed algorithms have achieved 96.33, 97.93, 96.67, 98.09 and 96.88% of accuracy with MSE of (0.001, 0.002, 0.001, 0.0032, 0.001). While the other algorithms fall behind in performance with MSE's of 0.26, 0.24, 0.30, 0.24 and 0.25, and accuracy of 95.08, 95.10, 94.67, 95.26, and 95.23%. The accuracy and MSE performances for all algorithms are shown in Fig. 2.

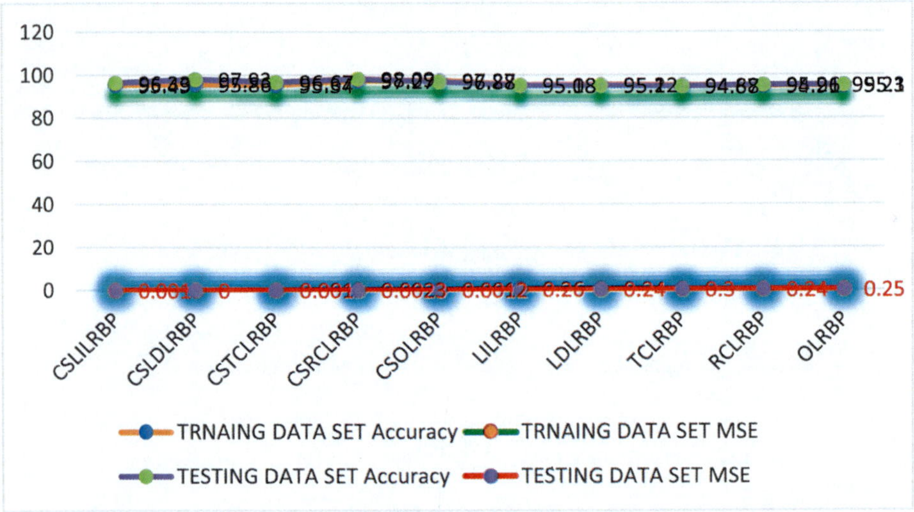

Fig. 2. Comparison of average MSE and accuracy for diabetes classification

5 Conclusion

This research, enhanced the learning process of BP algorithm with cuckoo search based variable learning rate. The conventional BP algorithms have some weakness such as slow convergence rate, local minima and stuck on local minima. This paper proposed an optimized CS algorithm, with variable learning rate to train the network. The proposed variable learning rate and cuckoo search algorithm speed up the slow convergence and solve the local minima problem of the back-propagation algorithm. The proposed CS variable learning rate BP algorithms are compared with traditional VLRBP algorithms. The proposed CSLILRBP, CSLDLRBP and CSOLRBP algorithms are compared with traditional algorithm such as LILRBP, LDLRBP, CLRBP, RCLRBP, and OLRBP. To check the performance of the proposed models breast cancer and diabetes benchmark classification problems are used. The simulation results show that proposed algorithm show high efficiency and enhanced performance.

Acknowledgements. "The authors would like to thanks the institute of Computer Sciences and Information Technology, Faculty of Management and Computer Sciences the University of Agriculture Peshawar Pakistan for support this research".

References

1. Karlik, B.: Machine learning algorithms for characterization of EMG signals. Int. J. Inf. Electron. Eng. **4**(3), 189 (2014)
2. İşeri, A., Karlık, B.: An artificial neural networks approach on automobile pricing. Expert Syst. Appl. **36**(2), 2155–2160 (2009)
3. Chiang, W.-Y.K., Zhang, D., Zhou, L.: Predicting and explaining patronage behavior toward web and traditional stores using neural networks: a comparative analysis with logistic regression. Decis. Support Syst. **41**(2), 514–531 (2006)
4. Hameed, A.A., Karlik, B., Salman, M.S.: Back-propagation algorithm with variable adaptive momentum. Knowl.-Based Syst. **114**, 79–87 (2016)
5. Ranganathan, V., Natarajan, S.: A new backpropagation algorithm without gradient descent. arXiv preprint arXiv:1802.00027 (2018)
6. Glorot, X., Bengio, Y.: Understanding the difficulty of training deep feedforward neural networks. In: Proceedings of the Thirteenth International Conference on Artificial Intelligence and Statistics (2010)
7. Lee, S.: Application of artificial neural networks in geoinformatics 2018, MDPI (2018)
8. Lee, S., Lee, M.-J., Jung, H.-S.: Data mining approaches for landslide susceptibility mapping in Umyeonsan, Seoul, South Korea. Appl. Sci. **7**(7), 683 (2017)
9. Karlik, B.: Soft computing methods in bioinformatics: a comprehensive review. Math. Comput. Appl. **18**(3), 176–197 (2013)
10. Samborska, I.A., et al.: Artificial neural networks and their application in biological and agricultural research. J. NanoPhotoBioSciences **2**, 14–30 (2014)
11. Karlik, B., Sarli Cemel, S.: Diagnosing diabetes from breath odor using artificial Neural Networks (2012)
12. Karlik, B.: Differentiating type of muscle movement via AR modeling and neural network classification. Turk. J. Electr. Eng. Comput. Sci. **7**(1–3), 45–52 (2000)
13. Sun, Y.J., Zheng, S., Miao, C.X., Li, J.: M, Improved BP neural network for transformer fault diagnosis. J. China Univ. Min. Technol. **17**(1), 138–142 (2007)
14. Nawi, N.M., Khan, A., Rehman, M.Z.: A new back-propagation neural network optimized with Cuckoo search algorithm. In: Murgante, B., et al. (eds.) ICCSA 2013. LNCS, vol. 7971, pp. 413–426. Springer, Heidelberg (2013). https://doi.org/10.1007/978-3-642-39637-3_33
15. Abdul Hamid, N., Mohd Nawi, N., Ghazali, R., Mohd Salleh, M.N.: Accelerating learning performance of back propagation algorithm by using adaptive gain together with adaptive momentum and adaptive learning rate on classification problems. In: Kim, T.-H., Adeli, H., Robles, R.J., Balitanas, M. (eds.) UCMA 2011. CCIS, vol. 151, pp. 559–570. Springer, Heidelberg (2011). https://doi.org/10.1007/978-3-642-20998-7_62
16. Khan, A., et al.: Chicken S-BP: an efficient chicken swarm based back-propagation algorithm. In: Herawan, T., Ghazali, R., Nawi, N.M., Deris, M.M. (eds.) SCDM 2016. AISC, vol. 549, pp. 122–129. Springer, Cham (2017). https://doi.org/10.1007/978-3-319-51281-5_13
17. Haykin, S.: Neural Network : A Comprehensive Foundation. Macmillan, New York (1994)
18. Madhiarasan, M., Deepa, S.: ELMAN neural network with modified grey wolf optimizer for enhanced wind speed forecasting. Circuits Syst. **7**(10), 2975–2995 (2016)
19. Nawi, N.M., Rehman, M.Z., Khan, A.: A new bat based back-propagation (BAT-BP) algorithm. In: Swiątek, J., Grzech, A., Swiątek, P., Tomczak, J.M. (eds.) Advances in Systems Science. AISC, vol. 240, pp. 395–404. Springer, Cham (2014). https://doi.org/10.1007/978-3-319-01857-7_38
20. Abbas, Q., Ahmad, F., Imran, M.: Variable learning rate based modification in backpropagation algorithm (mbpa) of Artificial Neural Network for data classification. Sci. Int. **28**(3) (2016)

21. Becker, S., Le Cun, Y.: Improving the convergence of back-propagation learning with second order methods. In: Proceedings of the 1988 connectionist models summer school. Morgan Kaufmann, San Matteo, CA (1988)
22. Deng, W.J., Chen, W.C., Pei, W.: Back-propagation neural network based importance-performance for determining critical service attributes. J. Expert Syst. Appl. **34**(2), 1–26 (2008)
23. Bi, W., Wang, X., Tang, Z., Tamura, H.: Avoiding the local minima problem in backpropagation algorithm with modified error function. IEICE Trans. Fundam. Electron. Commun. Comput. Sci. **E88-A**(12), 3645–3653 (2005)
24. Hamid, N.A., Nawi, N.M., Ghazali, R.: The effect of Adaptive Gain and adaptive Momentum in improving Training Time of Gradient Descent Back Propagation Algorithm on Classification problems. In: Proceeding of the International Conference on Advanced Science, Engineering and Information Technology, pp. 178–184 (2011)
25. Mohd Nawi, N., Ransing, R., Abdul Hamid, N.: BPGD-AG: a new improvement of back-propagation neural network learning algorithms with adaptive gain. J. Sci. Technol. **2**(2) (2011)
26. Smith, L.N.: Cyclical learning rates for training neural networks. In: 2017 IEEE Winter Conference on Applications of Computer Vision (WACV). IEEE (2017)
27. Peace, I.C., Uzoma, A.O., Ita, A.: Effect of learning rate on artificial neural network in machine learning. Int. J. Eng. Res. Technol. (IJERT) **4**(3), 359–363 (2015)
28. Yu, X.H., et al.: Dynamic learing rate optimization of the back propagation algorithm. IEEE Trans. Neural Network **6**, 669–677 (1995)
29. Ye, Y.C.: Application and Practice of the Neural Networks. Scholars Publication, Taiwan (2001)
30. Yu, C.-C., Liu, B.-D.: A backpropagation algorithm with adaptive learning rate and momentum coefficient. In: Proceedings of the 2002 International Joint Conference on Neural Networks. IJCNN 2002 (Cat. No. 02CH37290). IEEE (2002)
31. Rehman, M.Z., Nazri, M.N.: The effect of adaptive momentum in improving the accuracy of gradient descent back propagation algorithm on classification problems. CCIS J. Softw. Eng. Comput. Syst. **179**(6), 380–390 (2011)
32. Yuemei, X., Hong, Z.: Study on the improved BP algorithm and Application. In: Asia-Pacific Conference on Proceedings of the Information Processing. APCIP 2009, pp. 7–10 (2009)
33. Thota, L.S., Changalasetty, S.B.: Optimum learning rate for classification problem with MLP in data mining. Int. J. Adv. Eng. Technol. (IJAET) **6**(1), 35–44 (2013)
34. Nawi, N.M., khan, A., Rehman, M.Z.: CSBPRNN: a new hybridization technique using cuckoo search to train back propagation recurrent neural network. In: Herawan, T., Deris, M.M., Abawajy, J. (eds.) Proceedings of the First International Conference on Advanced Data and Information Engineering (DaEng-2013). LNEE, vol. 285, pp. 111–118. Springer, Singapore (2014). https://doi.org/10.1007/978-981-4585-18-7_13
35. Xin-She, Y., Deb, S.: Cuckoo Search via Levy flights. In: World Congress on Nature and Biologically Inspired Computing, pp. 210–214 (2009)
36. Yang, X.-S.: Nature-inspired Metaheuristic Algorithms. Luniver Press, Vancouver (2010)
37. Yang, X.S.: Engineering Optimization: An Introduction with Metaheuristic Application. Wiley.com, New York (2010)
38. Yang, X.-S.: A new metaheuristic bat-inspired algorithm. In: Pelta, D.A., Cruz, C. (eds.) Nature Inspired Cooperative Strategies for Optimization (NICSO 2010). SCI, vol. 284, pp. 65–74. Springer, Heidelberg (2014). https://doi.org/10.1007/978-3-642-12538-6_6
39. Lahmiri, S.: A comparative study of back propagation algorithms in financial prediction. Int. J. Comput. Sci. Eng. Appl. IJCSEA **1**(4), 15 (2011)
40. Nandy, S., Sarkar, P.P., Das, A.: Training a Feed-forward Neural Network with Artificial Bee Colony Based Backpropagation Method. arXiv preprint arXiv:1209.2548 (2012)

41. Nawi, N.M., Rehman, M., Khan, A.: Verifying the accuracy of GDAM algorithm on multiple classification problems. In: International Conference on Advances in Intelligent Systems in Bioinformatics (2013). Atlantis Press (2014)
42. Jin, W., et al.: The improvements of BP neural network learning algorithm. In: 5th International Conference on Signal Processing Proceedings 2000. WCCC-ICSP 2000, vol. 3, pp. 1647–1649 (2000)
43. Tuba, M., Subotic, M., Stanarevic, N.: Modified cuckoo search algorithm for unconstrained optimization problems. In: Proceedings of the 5th European Conference On European Computing Conference, pp. 263–268 (2011)
44. Walton, S., et al.: Modified cuckoo search: a new gradient free optimisation algorithm. Chaos, Solitons Fractals **44**(9), 710–718 (2011)
45. Jovanovic, R., Tuba, M., Brajevic, I.: Parallelization of the cuckoo search using CUDA architecture. Institute of Physics Recent Advances in Mathematics (2013)
46. Shawkat, N., Tusiy, S.I., Ahmed, M.A.: Advanced Cuckoo search algorithm for optimization problem. Int. J. Comput. Appl. **132**(2), 31–36 (2015)
47. Yang, X.-S., Deb, S.: Multiobjective cuckoo search for design optimization. Comput. Oper. Res. **40**(6), 1616–1624 (2013)
48. Wolberg, W.H., Mangasarian, O.L.: Multisurface method of pattern separation for medical diagnosis applied to breast cytology. National Academy of Sciences, pp. 9193–9196 (1990)
49. Smith, J.W., Everhart, J.E., Dickson, W.C., Knowler, W.C., Johannes, R.S.: Using the ADAP learning algorithm to forecast the onset of diabetes mellitus. In: Symposium on Computer Applications and Medical Care, pp. 261–265. IEEE Computer Society Press (1988)

Check for updates

A Metaheuristic Based Virtual Machine Allocation Technique Using Whale Optimization Algorithm in Cloud

Nadim Rana[1,2](✉), Shafie Abd Latiff[1], and Shafi'i Muhammad Abdulhamid[3]

[1] School of Computing, Faculty of Engineeirng, Universiti Teknologi Malaysia, Johor Bahru, Johor, Malaysia
nadimrana@jazanu.edu.sa

[2] College of Computer Science and Information Technology, Jazan University, Jazan, Kingdom of Saudi Arabia

[3] Department of Cyber Security Science, Federal University of Technology, Minna, Minna, Nigeria

Abstract. In modern datacenters in a cloud computing system, optimal resource allocation is always a tedious task due to the ever-changing environment. Millions of resources are allocated and deallocated on the datacenters in a fraction of the time using the virtual machine (VM). Thus, VM scheduling in cloud computing always remains a significant concern by cloud vendors to address the resource allocation problem. The critical concentration in this regard is optimally allocating the VM resources for task execution in minimum time and cost. Since scheduling is an NP-hard problem, the metaheuristic approach is proven to be significant and depicts better results in recent times. Therefore, this paper put forward an Improved Whale Optimization Algorithm (IWOA) based VM allocation technique for efficient allocation of resources in Infrastructure as a service (IaaS) cloud computing. The experimental results confirm that the proposed algorithm outperforms with GA, NSGAIII, and PSO in terms of makespan and cost minimization.

Keywords: Cloud computing · VM scheduling · Metaheuristic · Whale optimization algorithm

1 Introduction

Cloud computing is an on-demand computing model that provides distributed system services through the internet. The computing recourses are rationed to the clients over a virtualized network facility by cloud service providers [1, 2]. Cloud computing structure provides a stack of provisioning systems, in which the traditional ones being: Infrastructure as a Service (IaaS), which deals with the services involving virtual resources, such as servers, storage, compute nodes (processors) and network bandwidth. Amazon is one of the prominent cloud providers that offer IaaS for vendors and users. For instance, the service module Elastic Compute Cloud (EC2) and Simple Storage Service (S3) are offered by Amazon for virtual infrastructure management [1].

J. H. Abawajy et al. (Eds.): EATI 2020, LNNS 254, pp. 22–38, 2021.
https://doi.org/10.1007/978-3-030-80216-5_3

The second service model is called Platform as a Service (PaaS) that provides multiple flavors of operating systems and a lot of system tools and utilities as a service to the customers. It also allows computer programmers and engineers to develop and deploy their applications in cloud. Microsoft Azure is an example of PaaS [2].

The third model is the Software as a Service (SaaS), which permits suppliers to concede the client's access to authorized software. SaaS manages to utilize any application or administration through cloud. Google calendar is one of the applications that give coordinated effort on various applications, similar to event management, project management, etc. using a thin client. Salesforce is additionally a typical and well-known SaaS-based application for client relationship management (CRM) [2]. Each physical machine (PM) in cloud datacenter can hold several provisioned VMs which further consists of several virtual resources attached to it, such as CPUs, memory, storage, and network as shown in Fig. 1 [3].

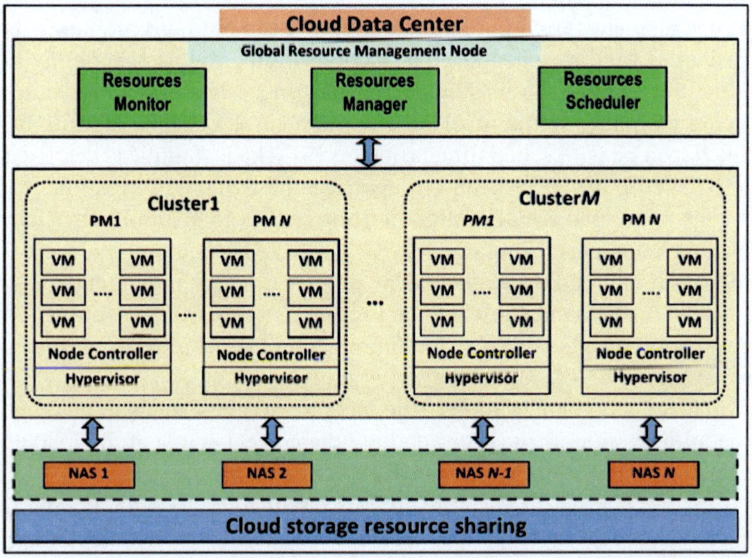

Fig. 1. Virtual resource model of datacenter [4]

The organization of the paper is as follows. Section 2 presents the background and previous works. The problem formulation and performance metrics for VM scheduling are discussed in Sect. 3. The outlines of the proposed IWOA algorithm is explained in Sect. 4. Section 5 discuss the experimental setup and statistical results. Finally, Sect. 6 concludes the work and recommends some future directions.

2 Background and Related Works

Virtual machine scheduling is a process of figuring out the resource movements to be performed on the datacenter. It is a request for the resources for an appropriate assignment of VMs and its resources to the PMs. The end goal of the VM scheduling is the maximum utilization of resources with minimum incurred cost. Hence, efficient VM scheduling always remains the critical concern of cloud providers and considered to be a primary feature of cloud computing [5, 6].

VM scheduling in cloud computing datacenter can be expressed as the distribution of a number of VMs to various PMs so that the amount of PMs can be minimized. In [7], the authors propose a VM assignment in cloud datacenter. The technique stressed upon the efficient VM provisioning to physical hosts to reduce the entire resource utilization and to reduce the numbers of PM required for specific usage. The outcome of the simulation shows the lowest provisioning of the VMs to the host machine. The proposed algorithm outperforms in view of its efficiency and scalability. In [8], the researchers combine the particle swarm optimization algorithm (PSO) with a variable-sized bin packing strategy to address VM placement problem. The best-fit bin packing scheme is used to find the optimal number of VMs to be fit on each PM to minimize energy consumption with improved resource utilization.

Virtual machine allocation in cloud computing is an NP-hard problem where thousands of resources need to schedule in a fraction of time by the scheduling algorithm. Assuming the fluctuation in user demand and historical data as a significant aspect of load balancing within the dynamic cloud environment, the genetic algorithm (GA) has been modified to cope with scheduling problems [9]. The technique is designed to attain better load balancing of the PMs and to decrease the dynamic migration of the VMs. The test results prove that the algorithm performs better to achieve better load balancing and low migration cost. Similarly, a PSO-based VM scheduling policy is presented for VM allocation in cloud datacenters. The proposed policy intelligently allocates the VMs to physical hosts to cut short the total resource waste and the numbers of servers used. Experimental results confirm the minimization in VM allocation time and better performance [10].

In a virtualized cloud environment, the incoming request tends to be dynamic. The typical scheduling system mends for handling the current status of the load and hardly anticipate the fluctuating workload on the servers which, usually leads to load imbalance problems. Considering that in [11], the authors modified the GA to cope with the load imbalance problem to archive the information of incoming loads on the datacenter. The algorithm proactively anticipates the incoming loads on the system generating multiple solutions and choose the leas affected one. The simulation graphs confirm that the algorithm reduces the load imbalance and migration cost of the VMs. In a similar study in [12], GA is integrated with greedy method to tackle task scheduling problems on cloud system. This metaheuristic-based method shown optimal mapping solutions for the tasks to the resources with minimum generation of iterations.

In [13] the authors present a load balancing aware hybrid metaheuristic algorithm combining ant colony optimization (ACO) with PSO to address the VM allocation problem. The ant colony optimization and particle swarm optimization (ACOPS) algorithm use previously-stored historical information on the server to predict the incoming workload to adjust the dynamic load. Moreover, the ACOPS denies the request that does not fulfill the scheduling requirements to decrease the computing time. The simulation results show that the presented technique outperforms when compare to the other benchmark techniques and also achieved reasonable load balancing. In [14] PSO is combined with opposition based learning (OBL) method to improve the initial solution generated by the proposed technique and avoid the local optima entrapment problem. The method is applied for resource allocation problem on the network datacenter and gained optimum performance in reducing execution time and load imbalance through its results. In [15], modified flower pollination algorithm (FPA) to cope with the VM allocation problem in cloud computing. The proposed scheme highlights one of the issues of green computing by handling the energy consumption on cloud datacenter. The simulation results show better quality solutions generated by the presented scheme and a significant reduction in power consumption on the servers.

Many works of literature have shown the capability of the metaheuristic algorithms in solving various Np-hard scheduling problems in cloud computing such as resource scheduling [16], task scheduling [17], workflow scheduling [18, 19], load balancing [20] and virtual machine scheduling [8] a few. This work utilizes the capability of the newly developed whale optimization algorithm (WOA) and modify it to tackle the VM scheduling problem in the virtual cloud system [22].

3 Problem Formulation of VM Scheduling

3.1 VM Model

Figure 2 demonstrates the scheduling mechanism of VMs and PMs. The set of PMs is denoted by $P = \{P_1, P_1, \ldots, P_N\}$ where N is referred to as the numbers of PMs on a datacenter (DC), whereas the set of VMs allocated on PM is denoted by $i = \{V_1, V_1, \ldots, V_M\}$ where M referred to as numbers of VMs. Suppose we need to assign virtual machine v_i on p_i at current state of the system, so to represent as a mapping solution set $S = \{S_1, S_1, \ldots, S_P\}$ is used after v_i is scheduled on each PM. Here S_i referred to as a set of mapping solutions corresponding to VM v_i mapped to the PM p_i.

3.2 The Expression of Load

The load of a PM can be achieved by increasing loads of the VMs currently running on it. In this regard, T is considered the best time span examined by archived data. Here, T is considered as the current time zone from the current time zone of historical monitoring data. As per the dynamic nature of PM load, we can divide time t into n time periods. Therefore we express $T = [(t_1 - t_0), [(t_2 - t_1) \ldots\ldots\ldots (t_n - t_{n-1})]]$.

As per the expression, $(t_k - t_{k-1})$ represents the time period k. Let us assume that the load of the VMs is somewhat stable in every period, then it may be defined for the load of VM no. i for the period k is $v(i, k)$. Hence, it may be concluded that, for the cycle T for the VM v_i, the average load on the PM p_i is expressed by Eq. (1):

$$\overline{v_\iota(\iota, \tau)} = \frac{1}{\tau} \sum_{k=1}^{n} v(i, k) \times (t_k - t_{k-1}) \tag{1}$$

As per the system configuration, the load of a PM can be computed by increasing loads of the VMs running on it. Thus the load on the PM p_i can be defined as Eq. (2).

$$\rho(i,T) = \sum_{j=1}^{m_i} \overline{v_i(j,\tau)} \tag{2}$$

The deployment of the current VM is v. Since the VM has already configured with its required resources information, thus we can compute the current load of the VM as v' according to the appropriate information. Therefore, when VM v is allocated to a PM, the load of every PM is expressed by Eq. (3):

$$\rho(i,\tau)' = \begin{cases} \rho(i,\tau) + v' & \text{After deploy } v \\ \rho(i,\tau) & \textit{Others} \end{cases} \tag{3}$$

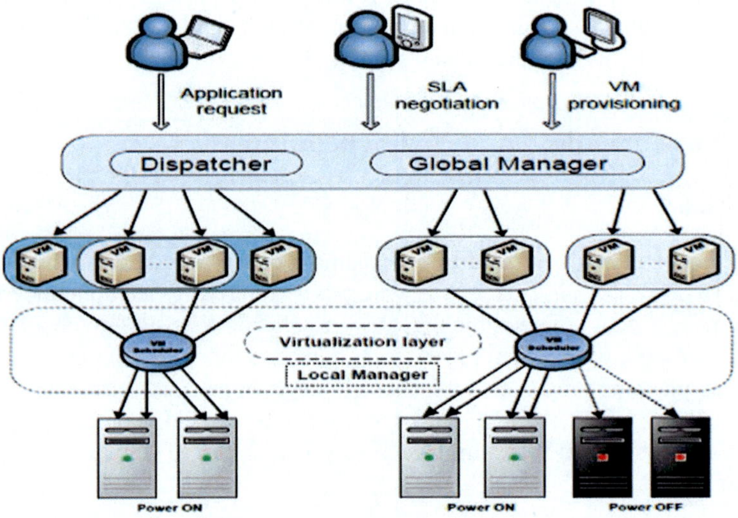

Fig. 2. VM scheduling overview [4]

Generally, when VM v is allocated to the PM p_i, a noticeable rise in the system load can be observed. So, we need to make an arrangement to load the balance of PMs by distributing the workload with under loaded PMs. Thus, the load variation of the PMs can be expressed by the following mathematical expressions. Where mapping solution S_i in time t period after VM v is allocated to PM p_i is expressed by Eqs. (4) and (5):

$$\sigma_i(\tau) = \sqrt{\frac{1}{\aleph}\sum_{i=1}^{\aleph}(p(\tau)' - p(i,\tau))'} \tag{4}$$

Where

$$\overline{\rho(\tau)'} = \frac{1}{\aleph}\sum_{i=1}^{\aleph}\rho(i,\tau)' \tag{5}$$

3.3 Performance Metrics

In this work, we considered the performance evaluation of the IWOA considering makespan and execution cost of the tasks in comparison to other algorithms.

Makespan: It is the maximum total execution time taken by the system to execute all tasks on all virtual machines. One of the critical objectives of the scheduling technique to minimize the makespan, as expressed by Eq. (6):

$$Makespan = min\{C_{i,j}\} \tag{6}$$

Where, Cij may define the actual time a resource j is expected to complete task Ti. Where resource denotes the virtual machine to which task is to map. Hence, $\sum Cj$ represents the total time that resource j completes all the tasks submitted to the virtual machines.

Execution Cost: It is the total cost taken by the VM instance per unit time by the service provider to execute the tasks of cloud users defined by Eq. (7):

$$Execution\ \ Cost = exe_{costj}(C_j * Time_j) \tag{7}$$

The execution cost exe_{costj} is the cost C_j of using a resource j by a cloud consumer per unit time $Timej$ for which the resource j is being utilized. Therefore, $\sum_j^m$ (Cj $*$ Timej) is the total execution cost for all resource j used for computing all tasks.

4 Improved Whale Optimization Algorithm (IWOA)

Whale Optimization Algorithm (WOA) is the recent incarnation of the natural phenomena motivated by the bubble-net hunting maneuvering of humpback whales. The swarm-based intelligent WOA algorithm is designed to solve complex engineering optimization problems [22]. Since the inception of WOA in 2016, it has been widely utilized to solve different optimization problems in multidisciplinary areas, such as electrical engineering, computer science, aeronautical and navigation systems, wireless networks, construction, and planning amongst few [4, 23–31]. The time-constraint based position of the individual whale is measured by three different functional processes: (i) shrinking encircling prey, (ii) bubble-net attacking method (exploitation), and (iii) search for prey (exploration).

4.1 Population Initialization

The WOA starts with the random generation of the population X_i, the unique hunting maneuver of the humpback whales is used to identify the target prey location. Once the search agents identify the target prey, they start to encircle them, moving around in 9 shapes, as shown in Fig. 3. Because the optimal position of the prey is not known in advance in the vast search space, the WOA presumably considers the target prey as the current best candidate solution $\vec{X}$ or near to optimum solution. All the possible effort is carried out to find the best search agent $\overrightarrow{X^*}$ in the exploration and exploitation phase by the WOA, whereas the other search agents update their positions near to the best search agent.

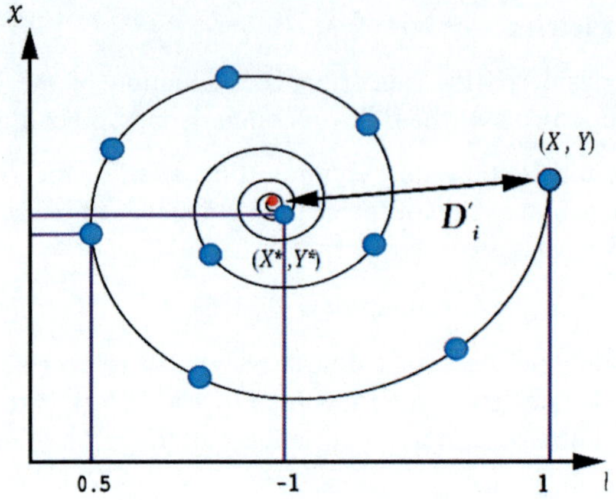

Fig. 3. Spiral updating position [24]

4.2 Encircling Prey

As soon as whales recognize the best search agent, the population of whales starts encircling them. Then the current positions of the search agents are updated as they get closer to the prey. The mathematical expressions of the encircling prey are defined as Eqs. (8), (9), (10) and (11), respectively:

$$\vec{D} = \left|\vec{C} \bullet \overrightarrow{X^*}(t) - \vec{X}(t)\right| \tag{8}$$

$$\vec{X}(t+1) = \overrightarrow{X^*}(t) - \vec{A} \bullet \vec{D} \tag{9}$$

$$\vec{A} = 2 \bullet \vec{a} \bullet \vec{r} - \vec{a} \tag{10}$$

$$\vec{C} = 2 \bullet \vec{r} \tag{11}$$

Where $\vec{D}$ is the distance between the best solution $\overrightarrow{X^*}(t)$ and the current solution $\vec{X}(t)$ with iterations t, whereas $\vec{r}$ is a random number equally distributed in the range of [0, 1]. $\vec{a}$ is the shrinking decreased linearity value reduced within the range of 2 to 0 throughout iterations, and $\bullet$ represent the element-wise multiplication.

4.3 Bubble-Net Attacking Method (Exploitation Phase)

The exploitation phase of the WOA shown in Fig. 4. A logarithmic spiral is applied to identify the equidistance between the whale position and the prey's position. After each movement, an update is performed between the whale positions in a helix shape in order to adjust to the current search space. The algorithm can choose the bubble-net hunting behavior in two ways (1) shrinking encircling and (2) spiral update based on the random selection probability. If the algorithm selects the first method for the solution update, then it applies the same method as decreasing value a in Eq. (8). Otherwise, it goes for the second method for solution update, and then the population of whales swim around the current best solution $\vec{X}$ so far, this can be expressed by Eqs. (12) and (13) [22].

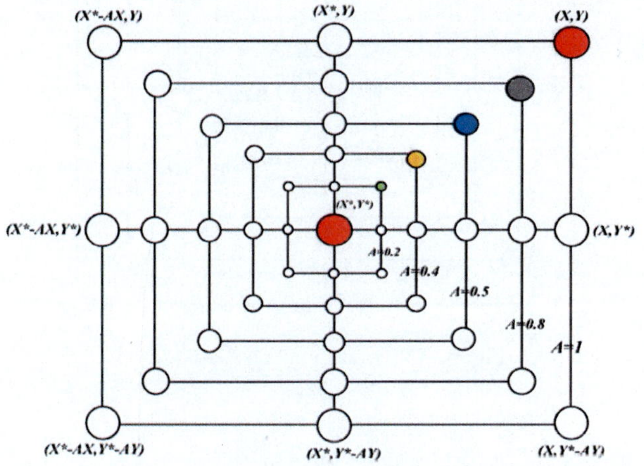

Fig. 4. Exploitation mechanism [22]

$$\vec{X}(t+1) = \overrightarrow{D'} \bullet e^{bk} \bullet \cos(2\pi k) + \overrightarrow{X^*}(t) \tag{12}$$

$$\overrightarrow{D'} = \left|\overrightarrow{X^*}(t) - \vec{X}(t)\right| \tag{13}$$

where k is the random variable number equally distributed in the range of $[-1, 1]$ and b represents a constant for describing the logarithmic spiral shape. As per the description of the WOA in [22], the algorithm can switch between the shrinking and the spiral path for solution update based on the random probability, which can be described as Eq. (14):

$$\vec{X}(t+1) = \begin{cases} \overrightarrow{X^*}(t) - \vec{A} \bullet \vec{D} & if \;\; p < 0.5 \\ \overrightarrow{D'} \bullet e^{bk} \bullet \cos(2\pi k) + \overrightarrow{X^*}(t) & if \;\; p \geq 0.5 \end{cases} \tag{14}$$

Where p is the probability of random number for the selection between the two methods.

4.4 Search for Prey (Exploration Phase)

In the exploration phase, IWOA uses $\vec{A}$ vector to search for the prey in the global search space. Where, the random value is used for vector $A > 1$ or $A < -1$, to force the search agent to move from the reference whale. The moving behavior of random vector $\vec{A}$ is shown in Fig. 5. The whales randomly search for the best prey $\overrightarrow{X^*}$ based on the random position $\overrightarrow{X_{rand}}$, not by the best solution $(\overrightarrow{X^*})$ as defined in Eqs. (15) and (16):

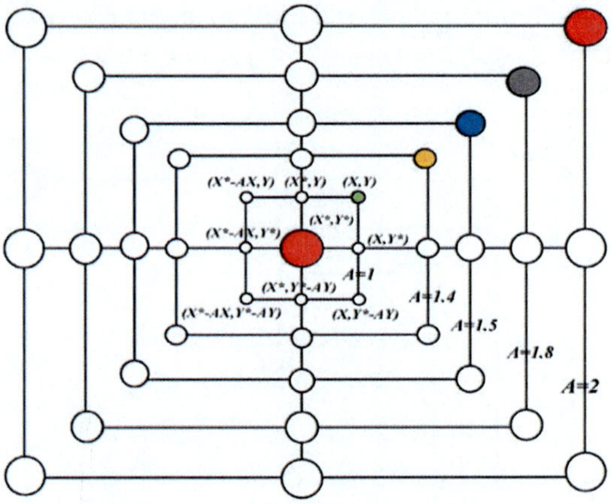

Fig. 5. Exploration mechanism [22]

$$\vec{X}(t+1) = \overrightarrow{X_{rand}} - A \bullet \overrightarrow{D'} \tag{15}$$

$$\overrightarrow{D'} = \left|\vec{C} \bullet \overrightarrow{X_{rand}} - \vec{X}(t)\right| \tag{16}$$

Where $\overrightarrow{X_{rand}}$ nominated as a randomly selected whale from currently generated population [22]. As suggested by [22], the parameters used for updating solutions are a, A, C and p. If the value of $p > 0.5$, the solution is updated by Eq. (12), if the value of $p < 0.5$, the solution is updated by Eq. (15) and (16). Moreover, it can also be updated by Eq. (8) and Eq. (9) based on the value of $|\vec{A}|$. The WOA continues to update the position of the solution until the set criteria are met. Once the termination condition is satisfied, it returns with X^* as the best solution found so far by the WOA.

Pseudo-code of Improved Whale Optimization Algorithm (IWOA)

```
Initialize the whales population X_i (i = 1,2,3, ..., n)
Calculate the fitness of each search agent
X* = the best search agent
while (t < Max_itr)
    for each search agent
        Update a, A, C, l, and p
            if1 (p < 0.5)
                if2 (|A|<1 )
                    Update the position of the current search agent by Eq. (9)
                else if2 (|A|≥1 )
                    Select a random search agent (X_rand)
                    Update the position of the current search agent by Eq. (15)
                end if2
            else if1 (p ≥ 0.5)
                    Update the position of the current search by Eq. (12)
            end if1
    end for
    Check if any search agent goes beyond the search space and amend it
    Calculate the fitness of each search agent
    Update X* if there is a better solution
    t = t + 1
end while
return X*
Begin
Input : Tasks T, virtual machines V
Output: Fitness value
    For (all VMs)
        Find the cost function of CPU by Eq. (6)
        Find the execution cost function by Eq. (7)
    End for
End
```

5 Experimental Setup

The experimental setup is defined to analyze the computational results generated from the proposed IWOA scheme. The simulation is executed on the personal computer with Intel Core i5-6200U with a 2.30 GHz processor with 4G RAM on Windows 10 OS. The IWOA scheme is implemented on JAVA based CloudSim 3.0.3 framework on the Eclipse IDE Luna 4.4.0 platform [32]. Two standard real workload traces are used, namely NASA and HPC2N offered by "Ake Sandgren, Bill Nitzberg and Victor Hazlewood" to evaluate the performance of the proposed algorithm. Three different benchmark techniques are used to compare the performance of the presented algorithm, those are GA [33], NSGAIII [34], and PSO [9], the parameter setting of these algorithms are shown in Table 1.

5.1 First Scenario

The experiential setup is formed in two scenarios. In the first scenario, we consider five cloud users with five cloud brokers with two datacenters. A total of thirty VMs are created using time-shared policy. Each VM has 0.5 GB of RAM, 10,000 MIPS of bandwidth, one CPU on each VM, and 10 GB of storage managed by Xen Virtual Machine Monitor (VMM) running on Linux operating system. Whereas, each host on the datacenter has 2 GB of RAM, 1,000,000 MIPS of processing power, 1 GB of storage with bandwidth of 10 GB/s. The cloudlets submitted to the system ranges from 200 to 1200, with cloudlet length 800,000 MIs each, and the file size is 600. The simulation in the first scenario is performed using HPC2N dataset with 527,371 cloudlets.

Table 1. Parameter setup for IWOA

Algorithms	Parameter	Value
GA	Population size	1000
	Max iteration	1000
	Crossover rate	0.5
	Mutation rate	0.1
NSGAIII	Population size	1000
	Max iteration	1000
	Crossover probability, Pc	0.8
	Mutation probability, Pm	0.1
PSO	Particle size	100
	Self recognition coefficients, $c1$, c2	2
	Uniform random number, $R1$	[0, 1]
	Max Iteration	1000
	Variable inertia weight (W)	0.9–0.4
IWODE	Population size, X	50
	Max iteration, Max_{itr}	1000
	Shrinking decreased linearly, $\vec{a}$	[2-0]

5.2 Second Scenario

Similarly, in the second scenario, ten cloud users with ten cloud brokers are created with five datacenters. A total of fifty VMs are created using space-shared policy with the same configuration for VM and cloudlet length mentioned in the above scenario. In the second scenario, the simulation is performed using NASA dataset with 14,794 cloudlets. The simulation is executed for 30 iterations for each technique and the average results are calculated for makespan and execution cost. In both the scenarios, simulation is performed using the same parameter values mentioned in Table 1 for all the techniques.

5.3 Result and Discussion

This section presents the statistical analysis of the results produced by IWOA for both scenarios for makespan and execution cost. The comparative results obtained from IWOA with respect to GA, NSGAIII, and PSO are shown in Figs. 6, 7, 8 and 9.

5.4 Statistical Analysis

In the first scenario in Fig. 6, the IWOA is executed using HPC2N workload starting from the minimum of 200 cloudlets, the average makespan time 2460.78 s for GA, 2375.72 s for NSGAIII, 2465.58 for PSO, and 2179.01 s taken by the proposed IWOA algorithm. When the same experiment performed on the maximum 1200 cloudlets, the average makespan time 8466.96 s for GA, 8174.47 s for NGSAIII, 8449.14 for PSO, and 7988.32 s taken by IWOA. Similarly, the simulation is done on 400, 600, 800, and 1000 cloudlets for 30 iterations each cycle, and the average is observed. In all conditions, the IWOA displays a significant reduction in makespan.

In the second scenario in Fig. 7, the simulation is implemented on the NASA workload with the same parameter settings shown in Table 1 for all the algorithms. When the experiment is performed on a minimum of 200 cloudlets, the makespan time taken is 170.68 s for GA, 147.97 s for NGSAIII, 198.91 for PSO, and 116.85 s taken by the proposed IWOA. When the experiment is performed on 1200 cloudlets, the makespan time is 1867.32 s for GA, 1677.44 s for NGSAIII, 1754.01 for PSO, and 1498.80 s taken by IWOA. Also, the simulation is done on 400, 600, 800, and 1000 cloudlets for 30 iterations each cycle and the average is observed. In all conditions, the IWOA displays a significant reduction in makespan.

Figures 8 and 9 illustrate the execution cost incurred by the proposed IWOA algorithm using HPC2N and NASA workloads consecutively. The simulation is done in both scenarios for execution cost, and the average results are observed after 30 iterations each cycle for the tasks from 200 to 1200. The results display a better performance of IWOA in execution cost reduction compared to other benchmark algorithms.

In the first scenario in Fig. 8, when the experiment is set to execute using HPC2N workload on a minimum of 200 cloudlets, the execution cost is 155.09 for GA, 32.09 for NGSAII, 198.76 for PSO, and 16.45 taken by the proposed IWOA algorithm. When performed the same experiment on a maximum of 1200 cloudlets, the execution cost is 369.32 for GA, 117.87 for NGSAIII, 432.10 for PSO, and 61.76 taken by IWOA. Likewise, the simulation is done on 400, 600, 800, and 1000 cloudlets for 30 iterations each cycle, and the average is observed. In all conditions, the IWOA displays a significant reduction in execution cost.

In the second scenario in Fig. 9, the simulation is implemented on the NASA workload with the same parameter setting for all the algorithms. When the experiment is performed on a minimum of 200 cloudlets, the execution cost is 0.36 for GA, 0.31 for NGSAIII, 0.98 for PSO, and 0.28 taken by IWOA.

When the experiment is performed on a maximum of 1200 cloudlets, the execution cost is 59.09 for GA, 16.44 for NSGAIII, 52.54 for PSO, and 7.43 taken by IWOA algorithm. Similarly, the simulation is done on 400, 600, 800, and 1000 cloudlets for 30 iterations each cycle, and the average is observed. In all conditions, the IWOA displays a

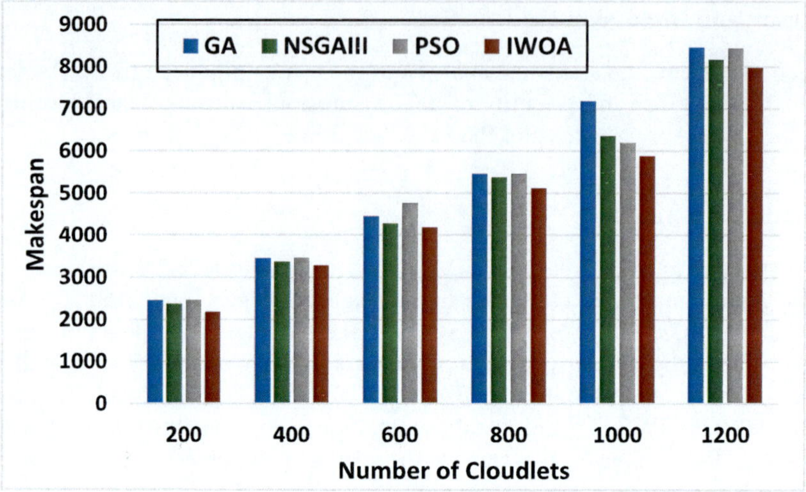

Fig. 6. Makespan time on HPC2N dataset

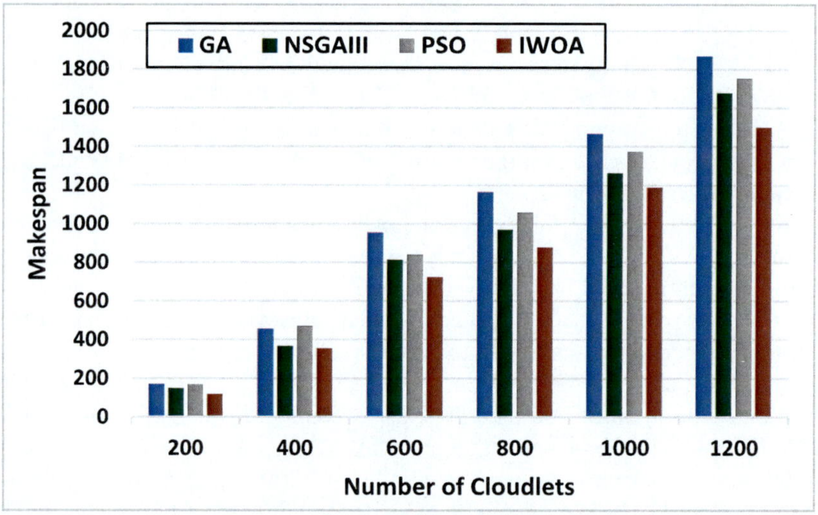

Fig. 7. Makespan time on NASA dataset

significant reduction in execution cost. With the above results, it can be iterated that the IWOA has noteworthy potential to generate optimal quality solutions for VM scheduling problem.

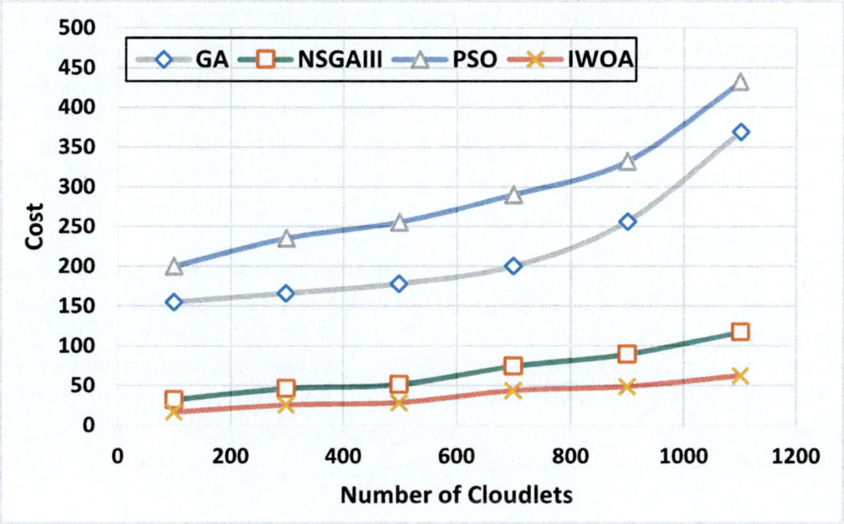

Fig. 8. Execution cost on HPC2N dataset

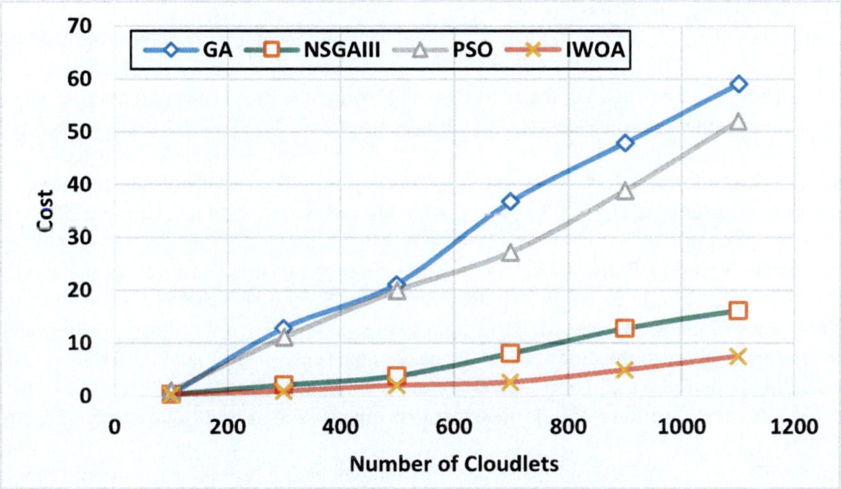

Fig. 9. Execution cost on NASA dataset

6 Conclusion and Future Work

The paper presents a metaheuristic-based Improved Whale Optimization Algorithm (IWOA) for makespan and cost minimization for virtual machine allocation in cloud computing. The algorithm is evaluated on a well-known CloudSim framework with two standard real workload traces HPC2N, and NASA offered by "Ake Sandgren, Bill

Nitzberg and Victor Hazlewood". The simulation outcomes display a superior performance by the proposed IWOA compared to the other benchmark metaheuristic algorithms such as GA, NSGAIII, and PSO for makespan and cost minimization. It is also observed that the proposed IWOA converges speedily towards finding optimal solutions compare to other algorithms and shows an acceptable balance between local and global search. From these observations, it can be deduced that the proposed IWOA provides better quality results in VM allocation and could perform well in solving other cloud scheduling problems.

In the future, we intend to use the WOA algorithm to solve other problems such as task scheduling, resource scheduling, load balancing and virtual machine placement in cloud computing. We also intend to hybridize the WOA with other local search based metaheuristic algorithms to tackle multi-objective performance metrics such as energy, utilization, throughput, and load balancing to name a few. We further wish to recommend using IWOA in other soft computing domains to test its validly like neural net systems, chaotic systems, and fuzzy optimization.

References

1. Buyya, R., Broberg, J., Goscinski, A.M.: Cloud Computing: Principles and Paradigms, vol. 87. Wiley, Hoboken (2010)
2. Foster, I., Zhao, Y., Raicu, I., Lu, S.: Cloud computing and grid computing 360-degree compared. In: 2008 Grid Computing Environments Workshop, pp. 1–10. IEEE Press (2008)
3. Beloglazov, A., Abawajy, J., Buyya, R.: Energy-aware resource allocation heuristics for efficient management of data centers for cloud computing. Future Gener. Comput. Syst. **28**, 755–768 (2012)
4. Rana, N., Abd Latiff, M.S.: A cloud-based conceptual framework for multi-objective virtual machine scheduling using whale optimization algorithm. Int. J. Innov. Comput. **8**(3), 53–58 (2018). https://doi.org/10.11113/ijic.v8n3.199
5. Ghiasi, H., Arani, M.G.: Smart virtual machine placement using learning automata to reduce power consumption in cloud data centers. SmartCR. **5**, 553–562 (2015)
6. Xu, F., Liu, F., Liu, L., Jin, H., Li, B., Li, B.: iAware: making live migration of virtual machines interference-aware in the cloud. IEEE Trans. Comput. **63**, 3012–3025 (2013)
7. Eden, M., Jain, R.: Washington University in St. Louis (2011)
8. Fatima, A., et al.: Virtual machine placement via bin packing in cloud data centers. Electronics **7**, 389 (2018)
9. Kumar, D., Raza, Z.: A PSO based VM resource scheduling model for cloud computing. In: 2015 IEEE International Conference on Computational Intelligence & Communication Technology, pp. 213–219. IEEE Press (2015)
10. Gondhi, N.K., Sharma, A.: Local search based ant colony optimization for scheduling in cloud computing. In: 2015 Second International Conference on Advances in Computing and Communication Engineering, pp. 432–436. IEEE Press (2015)
11. Hu, J., Gu, J., Sun, G., Zhao, T.: A scheduling strategy on load balancing of virtual machine resources in cloud computing environment. In: 2010 3rd International Symposium on Parallel Architectures, Algorithms and Programming, pp. 89–96. IEEE Press (2010)
12. Zhou, Z., Li, F., Zhu, H., Xie, H., Abawajy, J.H., Chowdhury, M.U.: An improved genetic algorithm using greedy strategy toward task scheduling optimization in cloud environments. Neural Comput. Appl. **32**(6), 1531–1541 (2019). https://doi.org/10.1007/s00521-019-04119-7

13. Cho, K.-M., Tsai, P.-W., Tsai, C.-W., Yang, C.-S.: A hybrid meta-heuristic algorithm for VM scheduling with load balancing in cloud computing. Neural Comput. Appl. **26**(6), 1297–1309 (2014). https://doi.org/10.1007/s00521-014-1804-9
14. Zhou, Z., Li, F., Abawajy, J.H., Gao, C.: Improved PSO algorithm integrated with opposition-based learning and tentative perception in networked data centres. IEEE Access **8**, 55872–55880 (2020)
15. Usman, M.J., et al.: Energy-efficient virtual machine allocation technique using flower pollination algorithm in cloud datacenter: a panacea to green computing. J. Bionic Eng. **16**(2), 354–366 (2019). https://doi.org/10.1007/s42235-019-0030-7
16. Singh, S., Chana, I.: A survey on resource scheduling in cloud computing: issues and challenges. J. Grid Comput. **14**(2), 217–264 (2016). https://doi.org/10.1007/s10723-015-9359-2
17. Singh, P., Dutta, M., Aggarwal, N.: A review of task scheduling based on meta-heuristics approach in cloud computing. Knowl. Inf. Syst. **52**(1), 1–51 (2017). https://doi.org/10.1007/s10115-017-1044-2
18. Masdari, M., ValiKardan, S., Shahi, Z., Azar, S.I.: Towards workflow scheduling in cloud computing: a comprehensive analysis. J. Netw. Comput. Appl. **66**, 64–82 (2016)
19. Shojafar, M., Canali, C., Lancellotti, R., Abawajy, J.: Adaptive computing-plus-communication optimization framework for multimedia processing in cloud systems. IEEE Trans. Cloud Comput. **8**(4), 1162–1175 (2016)
20. Milan, S.T., Rajabion, L., Ranjbar, H., Navimipoir, N.J.: Nature inspired meta-heuristic algorithms for solving the load-balancing problem in cloud environments. Comput. Oper. Res. **110**, 159–187 (2019)
21. Mirjalili, S., Lewis, A.: The whale optimization algorithm. Adv. Eng. Softw. **95**, 51–67 (2016)
22. Kaveh, A.: Sizing optimization of skeletal structures using the enhanced whale optimization algorithm. In: Kaveh, A. (ed.) Applications of metaheuristic optimization algorithms in civil engineering, pp. 47–69. Springer, Cham (2017). https://doi.org/10.1007/978-3-319-48012-1_4
23. Rana, N., Latiff, M.S.A., Abdulhamid, S.M., Chiroma, H.: Whale optimization algorithm: a systematic review of contemporary applications, modifications and developments. Neural Comput. Appl. **32**(20), 16245–16277 (2020). https://doi.org/10.1007/s00521-020-04849-z
24. Mirjalili, S., Mirjalili, S.M., Saremi, S., Mirjalili, S.: Whale optimization algorithm: theory, literature review, and application in designing photonic crystal filters. In: Mirjalili, S., Song Dong, J., Lewis, A. (eds.) Nature-Inspired Optimizers. SCI, vol. 811, pp. 219–238. Springer, Cham (2020). https://doi.org/10.1007/978-3-030-12127-3_13
25. Nazari-Heris, M., Mehdinejad, M., Mohammadi-Ivatloo, B., Babamalek-Gharehpetian, G.: Combined heat and power economic dispatch problem solution by implementation of whale optimization method. Neural Comput. Appl. **31**(2), 421–436 (2017). https://doi.org/10.1007/s00521-017-3074-9
26. Tubishat, M., Abushariah, M.A.M., Idris, N., Aljarah, I.: Improved whale optimization algorithm for feature selection in Arabic sentiment analysis. Appl. Intell. **49**(5), 1688–1707 (2018). https://doi.org/10.1007/s10489-018-1334-8
27. Chen, H., Xu, Y., Wang, M., Zhao, X.: A balanced whale optimization algorithm for constrained engineering design problems. Appl. Math. Model. **71**, 45–59 (2019)
28. Nasiri, J., Khiyabani, F.M.: A whale optimization algorithm (WOA) approach for clustering. Cogent Math. Stat. **5**(1), 1–13 (2018). https://doi.org/10.1080/25742558.2018.1483565
29. Huang, X., Wang, R., Zhao, X., Hu, K.: Aero-engine performance optimization based on whale optimization algorithm. In: 2017 36th Chinese Control Conference (CCC), pp. 11437–11441. IEEE Press (2017)
30. Jadhav, A.R., Shankar, T.: Whale optimization based energy-efficient cluster head selection algorithm for wireless sensor networks. arXiv preprint arXiv:1711.09389 (2017)

31. Buyya, R., Ranjan, R., Calheiros, R.N.: Modeling and simulation of scalable Cloud computing environments and the CloudSim toolkit: challenges and opportunities. In: 2009 International Conference on High Performance Computing & Simulation, pp. 1–11. IEEE Press (2009)
32. Abdulhamid, S.M., Abd Latiff, M.S., Abdul-Salaam, G., Madni, S.H.H.: Secure scientific applications scheduling technique for cloud computing environment using global league championship algorithm. PloS One **11**, e0158102 (2016)
33. Ye, X., Yin, Y., Lan, L.: Energy-efficient many-objective virtual machine placement optimization in a cloud computing environment. IEEE access. **5**, 16006–16020 (2017)

Data Sampling-Based Feature Selection Framework for Software Defect Prediction

Abdullateef O. Balogun[1(✉)], Fatimah B. Lafenwa-Balogun[1], Hammed A. Mojeed[1], Fatimah E. Usman-Hamza[1], Amos O. Bajeh[1], Victor E. Adeyemo[2], Kayode S. Adewole[1], and Rasheed G. Jimoh[1]

[1] Department of Computer Science, University of Ilorin, Ilorin 1515, Nigeria
{balogun.ao1,raji.fb,mojeed.ha,usman-hamza.fe,bajehamos, adewole.ks,jimoh_rasheed}@unilorin.edu.ng

[2] School of Built Environment, Engineering and Computing, Leeds Beckett University, Headingley Campus, Leeds LS6 3QS, UK
v.adeyemo5225@student.leedsbeckett.ac.uk

Abstract. High dimensionality and class imbalance are latent data quality problems that have a negative effect on the predictive capabilities of prediction models in software defect prediction (SDP). As a viable solution, data sampling and feature selection (FS) has been used to address the class imbalance and high dimensionality problem respectively. Most of the existing studies in SDP addressed these data quality problem individually which often affects the generalizability of such studies. Hence, this study proposed a novel framework based on correlation-based feature selection (CFS) and synthetic minority oversampling (SMOTE) methods for software defect prediction. CFS based on best-first search (BFS) method is used to handle feature selection while SMOTE sampling technique is used for the class imbalance. The proposed framework was developed with Bayesian Networks (BN) and Decision Tree (DT) classifiers on defects dataset from NASA repository. The experimental results showed that the proposed framework outperformed other experimental methods. The high accuracy (83.84%) and AUC (0.90) values of the proposed framework indicates its ability to differentiate between the defective and non-defective labels without bias. The proposed framework is therefore recommended for SDP as it handles both class imbalance and high dimensionality in SDP with high prediction performance.

Keywords: Software defect prediction · High dimensionality · Class imbalance · Feature selection · Data sampling

1 Introduction

Software Defect Prediction (SDP) is amongst the foremost beneficial practices of software testing in the software development life cycle (SDLC). SDP distinguishes the modules that are susceptible to defect and hence need further testing [1, 2]. This way, software testing tools and effort can be utilized efficiently without compromising the quality of the software. Besides, it is known that adequate software testing requires a

J. H. Abawajy et al. (Eds.): EATI 2020, LNNS 254, pp. 39–52, 2021.
https://doi.org/10.1007/978-3-030-80216-5_4

high amount of resources within the SDLC; hence an efficient way is required to address defects using the limited resources [1, 3, 4].

SDP is an efficient process to resolve defective software modules or components thereby ensuring high quality and reliable software systems. With this approach, the software modules which are predicted as defective are often tested thoroughly as compared to those modules which are predicted as non-defective [1, 2, 5, 6].

SDP models make use of data and details from software systems such as source code complexity, software development history, software cohesion and coupling to predict defect prone software modules or components. This information is numerically measured using software metrics to ascertain the degree of software quality and reliability [7, 8].

Machine learning methods (supervised and unsupervised) are broadly utilized for building SDP models using software metrics [7–10]. Using supervised machine learning methods, which are classification algorithms such as Naïve Bayes (NB), decision tree (DT), k-Nearest Neighbor (kNN), Bayesian Network (BN) etc., the model will be built by training the classification technique on the software metric data. The goal is to use the model with some level of accuracy and precision on new data [10, 11]. However, the prediction performance of the SDP models depends on the quality of the software metric dataset used for developing the models. That is, software data used for building SDP models affect the performance of SDP models. From existing studies, high dimensionality and class imbalance are some of primary data quality issues which impede the predictive performances of SDP models [12–16].

These software data are highly complex and skewed, which can be attributed to the class imbalance problem. An imbalanced data has an un-even depiction of its class labels with non-defective instances as the majority [15, 17]. Also, software metric data reflect the characteristics of software modules, however, the numbers of metrics generated are usually numerous. This is due to different types of software metric mechanisms used to determine the quality and reliability of a software system thereby generating a large number of metric values. Consequently, this leads to a high-dimensionality problem as many software metric values are produced. Besides, some of these features (metrics) may be more important to the class (defective or non-defective) than others [14, 18, 19].

The attention of researchers has been drawn to these issues and many studies have been carried out to address them [12, 14, 15, 17, 18, 20, 21]. However, most of the existing studies addressed the issue of class imbalance and high dimensionality separately. Developing a framework that will address both class imbalance and high dimensionality is imperative and conversely will be a significant contribution to empirical software engineering and body of knowledge.

This study, therefore, proposes a novel framework based on correlation-based feature selection (CFS) and synthetic minority oversampling (SMOTE) methods for software defect prediction. CFS based on best-first search (BFS) method is used to handle feature selection by selecting relevant and irredundant features while SMOTE sampling technique is used for balancing the datasets. BN and DT algorithms are implemented on the new pre-processed datasets to develop classifiers and their respective predictive performances are evaluated based on accuracy, Area Under the Curve (AUC) and f-measure.

In summary, the main contributions of this study are:

i. A novel framework based on CFS and SMOTE sampling methods for software defect prediction.
ii. Empirical validation of the importance of addressing both high dimensionality and class imbalance data quality issues in SDP.
iii. Empirically showed that CFS and SMOTE improved the prediction performances of classifiers in SDP.

The rest of the paper is outlined as follows. Section 2 presents a review of related works on class imbalance and high dimensionality in SDP. Section 3 describes the research approach employed in this study. Experimental results and analyses are discussed in Sect. 4. Section 5 concludes the study.

2 Related Work

Class imbalance and high dimensionality are part of the primary data quality problems in machine learning. Datasets used in the machine learning task (supervised or unsupervised) naturally exhibit these characteristics. Besides, class imbalance and high dimensionality both have an adverse effect on the prediction performance of prediction models. SDP is no exception to these data quality problems. Feature selection (FS) and data sampling methods are viable solutions to both high dimensionality and class imbalance problems and various methods have been proposed to mitigate these data quality problems.

Yu, Jiang and Zhang [15] empirically validated the prediction performance stability of prediction models using sampling methods in addressing class imbalance problem. An extension of their work by Balogun, Basri, Abdulkadir, Adeyemo, Imam and Bajeh [17] considered both undersampling (RUS) and oversampling (SMOTE) with varying imbalance ratio (IR). Their experimental results disclosed that SDP datasets suffer class imbalance problem and showed a negative impact on prediction models in SDP.

Jia [22] proposed a hybrid feature selection method (HFS) and a combination of various feature sorting technology. The values for every function were determined including chi-squared (CS), information gain (IG) and Pearson correlation, respectively. Their result showed that their strategy can produce optimal results using the same proportion of features. Aforementioned studies showed and empirically substantiate that class imbalance and high dimensionality affects the prediction performance of SDP models and their respective findings complement other existing studies [12, 14, 16, 17, 23, 24].

Hamdy and El-Laithy [25] applied SMOTE and an FS method to predict bug severity levels. SMOTE oversampling-technique was used for balancing the severity classes and the FS method to reduce the data set and selecting the foremost insightful features for classifier training. The experimental results showed that their approach outperforms cutting-studies in predicting minority severity classes which significantly improved the performance of minority classes. SMOTE decreased the efficiency of the majority class within the event of a highly imbalanced dataset like Eclipse.

Suryadi [26] also looked at reducing the effect of class inequality in prediction models by proposing a framework based on data sampling (Random Under Sampling (RUS) and

SMOTE) and particle swarm optimization (PSO) as FS method. Experimental findings show that the proposed model will boost the efficiency of the AUC (Area Under Curve) overall value of Naive Bayes. These studies showed that both high dimensionality and class imbalance can be addressed together. Perhaps other forms of FS and data sampling methods could be as effective or better than the proposed framework.

Consequently, this study proposes a novel framework for SDP based on SMOTE and CFS. SMOTE, a data oversampling method geared toward balancing the class labels (minority and majority). CFS based on the BFS method is used to screen out irrelevant and noisy features by calculating the similarity between features and classes and collective features. This study differs from the works of Hamdy and El-Laithy [25], Kuhn and Johnson [27], and Yohannese and Li [28] in terms of the classifiers, defect datasets, and evaluation metric used in this study.

3 Methodology

This section presents and discusses the SMOTE technique, CFS method, classification algorithms, software defect datasets, evaluation metrics and experimental framework used in this study.

3.1 Synthetic Minority Over-Sampling Technique (SMOTE)

SMOTE is a statistical method that generates synthetic instances for minority class labels without diminishing the size of majority labels. New instances are created in the neighbourhood of existing minority instances which are not same as the existing minority instances [25, 29, 30]. Consequently, this increases the population size of the minority class such that there exists no difference between the majority and minority instances.

Software defect data are known to suffer from significant class imbalance [12, 15, 17, 31]. As SDP intends to predict defective instances, the developed classification model must be able to significantly discriminate between defective and non-defective instances without bias. SMOTE algorithm is presented in Algorithm 1.

Algorithm 1. SMOTE Algorithm

Input: Minority set: M; Amount of Synthetic dataset: Q; integer K (number of nearest neighbours).

```
S = Ø
for all x ∈ to M do
    determine the k-neighbourhood of X {Creation of synthetic instances}
    for i=1: Q do
        randomly select X^r ∈ N_k(X)
        s^i = x + U_i(0,1) · (X^ri − X)
        S = S ∪ s^i)
    end for
end for
return S
```

3.2 Correlation Feature Selection (CFS) Method

Correlation feature selection (CFS) method assesses the correlation between a certain feature and a class label and at the same time, the inter-correlation amongst features of a dataset [7, 32]. CFS is used with a search method (heuristic or metaheuristic algorithm) to select important features from a dataset. The application and choice of search method depend on the number of features in a dataset. Search methods such as best-first search (BFS) can deal with high-dimensional datasets. In this study, CFS with BFS is used. We aimed to select the optimal subset of features while still maintaining high or better classification performance. Details on CFS methods can be gotten from [33–35].

3.3 Classification Algorithms

In this study, Decision Tree (DT) and Bayesian Network (BN) algorithms are used as the base-line prediction models (see Table 1). The ensuing prediction models will be analyzed for the efficacy and efficiency of the proposed framework. Besides, DT and BN classifiers have been widely applied in numerous SDP studies with high predictive performance and they have been reported to be stable with imbalance datasets [8, 15, 36, 37]. Table 1 shows DT and BN classifiers with their respective parameter setting as used in this study.

Table 1. Classification algorithms

Classifier	Parameter setting
Decision Tree (DT)	ConfidenceFactor = 0.25; MinObj = 2
Bayesian Network (BN)	SimpleEstimator = alpha (0.25); SearchMethod = hillclimbing; MaxNoParents = 1

3.4 Software Defect Datasets

The software defect datasets used in this study were selected from the NASA software repository [36]. These datasets have a varying number of features and modules which makes them suitable for our study. Besides they have been widely used by researchers for the same process [13–17]. Shepperd, Song, Sun and Mair [38] version of the NASA corpus was used to build the SDP models. A detailed summary of the datasets is presented in Table 2.

Table 2. Summary of selected software defect datasets

Datasets	# of features	# of modules	# of defective Modules	# of non-defective modules
KC1	22	1162	294	868
KC3	40	194	36	158
MC2	40	124	44	80
PC1	38	679	55	624
PC3	38	1053	130	923
PC4	38	1270	176	1094
PC5	39	1694	458	1236

3.5 Performance Evaluation Metrics

For evaluating the predictive performances of the respective SDP models, Accuracy, F-Measure and Area under the curve (AUC) are selected. These metrics are used to assess the predictive performance of SDP models based on the impact of the proposed framework. Also, accuracy, f-measure and AUC have been widely used in existing SDP studies [17, 23, 39].

i. Accuracy is the number or percentage of correctly classified instances to the total sum of instances.

$$\text{Accuracy} = \frac{\text{TP} + \text{TN}}{\text{TP} + \text{FP} + \text{FN} + \text{TN}} \tag{1}$$

ii. F-Measure is defined as the weighted harmonic mean of the test's precision and recall

$$\text{F} - \text{Measure} = 2 \times (\text{Precision} \times \text{Recall}/\text{Precision} + \text{Recall}) \tag{2}$$

iii. The Area under Curve (AUC) shows the trade-off between TP and FP. It provides an aggregate measure of performance across all possible classification thresholds.

Where $Precision = \frac{TP}{TP+FP}$, $Recall = \frac{TP}{TP+FN}$, TP = Correct Classification, FP = Incorrect Classification, TN = Correct Misclassification, and FN = Incorrect Misclassification.

3.6 Experimental Framework

Figure 1 presents the experimental framework of the proposed method in this study. To empirically assess the efficacy of the proposed method on SDP models, the experimental framework is used on 7 defect datasets (see Table 2) with DT and BN classifiers (see Table 1). Due to variability as reported in some existing studies [23, 40, 41], cross-validation (CV) is used for the experimental evaluation method [40]. Specifically, k-fold (where k = 10) CV is used in this study. The 10-fold CV has been used in existing studies to build SDP models with low bias and variance [27, 40].

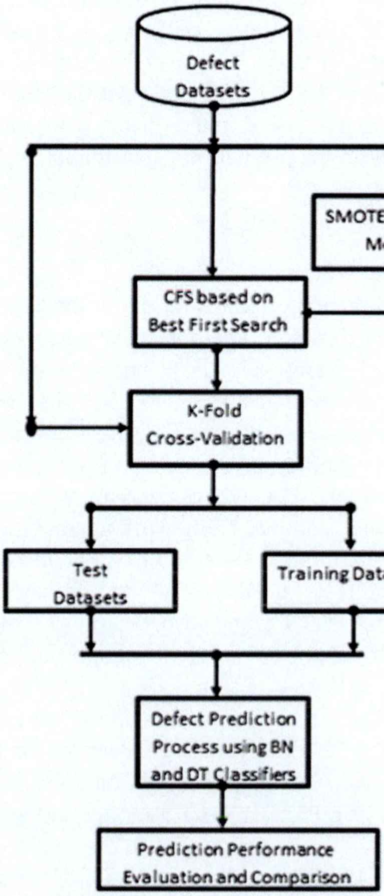

Fig. 1. Experimental framework

As depicted in Fig. 1, SMOTE sampling technique is used to balance the class labels in each dataset. In this study, equal representation percentage was used for both class labels (i.e. 50% defective label and 50% non-defective labels) as recommended by [15] and [17]. Besides, this will give more credibility to the resulting SDP models in terms of predicting any of the class labels (defective or non-defective). Thereafter, each newly generated (balanced) datasets is evaluated based on k-fold cross-validation which divides each dataset into training and testing. This procedure is to guide against the inherent error made in some existing studies by pre-processing the whole dataset instead of only the training dataset [36, 41]. In the study, 10-fold (i.e. k = 10) cross-validation is used for model evaluation. CFS based on BFS is applied to each training dataset. Consequently, each CFS generates a subset of features with high predictive capabilities. Our choice of SMOTE as sampling technique and CFS for feature selection is to due to their respective wide acceptance in the literature [14, 15, 17, 25, 39, 41].

Consequently, the reduced dataset is processed by BN and DT classifiers and the respective prediction performances of the ensuing SDP models are assessed using accuracy, f-measure and AUC. For fair comparative analysis, SDP models based on BN and DT with and without SMOTE and CFS (–i- BN, –ii- BN and CFS, -iii- BN and SMOTE (i.e BN+SM), -iv- DT, v- DT and CFS, -vi- DT and SMOTE (i.e. DT+SM)) were also developed and assessed accordingly. The essence of this is to empirically validate the effect of both SMOTE and CFS on the prediction performance of SDP models.

4 Results

In this section, the results from the experiments based on the experimental framework (see Fig. 1) were presented and discussed. The predictive performance of the proposed framework based on BN and DT were evaluated using accuracy, f-measure and AUC values. All experiments were performed using the WEKA machine learning tool [42] and Scott-KnottESD was used for statistical tests [23, 41, 43, 44].

Tables 3, 4 and 5 present the accuracy, AUC and F-measure results of each variant of the proposed framework respectively. This was done to allow fair comparative analysis of prediction performances of the proposed framework against the baseline classifiers. The developed prediction models include –i- BN and CFS, -ii- BN and SMOTE (i.e. BN+SM), -iii- BN, CFS and SMOTE (i.e. BN+CFS+SM), -iv- DT and CFS, -v- DT and SMOTE (i.e. DT+SM) and –vi- DT, CFS and SMOTE (i.e. DT+CFS+SM). Thus, six (6) variants of SDP models with emphasis on the novel approaches (i.e. BN+CFS+SM and DT+CFS+SM).

As presented in Tables 3, 4 and 5, BN classifier produced an average accuracy of 70.7%, average AUC value of 0.72 and average F-Measure value of 0.73 across all studied datasets. Besides, BN had its peak accuracy value (77.83%) on the KC3 and its lowest accuracy value (67%) on PC5. BN+CFS model had an average accuracy of 75.17%, average AUC value of 0.73 and average F-Measure value of 0.76 with its peak accuracy value (83.78%) on PC4 and lowest accuracy (68.53%) on PC5. On the other hand, DT classifier had an average accuracy of 79.54%, average AUC 0.64 and average F-Measure score of 0.78 across all studies dataset. DT recorded its peak accuracy value (91.5%) on PC1 dataset and lowest accuracy value (60.5%) on MC2. DT based on CFS (DT+CFS) model produced an average accuracy of 81.40%, average AUC value of 0.65 and average F-Measure value of 0.78 with its peak accuracy value (91.46%) on PC1 and lowest accuracy value (65.32%) on MC2. BN+CFS and DT+CFS had+6.32% and+2.33% increase in average accuracy value over BN and DT respectively. From the above analysis, it was observed that CFS does not only reduce features of the studied datasets but also improved the prediction performances of BN and DT which correlates with the findings of some existing studies [14, 23, 39, 41, 45].

Table 3. Accuracy results of experimented prediction models based on BN and DT

Dataset	BN (%)	BN+CFS (%)	BN+SM (%)	BN+CFS+SM (%)	DT (%)	DT+CFS (%)	DT+SM (%)	DT+CFS+SM (%)
KC1	68.33	69.53	72.45	**77.46**	74.18	75.47	79.7	**81.90**
KC3	77.83	81.96	75.87	**82.54**	79.4	81.44	82.2	**82.21**
MC2	70.16	67.74	70.44	**71.07**	60.5	65.32	64.2	**66.67**
PC1	71.13	80.11	89.66	**92.78**	**91.5**	91.46	91.26	90.63
PC3	67.61	74.55	88.29	**92.03**	84.71	87.46	**88**	87.64
PC4	72.83	83.78	86.1	**91.04**	86.93	87.24	**91.31**	90.76
PC5	67	68.53	77.21	**79.93**	74	74.79	**79.3**	77.98
Average	70.70	75.17	80.00	**83.84**	79.54	81.40	82.78	**83.30**

Table 4. AUC results of experimented prediction models based on BN and DT

Dataset	BN	BN+CFS	BN+SM	BN+CFS+SM	DT	DT+CFS	DT+SM	DT+CFS+SM
KC1	0.681	0.673	0.8	**0.845**	0.604	0.634	0.807	**0.825**
KC3	0.584	0.59	0.849	**0.879**	0.653	0.585	0.858	**0.873**
MC2	0.614	0.588	0.729	**0.776**	0.589	0.569	**0.657**	0.645
PC1	0.811	0.846	0.969	**0.984**	0.598	0.631	**0.927**	0.916
PC3	0.779	0.807	0.966	**0.973**	0.591	0.612	0.902	**0.909**
PC4	0.81	0.869	0.954	**0.977**	0.789	0.829	0.915	**0.928**
PC5	0.741	0.751	0.864	**0.89**	0.673	0.695	0.816	**0.819**
Average	0.72	0.73	0.88	**0.90**	0.64	0.65	0.84	**0.85**

Table 5. F-measure values of experimented prediction models based on BN and DT

Dataset	BN	BN+CFS	BN+SM	BN+CFS+SM	DT	DT+CFS	DT+SM	DT+CFS+SM
KC1	0.698	0.701	0.724	**0.774**	0.717	0.713	0.797	**0.819**
KC3	0.775	0.803	0.756	**0.825**	0.783	0.796	**0.821**	**0.821**
MC2	0.693	0.656	0.7	**0.709**	0.608	0.632	0.641	**0.665**
PC1	0.776	0.838	0.896	**0.928**	0.901	0.9	**0.913**	0.906
PC3	0.731	0.784	0.883	**0.92**	0.839	0.82	**0.88**	0.876
PC4	0.767	0.846	0.861	**0.91**	0.869	0.847	**0.913**	0.908
PC5	0.687	0.7	0.772	**0.8**	0.737	0.739	**0.793**	0.78
Average	0.73	0.76	0.80	**0.84**	0.78	0.78	0.82	**0.83**

Furthermore, it was also observed that BN and DT based on SMOTE models (BN+SM (80%, 0.88, 0.8) and DT+SM (82.78%, 0.84, 0.82)) were both superior to base classifier BN (70.7%, 0.72, 0.73) and DT (79.54%, 0.64, 0.78) respectively in terms of the average accuracy, average AUC and average F-measure values. This finding indicates that the removal of class imbalance in datasets can improve the predictive performance of SDP models [12, 15, 17, 25].

Prediction performance of BN and DT models based on the proposed framework (i.e. BN+CFS+SM and DT+CFS+SM) outperformed other experimented methods (BN, BN+CFS, BN+SM, DT, DT+CFS, DT+SM). BN+CFS+SM and DT+CFS+SM had an average accuracy, average AUV and average F-measure values of (83.84%, 0.9, 0.84) and (82.78%, 0.85, 0.83) respectively. From the experimental results, we observed +18.58% and +4.07% increase in the average accuracy value of BN+CFS+SM and DT+CFS+SM over BN and DT. On average AUC and F-measure values, BN+CFS+SM and DT+CFS+SM had+25%,+32.51% and 15.6%, 6.41% increase over the BN and DT. These analyses indicate that the proposed framework (CFS and SMOTE) had a positive effect on the prediction performances of BN and DT.

More importantly, we also observed +11.53% and +4.80% increase in the average accuracy value of BN+CFS+SM over BN+CFS and BN+SM respectively, +2.33% and + 0.63% increase in average accuracy value of DT+CFS+SM over DT+CFS and DT+SM. On average AUC, we recorded +23.29% and +2.73% increase for BN+CFS+SM over BN+CFS and BN+SM respectively, +30.77% and 1.20% increase for DT+CFS+SM over DT+CFS and DT+SM respectively. And lastly on average F-measure, percentage increase of +10.53% and +5.00% were recorded for BN+CFS+SM over BN+CFS and BN+SM respectively, +6.40% and +1.22% were recorded for DT+CFS+SM over DT+CFS and DT+SM respectively. These recorded percentage increase in accuracy, AUC and F-measure for the proposed approaches validates the superiority in the prediction performance of SDP models by applying both FS and sampling techniques rather than applying both separately. Noticeably, the higher percentage increase recorded by our approach over BN+CFS and DT+CFS than over BN+SM and DT+SM is an indication that SM contributes more to the increase in performance than CFS. Thus, our findings revealed that class imbalance impedes the performance of SDP models far more than high dimensionality does.

The high AUC values of BN+CFS+SM and DT+CFS+SM indicates the ability of the methods to differentiate between the class labels (defective or non-defective) without bias to a particular class label.

For the statistical test, Scott-KnottESD statistical rank test which is a mean comparison approach that uses hierarchical clustering to separate mean values into statistically distinct clusters with non-negligible mean differences was conducted [43, 44]. Figure 2 presents the Scott-KnottESD Rank test of experimented prediction models based on accuracy, AUC and F-measure. Based on average accuracy and average F-measure values, BN+CFS+SM and DT+CFS+SM ranked superior to other methods. While on average AUC, BN+CFS+SM ranked superior to amongst all other methods. These experimental results and findings empirically validate the positive effect of SMOTE and CFS on BN and DT.

Consequently, it is, therefore, recommended to address high dimensionality and class imbalance in SDP by deploying applicable data sampling and feature selection methods.

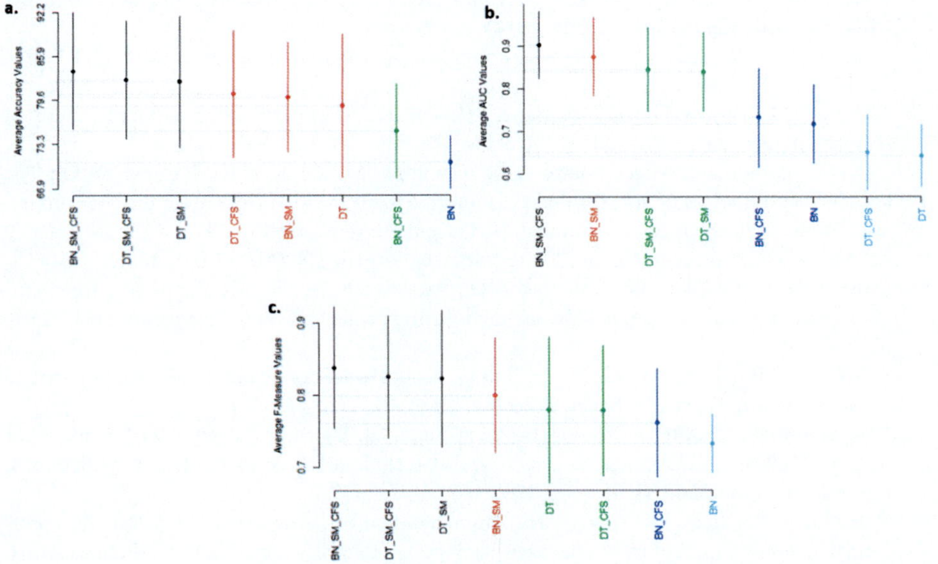

Fig. 2. Scott-KnottESD statistical rank test of experimented prediction models on the defect datasets (a) based on average accuracy values (b) based on average AUC values (c) based on average F-Measure values

5 Conclusions

This study proposes a novel SDP framework based on SMOTE sampling and CFS methods. The proposed framework was used with BN and DT classifiers on 7 defect datasets from NASA repository. The main essence of this study is to address high dimensionality and class imbalance which are primary data quality problems in SDP. The experimental results showed a reduction of features, that is, using FS (in this case CFS) in SDP identifies and selects the important features and produces an SDP model with better prediction performance. Furthermore, using an over-sampling method (i.e. SMOTE) on software defect datasets give a better representation of the dataset by addressing the latent class imbalance and subsequently develops a more generalizable prediction model with high prediction performance. The proposed framework was able to improve the prediction performance of BN and DT by using SMOTE and CFS based on BFS to address the latent class imbalance and high dimensionality problems in SDP. Results indicate better predictive performance when both techniques are applied together rather than individually. We, therefore, recommend that class imbalance and high dimensionality should be addressed when developing or deploying SDP models.

For future works, we intend to explore and investigate the occurrence and effect of other possible data quality problems on the prediction performance of SDP models. Besides, other forms of data sampling and FS methods are worth investigating.

References

1. Kamei, Y., Shihab, E.: Defect prediction: accomplishments and future challenges. In: IEEE 23rd International Conference on Software Analysis, Evolution, and Reengineering (SANER), vol. 5, pp. 33–45. IEEE (2016)
2. Li, Z., Jing, X.-Y., Zhu, X.: Progress on approaches to software defect prediction. IET Softw. **12**, 161–175 (2018)
3. Mahmood, Z., Bowes, D., Hall, T., Lane, P.C., Petrić, J.: Reproducibility and replicability of software defect prediction studies. Inf. Softw. Technol. **99**, 148–163 (2018)
4. Basri, S., Almomani, M.A., Imam, A.A., Thangiah, M., Gilal, A.R., Balogun, A.O.: The organisational factors of software process improvement in small software industry: comparative study. In: Saeed, F., Mohammed, F., Gazem, N. (eds.) IRICT 2019. AISC, vol. 1073, pp. 1132–1143. Springer, Cham (2020). https://doi.org/10.1007/978-3-030-33582-3_106
5. Mojeed, H.A., Bajeh, A.O., Balogun, A.O., Adeleke, H.O.: Memetic approach for multi-objective overtime planning in software engineering projects. J. Eng. Sci. Technol. **14**, 3213–3233 (2019)
6. Balogun, A., Bajeh, A., Mojeed, H., Akintola, A.: Software defect prediction: a multi-criteria decision-making approach. Niger. J. Technol. Res. **15**, 35–42 (2020)
7. Usman-Hamza, F., Atte, A., Balogun, A., Mojeed, H., Bajeh, A., Adeyemo, V.: Impact of feature selection on classification via clustering techniques in software defect prediction. J. Comput. Sci. Appl. **26** (2019)
8. Balogun, A., Oladele, R., Mojeed, H., Amin-Balogun, B., Adeyemo, V.E., Aro, T.O.: Performance analysis of selected clustering techniques for software defects prediction. Afr. J. Comput. ICT **12**, 30–42 (2019)
9. Li, J., He, P., Zhu, J., Lyu, M.R.: Software defect prediction via convolutional neural network. In: 2017 IEEE International Conference on Software Quality, Reliability and Security (QRS), pp. 318–328. IEEE (2017)
10. Bashir, K., Li, T., Yohannese, C.W., Mahama, Y.: Enhancing software defect prediction using supervised-learning based framework. In: 2017 12th International Conference on Intelligent Systems and Knowledge Engineering (ISKE), pp. 1–6. IEEE (2017)
11. Mabayoje, M.A., Balogun, A.O., Jibril, H.A., Atoyebi, J.O., Mojeed, H.A., Adeyemo, V.E.: Parameter tuning in KNN for software defect prediction: an empirical analysis. Jurnal Teknologi dan Sistem Komputer **7**, 121–126 (2019)
12. Chen, L., Fang, B., Shang, Z., Tang, Y.: Tackling class overlap and imbalance problems in software defect prediction. Softw. Qual. J. **26**(1), 97–125 (2016). https://doi.org/10.1007/s11219-016-9342-6
13. Tong, H., Liu, B., Wang, S.: Software defect prediction using stacked denoising autoencoders and two-stage ensemble learning. Inf. Softw. Technol. **96**, 94–111 (2018)
14. Balogun, A.O., Basri, S., Abdulkadir, S.J., Hashim, A.S.: Performance analysis of feature selection methods in software defect prediction: a search method approach. Appl. Sci. **9**, 2764 (2019)
15. Yu, Q., Jiang, S., Zhang, Y.: The performance stability of defect prediction models with class imbalance: an empirical study. IEICE Trans. Inf. Syst. **100**, 265–272 (2017)
16. Iqbal, A., Aftab, S.: A classification framework for software defect prediction using multi-filter feature selection technique and MLP. Int. J. Mod. Educ. Comput. Sci. **12** (2020)
17. Balogun, A.O., Basri, S., Abdulkadir, S.J., Adeyemo, V.E., Imam, A.A., Bajeh, A.O.: Software defect prediction: analysis of class imbalance and performance stability. J. Eng. Sci. Technol. **14**, 3294–3308 (2019)
18. Oluwagbemiga, B.A., Shuib, B., Abdulkadir, S.J., Sobri, A.: A hybrid multi-filter wrapper feature selection method for software defect predictors. Int. J Sup. Chain. Mgt **8**, 9–16 (2019)

19. Bajeh, A.O., Oluwatosin, O.-J., Basri, S., Akintola, A.G., Balogun, A.O.: Object-oriented measures as testability indicators: an empirical study. J. Eng. Sci. Technol. **15**, 1092–1108 (2020)
20. Yang, X., Lo, D., Xia, X., Sun, J.: TLEL: a two-layer ensemble learning approach for just-in-time defect prediction. Inf. Softw. Technol. **87**, 206–220 (2017)
21. Akintola, A.G., Balogun, A.O., Lafenwa, F., Mojeed, H.A.: Comparative analysis of selected heterogeneous classifiers for software defects prediction using filter-based feature selection methods. FUOYE J. Eng. Technol. **3**, 134–137 (2018)
22. Jia, L.: A hybrid feature selection method for software defect prediction. In: IOP Conference Series: Materials Science and Engineering, vol. 394, p. 032035. IOP Publishing (2018)
23. Ghotra, B., McIntosh, S., Hassan, A.E.: A large-scale study of the impact of feature selection techniques on defect classification models. In: 2017 IEEE/ACM 14th International Conference on Mining Software Repositories (MSR), pp. 146–157. IEEE (2017)
24. Kondo, M., Bezemer, C.-P., Kamei, Y., Hassan, A.E., Mizuno, O.: The impact of feature reduction techniques on defect prediction models. Empirical Softw. Eng. **24**(4), 1925–1963 (2019). https://doi.org/10.1007/s10664-018-9679-5
25. Hamdy, A., El-, A.: SMOTE and feature selection for more effective bug severity prediction. Int. J. Software Eng. Knowl. Eng. **29**, 897–919 (2019)
26. Suryadi, A.: Integration of feature selection with data level approach for software defect prediction. SinkrOn **4**, 51–57 (2019)
27. Kuhn, M., Johnson, K.: Applied Predictive Modeling. Springer, New York (2013). https://doi.org/10.1007/978-1-4614-6849-3
28. Yohannese, C.W., Li, T.: A combined-learning based framework for improved software fault prediction. Int. J. Comput. Intell. Syst. **10**, 647–662 (2017)
29. Kong, J., Rios, T., Kowalczyk, W., Menzel, S., Bäck, T.: On the performance of oversampling techniques for class imbalance problems. In: Lauw, H.W., Wong, R.-W., Ntoulas, A., Lim, E.-P., Ng, S.-K., Pan, S.J. (eds.) PAKDD 2020. LNCS (LNAI), vol. 12085, pp. 84–96. Springer, Cham (2020). https://doi.org/10.1007/978-3-030-47436-2_7
30. Gonzalez-, D., et al.: Synthetic minority oversampling technique for optimizing classification tasks in botnet and intrusion-detection-system datasets. Appl. Sci. **10**, 794 (2020)
31. Wang, S., Yao, X.: Using class imbalance learning for software defect prediction. IEEE Trans. Reliab. **62**, 434–443 (2013)
32. Mabayoje, M.A., Balogun, A.O., Bajeh, A.O., Musa, B.A.: Software defect prediction: effect of feature selection and ensemble methods. FUW Trends Sci. Technol. J. **3**, 518–522 (2018)
33. Sumaiya, I., Lavanya, K.: Credit card fraud detection using correlation-based feature extraction and ensemble of learners. In: Singh , G., Chaudhari, N.S., Barbosa, J.L.V., Aghwariya, M.K. (eds.) International Conference on Intelligent Computing and Smart Communication 2019. AIS, pp. 7–18. Springer, Singapore (2020). https://doi.org/10.1007/978-981-15-0633-8_2
34. Sharma, S., Jain, A.: An empirical evaluation of correlation based feature selection for tweet sentiment classification. In: Gunjan, V.K., Senatore, S., Kumar, A., Gao, X.-Z., Merugu, S. (eds.) Advances in Cybernetics, Cognition, and Machine Learning for Communication Technologies. LNEE, vol. 643, pp. 199–208. Springer, Singapore (2020). https://doi.org/10.1007/978-981-15-3125-5_22
35. Tripathi, D., Manoj, I., Raja Prasanth, G., Neeraja, K., Varma, M.K., Ramachandra Reddy, B.: Survey on classification and feature selection approaches for disease diagnosis. In: Venkata Krishna, P., Obaidat, M.S. (eds.) Emerging Research in Data Engineering Systems and Computer Communications. AISC, vol. 1054, pp. 567–576. Springer, Singapore (2020). https://doi.org/10.1007/978-981-15-0135-7_52
36. Menzies, T., Greenwald, J., Frank, A.: Data mining static code attributes to learn defect predictors. IEEE Trans. Software Eng. **33**, 2–13 (2007)

37. Lessmann, S., Baesens, B., Mues, C., Pietsch, S.: Benchmarking classification models for software defect prediction: a proposed framework and novel findings. IEEE Trans. Software Eng. **34**, 485–496 (2008)
38. Shepperd, M., Song, Q., Sun, Z., Mair, C.: Data quality: some comments on the nasa software defect datasets. IEEE Trans. Software Eng. **39**, 1208–1215 (2013)
39. Rathore, S.S., Gupta, A.: A comparative study of feature-ranking and feature-subset selection techniques for improved fault prediction. In: Proceedings of the 7th India Software Engineering Conference, p. 7. ACM (2014)
40. James, G., Witten, D., Hastie, T., Tibshirani, R.: An introduction to statistical learning. Springer, New York (2013). https://doi.org/10.1007/978-1-4614-7138-7
41. Xu, Z., Liu, J., Yang, Z., An, G., Jia, X.: The impact of feature selection on defect prediction performance: an empirical comparison. In: 2016 IEEE 27th International Symposium on Software Reliability Engineering (ISSRE), pp. 309–320. IEEE (2016)
42. Hall, M., Frank, E., Holmes, G., Pfahringer, B., Reutemann, P., Witten, I.H.: The WEKA data mining software: an update. ACM Sig. Exp. **11**, 10–18 (2009)
43. Tantithamthavorn, C., McIntosh, S., Hassan, A.E., Matsumoto, K.: Comments on "researcher bias: the use of machine learning in software defect prediction." IEEE Trans. Software Eng. **42**, 1092–1094 (2016)
44. Tantithamthavorn, C., McIntosh, S., Hassan, A.E., Matsumoto, K.: The impact of automated parameter optimization on defect prediction models. IEEE Trans S.E. 45, 683–711 (2018)
45. Al-Tashi, Q., Abdulkadir, S.J., Rais, H.M., Mirjalili, S., Alhussian, H.: Binary optimization using hybrid grey wolf optimization for feature selection. IEEE Access **7**, 39496–39508 (2019)

A Reliable Hybrid Software Development Model: CRUP (Crystal Clear & RUP)

Taghi Javdani Gandomani[1](✉), Mohammadreza Mollahoseini Ardakani[2], and Maryam Shahzeydi[2]

[1] Department of Computer Science, Shahrekord University, Shahrekord, Iran
javdani@sku.ac.ir
[2] Department of Computer Engineering, Maybod Branch, Islamic Azad University, Maybod, Iran

Abstract. The turning point of software development dates back to 2001 when some specialists published the Agile Manifesto. After that, many companies are moving to Agile methods. However, it seems that in many cases, software teams and companies prefer to use a combination of Agile and plan-driven methods together to take advantage of both discipline and agility simultaneously. In this viewpoint, RUP as a traditional method combined with one or two Agile methods to solve the inefficiency of software development methods against new and changing customer requirements. However, combining such methods may lead to organizational overheads and challenges as well. In this study, having reviewed the most recent attempts to reach hybrid models, a new hybrid model will be presented based on RUP and Crystal Clear. Then, the new model employed in a Case Study. The results showed that this hybrid model has considerable capabilities in small and medium projects compared to its parents' methods.

Keywords: Agile software development · Agile methods · Hybrid software methods · RUP · Crystal family · Crystal clear

1 Introduction

Software methodologies have been focused since the software industry tried to fit the gap of lack of a well-defined development process. Software methodologies gained more attention, mainly after the software industry felt to improve the development process by following proper disciplines (Awad 2005). In the software industry, Plan-Driven methods routed back to 1970, have been used for several years. These methods, which today are known as traditional methods, were following the Waterfall model for software development (Pressman 2009).

These methods were widely used until 2000. However, the advancements in technology expected new and changing needs in a short period whereby these methods were not able to provide these abilities (Turner and Boehm 2004). Therefore, some software methodologists created Agile manifesto focusing on user involvement and people collaboration in software development. The manifesto led to the formal introduction of several

J. H. Abawajy et al. (Eds.): EATI 2020, LNNS 254, pp. 53–69, 2021.
https://doi.org/10.1007/978-3-030-80216-5_5

lightweight methods, currently known as Agile methods. (Gandomani and Nafchi 2016; Gandomani et al. 2013; Jovanović et al. 2020; Neto et al. 2019). However, soon it was found that Agile methods cannot easily replace traditional methodologies. Indeed, Agile methods were not the best to be used Agile methodologies in all organizations and for all kinds of projects (Boehm 2002). Also, sometimes, adopting Agile methods is not easy as expected in the practice (Gandomani and Nafchi 2016; Gandomani et al. 2014).

To have more flexibility and better efficiency of available resources in software organizations and teams, as well as improvement of software development, researchers focused on a new concept "hybrid methodologies." In this strategy, the traditional and Agile methods are used simultaneously in such a way that the strengths of each method are involved in the target combination, and attempts are made to reduce weaknesses (Ahmad et al. 2014; Bashir and Qureshi 2012; Castilla 2014; Cho 2009; Gill et al. 2016; Rasool et al. 2013; Smoczyńska et al. 2018).

RUP[1] is the most disciplined methodology among the Plan-Driven methodologies (Ahmad et al. 2014). Therefore, this method has been used as a representative of traditional methods and combined with Agile methods in several studies. On the other hand, Scrum and XP[2] methods have been used mostly in hybrid methods. However, most of the hybrid methods have usually been defined and explained in theory only and rarely have been employed in real environments. Crystal Family methods are the most practical Agile methodologies to consider project criticality and team size (Abrahamsson et al. 2002; Alsitar Cockburn 2002; Shaw et al. 2019), none of which has been used in the previous studies. So, it seems that RUP and Crystal can produce a hybrid software development model by appropriate synergy.

In small and medium software projects that need considerable discipline in the development process, employing RUP is a good choice. However, employing Agile methods also provides software teams with notable advantages. Crystal methods also offer great values for the small and medium (non-critical) software project. This study aimed to provide a software process model by combining these methods. So, a combination of RUP and one of the Crystal methodologies, Crystal Clear, called CRUP, has been proposed and verified in this study to be used in small up to medium companies. The capabilities of this hybrid method are also proved in this study by using the hybrid model in a Case Study and applying it in real projects.

The rest of this paper is organized as follows: a short introduction to RUP and Crystal Clear methods been provided in Sect. 2, followed by addressing the main related work in Sect. 3. Section 4 describes the adopted research design. Then, the proposed hybrid method is explained in Sect. 5, followed by its application in a Case Study to evaluate its performance in Sect. 6. Finally, the last sections provide the limitations and conclusions relatively.

2 Background

In this section, a brief introduction to RUP and Crystal methods are presented, and the famous hybrid methodologies proposed in the previous studies are introduced, as well.

[1] Rational Unified Process.

[2] Extreme Programming.

2.1 RUP

RUP can be summarized in several sentences; RUP presents an incremental-iterative approach on an architectural framework for software development. This approach is suitable for large-scale projects because of its excessive focus on analysis and design processes (Ahmad et al. 2014). Each development "Process" is well defined in RUP. It means that the components of a process, that is, who, what, how, and when are well determined by the roles, activities, findings, and workflows in a two-dimensional structure formed by phases and disciplines, as shown in Fig. 1 (Ambler 2005; Cho 2009). Some of the advantages of this method include full documentation at all development stages, applicable in a wide range of projects, highly disciplined, and high capability in reusing the codes (Ambler et al. 2005; Shaw et al. 2019).

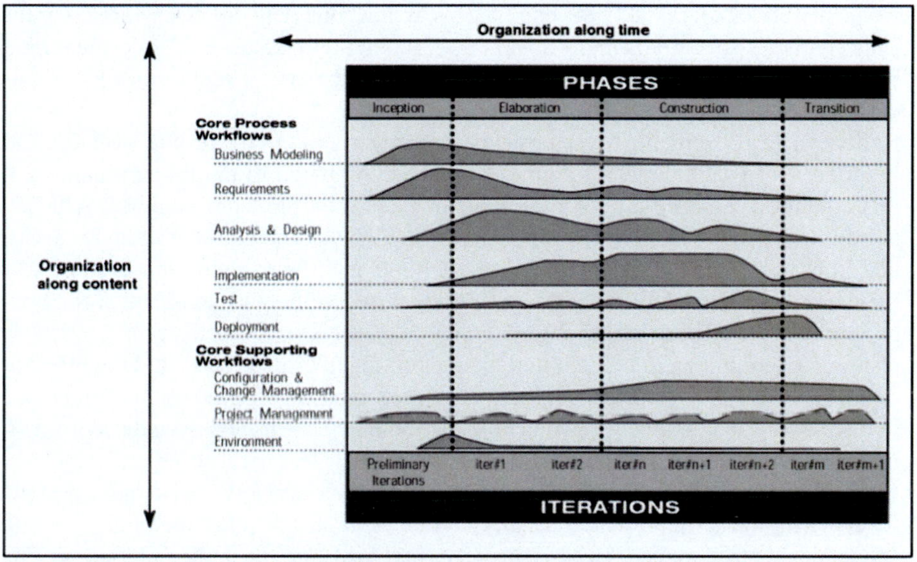

Fig. 1. RUP process model (Ambler 2005)

RUP disadvantages include high complexities, which lead to difficulties in understanding and learning the processes and adequate and accurate implementation of the processes, except for specialists and skilled people. The other disadvantage of RUP is its weak sociological aspect. Also, RUP has problems regarding non-organizational development, and despite separate disciplines defined in this method, there may be irregularities in the team (Ambler et al. 2005).

2.2 Crystal Family Methodologies

Alistair Cockburn was the innovator of Crustal Family methodologies. He believes that one method is not applicable to all software projects in different scales (Boehm 2007; Cockburn 2002; Cohen et al. 2004). Thus, he defined a set of Agile methods, not a unique method. These methods are lightweight, human-powered, and flexible in their core. The importance of the criticality and size of software projects are the main metrics for consideration of the required methodology. According to the size of the teams involved in the project and the level of the project's criticality, a spectrum of clear, yellow, orange, red, and darker colors are considered as Crystal methodologies.

Since this study aimed at creating a hybrid model of Crystal Clear and RUP, the major characteristics of this method provides in the following.

Crystal Clear

The below provides the main characteristics of Crystal Clear, as the most Agile method in the Crystal Family.

- Frequent delivery: The most important characteristic of Crystal Clear is its focus on frequent delivery. Crystal Clear emphasizes at least one delivery every three months. This characteristic entails many advantages, including the ability to get critical feedback quickly, end-user collaboration, focusing on the whole team, not individuals, etc.
- Reflective Improvement. Most projects lead to disappointing results over the first iteration and have delays in iterative deliveries compared to the initial planning. In this condition, holding a reflection workshop between the team members will be a helpful technique. In this workshop, that is usually held after each iteration, for a short period of time, things that are not correctly done, and the methods and strategies for them are discussed deeply. Therefore, by reflecting on the problems, the project will be improved efficiently and timely.
- Osmotic Communication: Team members work together in a room, and the information flow or questions and answers are active are streaming like the background music in the place. Each member uses this information flow unconsciously and according to his/her needs. The level of disorders and disturbance is less in this flow, and it includes more advantages. Osmotic communication and iterative delivery lead to quick and rich feedback in the project, which usually can be seen less in other methods. Personal safety, concentration, and accessibility to expert users are other characteristics of the Crystal Clear Method.

Apart from the characteristics mentioned above, Crystal Clear includes various strategies for software development, including full investigation, information reflection, walking skeleton, early win, and incremental re-architecture. Also, like other Agile methods, this methodology includes techniques to improve software development, including methodology formation, project estimation, miniature process, using a burndown chart, daily stand up meetings, etc.

2.3 RUP Versus Crystal Clear

Although RUP has recommended its advocators to change the methodology set up when needed, it has presented no theories in this regard; while this transformation is presented in a theoretical format and a set of principles in the Crystal Clear method.

The pace of methodology transformation is very higher and in a few days in Crystal Clear. But, this takes several months in RUP, and consequently, many companies accept RUP in its full mode instead of applying methodology change. This will, in turn, double inefficiency because a heavy and primarily full methodology is applied to the project instead of a light methodology, which is often appropriate for most software projects.

The procedure in RUP is general-to-specific. That is, to achieve the desired form of the methodology, the whole methodology should be taken into consideration, and unnecessary parts should be eliminated to reach the desired outcome. This procedure is vice versa in Crystal Clear. That is addressed as "stretch to fit" in the Crystal family, in which each project starts with small activities and processes, and more items are added if necessary to find the best choice of methodology in practice (Cockburn 2004).

3 Hybrid Methods

Several attempts have been paid on proposing hybrid software methodologies to take advantage of their potential benefits. The followings are the most notable ones.

3.1 XSR Model

RUP, XP, and Scrum methods have been used in this hybrid model, which was proposed by Ahmad et al. (2014). Indeed, XSR is a model rather than a methodology. This model provides various important characteristics that could make it a significant model of Scrum, XP, and RUP together in practice.

This model has tried to involve the main capabilities (Ahmad et al. 2014). Also, it claims that it solved the incomplete software development cycle, as a common challenge in almost all Agile methods. For example, Scrum is focused on managerial methods and lacks the architectural aspect, which has been overcome in XSR by combining it with RUP (Ahmad et al. 2014). Moreover, the coincidence of the beginning of the important development activities (e.g., test) with the initial phases like the start and establishment of the roles has been expressed clearly (Ahmad et al. 2014).

XSR *believes* that using comprehensive and straightforward names in methodologies instead of vague and strange names like "sprint" is better and will lead to a better understanding (Ahmad et al. 2014). Without elimination or substitution of any of the Agile practices, this model attempts to simplify and expand them. In other words, instead of saying that "RUP is used in this part of the project, " "now the performance will be based on XP method," or "Scrum method recommends this action in this project, " it simply states that "according to XSR, this should be done." In XSR, the focus is on iterations that, despite being simple and step by step, include Agile principles. The XSR life cycle includes the following important characteristics (Ahmad et al. 2014):

- Product delivery guarantee: XSR's life cycle follows Scrum patterns and explicitly shows all the steps from beginning to the end when the product is delivered to the customers (Ahmad et al. 2014).
- Explicit phases: Recognition, construction, and transition phases that include C3[3] agility principles.

[3] Coordinate Collaborate Conclude.

- Accurate and clear milestones: This model has different types of milestones that play an important role in project success and finally lead in the elimination of risks of releasing software products (Ahmad et al. 2014). As was mentioned earlier, this model is very general and wide, and using it requires developers to be fully proficient in all three basic methodologies.

3.2 eXRUP Model

This hybrid model was proposed in 2013 by Rasool et al. (2013). The main objective of proposing this model was achieving optimal software development in small and medium projects. RUP and XP methods are involved in this model by focusing on the advantages of these methods and overcoming their shortcomings and limitations (Rasool et al. 2013). The important feature of this method is that despite the combination of RUP and XP, it is still lightweight and is easily understandable by software developers and stakeholders (Rasool et al. 2013). This model includes the following five phases:

- Initialization Phase: Including the two activities of requirements gathering and project planning. In the former activity, there is a close and strong collaboration between developers and customers. In the later, operational and non-operational requirements are met.
- Evolution Phase: This is the first iterative phase in eXRUP that begins with risk analysis and management, and followed by design-related activities.
- Production Phase: As an iterative phase, this phase tries to ensure the development of test cases, modules, and subsystems according to the defined user stores as well as validation of them.
- Maintenance Phase: This is the final phase in the eXRUP iterations cycle, which focuses on managing the delivered systems and take care of it due to maintaining it in the running mode.
- Release Phase: This is the last phase in the development process of this model and includes three activities, namely product deployment, training users, and alpha system testing.

3.3 Combining RUP and SCRUM

In an academic study in 2014, Dalia Castilla (Castilla 2014) introduced this combination. She believes that these two methods are appropriate to be combined in many aspects, and they are complementary. She studied a Case Study for the Lobbyist Registration and Tracking System in Jacksonville by the use of this hybrid model (Castilla 2014).

In 2009, Cho (2009) suggested a hybrid model of Scrum and RUP resolve the shortcomings of Agile methods in big organizations. He assigned the same four phases of RUP in his model. However, he reduced the nine disciplines in the RUP to seven disciplines in his model to facilitate its applicability. The focus in the first phase was on building the business model. The focus in the "Elaboration" phase was on analysis and design. In the "Construction" phase, the focus was on implementation and testing the system, and in the "Transition" phase, the focus was on configuration and deployment. The researcher has attempted to observe Scrum principles in each phase. Therefore, he has put these method's activities as the center of his model. For example, construction of the business model in the first phase can be done based on the Scrum framework with several sprints, or the seven disciplines of the model can be stated in the form of daily Scrum meetings (Cho 2009).

3.4 Combining RUP, Scrum, and XP

In a study in 2012, Bashir and Qureshi (2012) combined the three methods of RUP, XP, and Scrum. They formed a development team including six developers, for a Case Study and trained them with XP practices such as Pair programming, simple designing, customer involvement, etc. and introduced RUP principles and disciplined to them. This method consists of three major phases and six logical activities.

4 Research Methodology

This study carried out as a Case Study as a subset of qualitative studies. Different steps in a Case Study and a qualitative study, in general, include research design, review of the related literature, data collection, results, and conclusion (Yin 2013, 2015). Details of the research steps are presented in the following.

4.1 Case Study

To achieve the objectives of this study, a Case Study was selected based on the required conditions. The Case Study company consisted of the required software experts. Most of the experts had BSc in software engineering, and some others had MSc. in software engineering or information technology. They had enough experience in using the RUP method and Agile methods. Also, the authors held a workshop to introduce the Crystal Clear process. It should be noted that the team members showed their passion for participating in novel research work.

4.2 Data Collection

According to Yin (2013), it should be assured that the collected data in the data collection procedure are valid and reliable enough to be used in the future and scientific applications. To this end, the author's interviewed the development team members who were well aware of the significance of the study. Also, receiving feedback from end-users and customers, and direct observation was adopted when necessary.

5 The Proposed Hybrid Methodology: CRUP

Different parts of the proposed hybrid model will be presented in this section. Adapted from RUP phases and then merging and summarizing, the main framework of the model is comprised of three phases. Also, RUP disciplines are aggregated and summarized with Crystal Clear strategies to be used in small projects. However, the following Agile

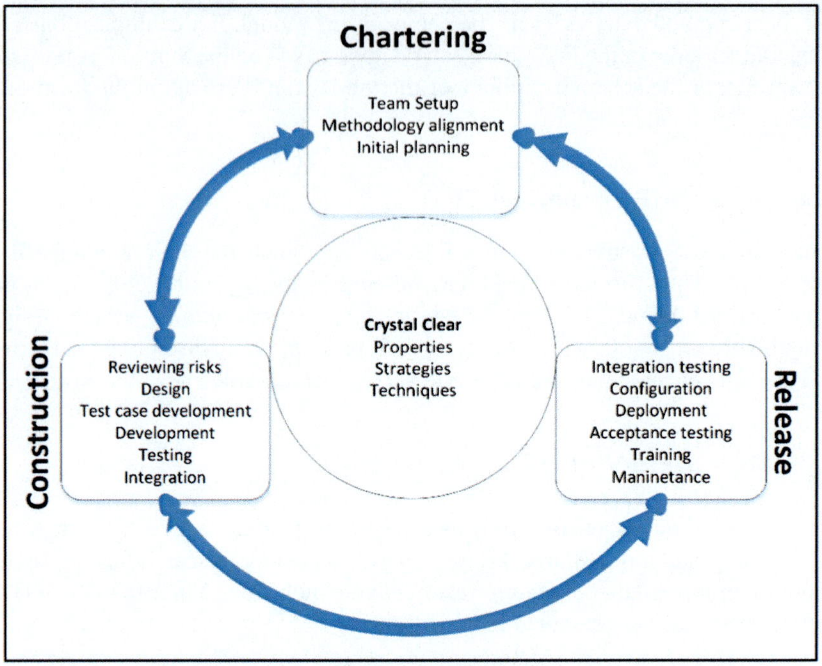

Fig. 2. The proposed model

principles of the Crystal method are ruling over all the development processes from beginning to end. Therefore, it will be hypothetically in the center of the model. Figure 2 shows the proposed model.

The proposed hybrid method is explained in detail in this section. For the benefit of the readers, the details are explained phase by phase.

5.1 Phase 1: Chartering Phase

The first phase in the proposed model is the chartering phase. This name is derived from the first software cycle in Crystal Clear methodology. In this phase, the production of a software package or its new version begins by the project manager or the executive support system. This phase mainly includes team setup, methodology alignment, and initial planning.

5.2 Phase 2: Construction Phase

This phase is the beginning of the incremental-iterative activity to develop the system. The activities in this phase include:

- Reviewing Risks: The list of risks should be repeatedly reviewed and edited. A number of the most important risks should be determined at the beginning of each iteration and resolve them in the rest of the current iteration. The executive supporter is responsible for this job in this model. Inspired by the Crystal Clear method, the list of risks will be installed on a panel or screen so that all development team members can see it. This depends on the project type and might not apply to some simple projects.

- Design: Designing in this model is done by the use of UML modeling language and by senior designers. The senior designer is the role given to a skilled and experienced individual in software development.
- Test case development: Creating test cases is an activity that is done before system development. Test case includes a set of conditions, and passing them indicated a successful evaluation of developed systems. Developers are responsible for this test case creation in this model.
- Development: This activity is somehow the heart of the process. Here, development means coding to meet the end-users' requirements. Most of the time of the project is spent in this stage. Therefore, in this model, a good development process requires high skilled developers.
- Testing: Software is tested by the developer at this stage, and afterward, the list of software errors is completed to be resolved in the current iteration.
- Integration: Development in this model is based on iterations and increments, and like other similar models, after each iteration, integration of the new product increment to previously developed part is done.

5.3 Phase 3: Release

Product release includes going to the customer's location, testing integration, configuring, installing the software, training the users, testing the software by end-users, developing a system guide, and finally supporting the system.

- Integration Testing: Component testing is done in the construction phase. However, after the integration of the components and development of the whole system, it should be checked that no error in integration is threatening the whole system. Therefore, an activity of integration testing is required at the beginning of the release phase. The executive supporter is responsible for it, and it results in a report called "load test."
- Configuration: For the successful deployment of a developed system, it should be first configured. The parts that need to be investigated in this configuration include hardware, software (operating system, required infra-structure software to run the developed system, etc.), individuals, data, etc. The executive supporter is responsible for this activity in the proposed model.
- Deployment: It includes installing the new system along with all its accessories on customers' systems such that the aforementioned system will be usable. Since the objective projects of this model are small and medium, the executive supporter is responsible for it.
- Acceptance Testing: Having established the system and before entering the actual data, the final users should test the delivered system with hypothetical data to make sure of its correct function and test its outputs. This phase is called the acceptance test and the development team, or its representative is present at the customer's location at this time. Expert users are responsible for this activity in this model.

- Training: If the outcome of the acceptance test is satisfactory, the users are trained on how to use the system, and they will be provided with a system user manual.
- Maintenance: Supporting and maintenance activities of the developed system should be considered at this stage.

6 Model Evaluation

A Case Study has been selected for the evaluation of the proposed hybrid method. This may result in a better understanding of the applicability of this method in real environments.

6.1 Case Study Specifications

In order to evaluate the proposed model, it is evaluated in a controlled Case Study. Three teams with four members were selected to develop a single system. To develop the aforementioned system, all three teams were in the same environment with the same tools, including:

- Visual Studio 2012
- MS SQL Server2012
- MS Visio
- Telerik Reporting

The primary author was playing the role of the project manager or executive supporter in all three teams. The determined time period to develop each team was two months. The number of iterations was the same in each release, and a briefing-training course was held at the beginning of each iteration. Each team is called with regard to the method used to develop the system. Therefore, the teams are called RUP, Crystal, and CRUP.

It should be noted that the team's atmosphere was good in all three projects. Team members knew that the researchers were trying to propose a new method to improve the development process. But, all the development process was carried out professionally to show the reality of their adopted methodology.

6.2 Evaluation Criteria

Evaluation metrics in software engineering are usually classified into three different groups; product, process, and project metrics (Kan 2002). Product metrics describe characteristics such as size, complexity, design features, efficacy, and quality level. Process metrics are used to monitor, control, and improvement of the software development and evolution process. Project metrics include the number of software developers, the significance of human relationships on software development life cycle, cost, scheduling, and productivity. Some of the criteria are used in different categories. Quality metric, for example, is both a process metric and a project metric (Kan 2002).

Software development processes have been considered in three operational teams of RUP, Crystal, and CRUP. Meanwhile, interviews were done with development team members and customers. The first author worked directly with the teams and monitored all the development activities carefully and observed the predefined data himself. Table 1 shows the collected data in the three teams. "R" stands for release, and "Total" indicates the total of each item. In the Case Study, four releases (R1, R2, R3, and R4) were required to complete the project.

Table 1. The data collected in the Case Study

Method	RUP					Crystal					CRUP				
Parameter/Release	R1	R2	R3	R4	**Total**	R1	R2	R3	R4	**Total**	R1	R2	R3	R4	**Total**
Project finish time (week)	3	2	1.5	1	**7.5**	2.7	1.7	1.3	0.8	**6.5**	2.5	1.5	1.1	0.6	**5.7**
Actual spent time (man-hour)	340	220	170	110	**840**	310	180	135	66	**691**	240	130	100	50	**520**
Average number of programmers' errors per release	15	11	9	5	**40**	11	8	6	3	**28**	8	6	4	2	**20**
Number of requests for re-work	5	3	2	1	**11**	2	2	1	0	**5**	2	1	1	0	**4**
Number of errors before release	21	16	15	10	**62**	17	16	15	5	**53**	8	5	3	1	**17**
Requests for change after release	5	3	2	1	**11**	3	2	1	1	**7**	2	1	1	0	**4**
Number of deficiencies after release	5	3	2	1	**11**	3	2	1	1	**7**	2	1	1	0	**4**
Time spent on corrections after release (hour)	48	36	30	21	**135**	39	30	24	18	**111**	30	24	17	0	**71**
Number of integrations	11	6	5	3	**25**	28	16	14	7	**65**	24	9	6	3	**42**
Number of user requests	12	9	5	2	**28**	12	9	5	2	**28**	12	9	5	2	**28**
Number of classes	31	22	14	8	**75**	33	23	15	9	**80**	32	22	15	8	**77**
Number of Lines of Codes (×1k)	4.1	2.8	1.9	0.9	**9.7**	4.1	2.8	1.9	0.9	**9.7**	4.1	2.8	1.9	0.9	**9.7**
Using side-by-side programming (percentage)	0	0	0	0	**0**	100	100	100	100	**100**	100	100	100	100	**100**
Customer participation percentage	10	10	10	10	**10**	5	5	5	5	**5**	5	5	5	5	**5**
Productivity (ratio of KLOC to time (week))	12.0	12.7	11.1	8.2	**11.55**	13.2	15.6	14.1	13.6	**14.04**	17.1	21.5	19	18	**18.65**
Number of user interface	5	2	4	3	**14**	5	2	4	3	**14**	5	2	4	3	**14**
Number of team members	4	4	4	4	**4**	4	4	4	4	**4**	4	4	4	4	**4**

The metrics for evaluation of the proposed hybrid method mainly were selected to measure the process quality and efficiency. These metrics derived from the most famous studies in this regard, including (Dybå et al. 2004; Ebert et al. 2005; Haug et al. 2011;

Kan 2002; O'Regan 2010, 2012; Stamelos 2007; von Wangenheim et al. 2010; Wong and Cukic 2011).

- Project Time (Week): Fig. 3 shows the project finish time. This time is shown in R1 up to R4 iterations. The diagram shows that using RUP and Crystal has improved the project finish time as well as project delivery time.
- Actual Team Performance (Man-Hour): Predicted time is the number of working weeks multiplied by the number of team members multiplied by 40, which is the number of working hours in a week. However, since the project manager and customer representative have had a common role in all the three teams in this project and have performed in parallel, the real-time would be less than the predicted time. The real-time spent by the CRUP development team was less than other groups.
- Number of Programmers' Errors in a Time Interval: The programming errors in the hybrid team have been fewer, mainly due to taking advantage of strengths in Crystal and RUP methods. Figure 4 shows the average number of programmers' errors in each release and total errors.

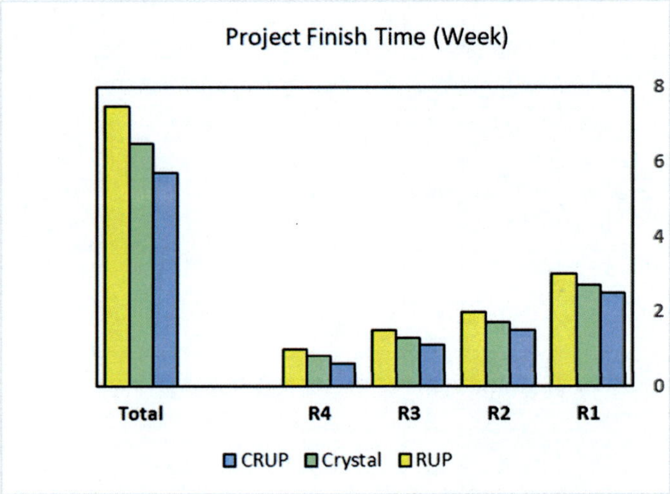

Fig. 3. Project time in the CRUP, RUP, and Crystal Clear

- Number of Requests for Re-Work: The number of requests for re-working has been equal to the number of deficiencies after release, or they have a direct relationship. Therefore, considering Fig. 5 that shows the number of deficiencies after software release and also considering the results listed in Table 1, the improvement in the performance of the hybrid model can be seen in this parameter.
- Number of Detected Errors Before Release: According to the hybrid model architecture that was presented in previous sections, there is a test process before releasing any subsystems. In this process, one who is responsible for testing examines the system and writes down its probable errors so that developers can take necessary measures to overcome them before software release.

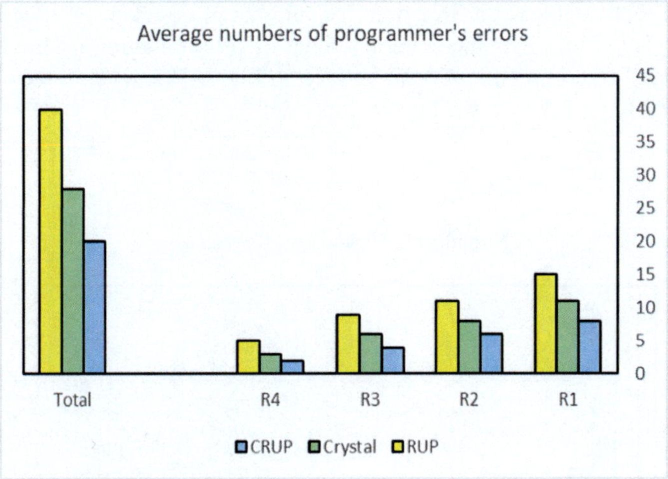

Fig. 4. The average number of the observed errors in the three methods

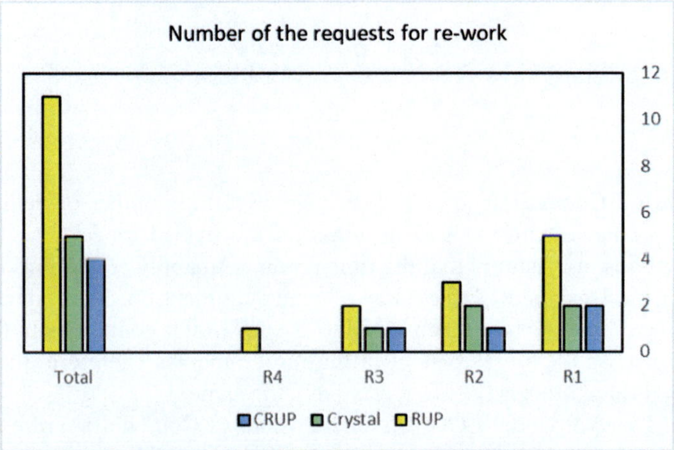

Fig. 5. Number of the requests for re-work

- Number of Requests for Change after Release: This parameter shows that the customer representative has not had a good performance regarding clearly expressing user requirements to the development team and/or the customers have had difficulties in understanding the system and its functionalities. Either way, their expected work has not been done, and they request a change in the presented system. According to the data presented in the table, the superiority of the hybrid model can be seen in this parameter.

- Number of Deficiencies after Release: The number of deficiencies after release is the parameter that determines the accuracy level of the evaluation by the development team of users' requirements. As can be seen in Fig. 6, this level in the hybrid model is considerably higher than the other two models.

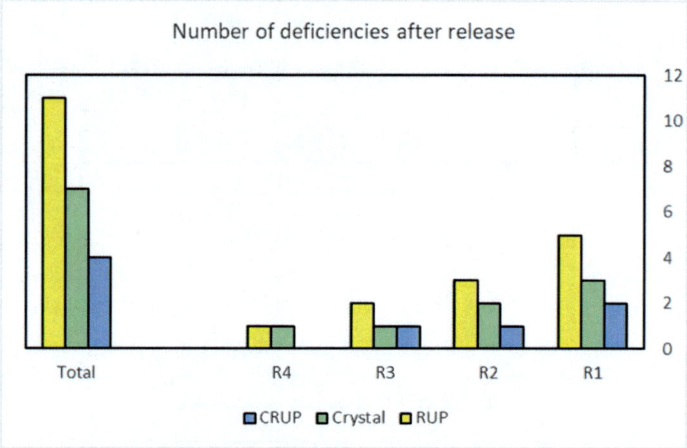

Fig. 6. Number of deficiencies after publishing in three methods

- Time Spent on Corrections after Release: Because the number of deficiencies and requests for changes after release is fewer in the hybrid model than in RUP and Crystal methods, it is natural that the time required to apply corrections is less too.
- Number of the Designed Classes: For a better judgment on the productivity of the hybrid model, attempts have been made to create similar conditions in the designed classes to discuss the target system in the three models. Therefore, the number of classes in all three models has been relatively the same.
- Number of Lines of Code (LOC): Since the Lines of Code plays a direct role in the estimation of productivity, all the three development teams are required to observe the equality in the number of the coded lines. This number has been 10000 lines after examination and third, publish which has been considered to be 500 lines in the whole project with an approximation of 100 lines per publish and final approximation.
- Number of Lines of Code: The number of integrations is fewer in RUP than the Crystal method that has defined the length of iterations in a limited time because the duration is higher in RUP. No specific time has been determined for time durations in the hybrid method, and the aim has been to reach the maximum efficiency in development.
- Productivity: Productivity has been higher in the hybrid method than the other two methods. Figure 7 presents a comparison of the productivity level in the three models.

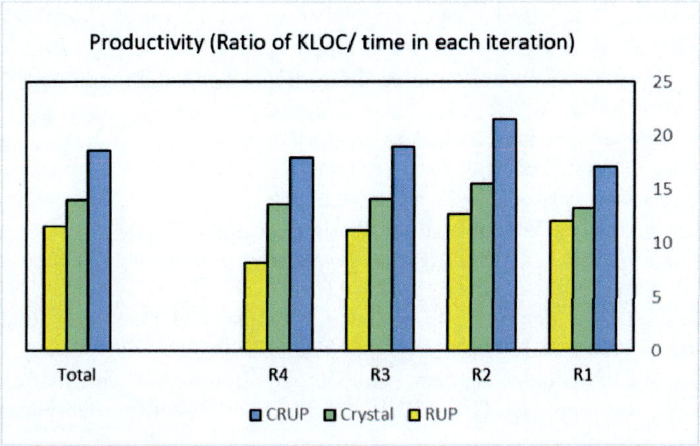

Fig. 7. Comparison of team productivity in the three methods throughout the Case Study

7 Limitations

The following limitations existed in the research stages and implementation of the Case Study:

- Limited resources (books and papers) on Crystal methodology.
- Limitation of Case studies on hybrid methodologies, especially those using Crystal family members.
- Limited access to experienced human resources in applying Agile methods.

8 Conclusion

This paper was carried out to propose a hybrid methodology in the software industry. Using Agile methods should be quickly put in the plan of software organizations. However, the proper time needed so that the organization reaches a suitable maturity to adapt to Agile methods. Furthermore, in many cases, software teams and companies need to use both Agile and non-Agile methods together. Therefore, there is a need to use hybrid methods and models. Considering the project size, these hybrid models attempt to periodically make an efficient and productive combination of the strengths of the traditional and Agile methods primarily to overcome the limitations of each model by the capabilities of the other model or models as a complementary role. This study proposed a hybrid model by combining Crystal Clear and RUP. This model examined in a Case Study. The results showed the validity and applicability of the proposed model. Using this model leads to achieving better team productivity, project duration, and improvement of project delivery.

References

Abrahamsson, P., Ronkainen, J., Warsta, J.: Agile software development methods: review and analysis. VTT Publ. **478**, 1–112 (2002)

Ahmad, G., Soomro, T.R., Brohi, M.N.: XSR: novel hybrid software development model (integrating XP, scrum & RUP). Int. J. Soft Comput. Eng. (IJSCE) **2**(3), 126–130 (2014)

Ambler, S.W.: A manager's introduction to the Rational Unified Process (RUP). Version. 4 December, 2005 (2005)

Ambler, S.W., Nalbone, J., Vizdos, M.J.: The Enterprise Unified Process: Extending the Rational Unified Process. Prentice Hall, Hoboken (2005)

Awad, M.: Comparison between agile and traditional software development methodologies. Honours program thesis, The University of Western Australia (2005)

Bashir, M.S., Qureshi, M.R.J.: Hybrid software development approach for small to medium scale projects: RUP XP & scrum. Cell **966**, 536474921 (2012)

Boehm, B.: Get ready for agile methods, with care. Computer **35**(1), 64–69 (2002). https://doi.org/10.1109/2.976920

Boehm, B.: A survey of agile development methodologies. Laurie Williams (2007)

Castilla, D.: A Hybrid Approach Using RUP and Scrum as a Software Development Strategy (2014)

Cho, J.: A hybrid software development method for large-scale projects: rational unified process with scrum. J. Issues Inf. Syst. **5**(2), 340–348 (2009)

Cockburn, A.: Agile Software Development. Pearson Education Inc., Boston (2002)

Cockburn, A.: Crystal clear: A Human-Powered Methodology for Small Teams. Pearson Education, London (2004)

Cohen, D., Lindvall, M., Costa, P.: An introduction to Agile methods. Adv. Comput. **62**, 1–66 (2004). https://doi.org/10.1016/S0065-2458(03)62001-2

Dybå, T., Dingsøyr, T., Moe, N.B.: Process Improvement in Practice: A Handbook for IT Companies, vol. 9. Springer, New York (2004). https://doi.org/10.1007/b116193

Ebert, C., Dumke, R., Bundschuh, M., Schmietendorf, A.: Best Practices in Software Measurement: How to Use Metrics to Improve Project and Process Performance: Springer, Heidelberg (2005). https://doi.org/10.1007/b138013

Gandomani, T.J., Nafchi, M.Z.: Agile transition and adoption human-related challenges and issues: a Grounded Theory approach. Comput. Hum. Behav. **62**, 257–266 (2016)

Gandomani, T.J., Zulzalil, H., Abdul Ghani, A.A., Sultan, A.B.M., Sharif, K.Y.: How human aspects impress Agile software development transition and adoption. Int. J. Softw. Eng. Appl. **8**(1), 129–148 (2014). https://doi.org/10.14257/ijseia.2014.8.1.12

Gandomani, T.J., Zulzalil, H., Ghani, A.A.A., Sultan, A.M., Nafchi, M.Z.: Obstacles to moving to agile software development; at a glance. J. Comput. Sci. **9**(5), 620–625 (2013). https://doi.org/10.3844/jcssp.2013.620.625

Gill, A.Q., Henderson-Sellers, B., Niazi, M.: Scaling for agility: a reference model for hybrid traditional-agile software development methodologies. Inf. Syst. Front. **20**(2), 315–341 (2016). https://doi.org/10.1007/s10796-016-9672-8

Haug, M., Olsen, E.W., Bergman, L.: Software Process Improvement: Metrics, Measurement, and Process Modelling: Software Best Practice, vol. 4. Springer, Heidelberg (2011). https://doi.org/10.1007/978-3-642-56618-9

Jovanović, M., Mesquida, A.-L., Mas, A., Colomo-Palacios, R.: Agile transition and adoption frameworks, issues and factors: a systematic mapping. IEEE Access **8**, 15711–15735 (2020)

Kan, S.H.: Metrics and Models in Software Quality Engineering. Addison-Wesley Longman Publishing Co., Inc. (2002)

Neto, G.T.G., Santos, W.B., Fagundes, R.A., Margaria, T.: Towards an understanding of value creation in agile software development. Paper Presented at the Proceedings of the XV Brazilian Symposium on Information Systems (2019)

O'Regan, G.: Introduction to Software Process Improvement: Springer, London (2010). https://doi.org/10.1007/978-0-85729-172-1

O'Regan, G.: A Practical Approach to Software Quality. Springer, New York (2012). https://doi.org/10.1007/978-0-387-22454-1

Pressman, R.S.: Software Engineering: A Practitioner's Approach. 7th edn. McGraw-Hill Science/Engineering/Math, New York (2009)

Rasool, G., Aftab, S., Hussain, S., Streitferdt, D.: eXRUP: a hybrid software development model for small to medium scale projects. J. Softw. Eng. Appl. **6**(09), 446 (2013)

Shaw, S., Johnson, T., Garcia, K., Greene, S.: Prototyping Rational Unified Process. Recent advances in software engineering and computer science 6(1) (2019)

Smoczyńska, A., Pawlak, M., Poniszewska-Marańda, A.: Hybrid agile method for management of software creation. Paper Presented at the KKIO Software Engineering Conference (2018)

Stamelos, I.G.: Agile software development quality assurance. IGI Global (2007)

Turner, R., Boehm, B.: Balancing Agility and Discipline: A Guide for the Perplexed, 1st edn. Addison-Wesley/Pearson Education, Boston (2004)

von Wangenheim, C.G., Hauck, J.C.R., Salviano, C.F., von Wangenheim, A.: Systematic literature review of software process capability/maturity models. Paper Presented at the Proceedings of International Conference on Software Process Improvement and Capabity Determination (SPICE), Pisa, Italy (2010)

Wong, W.E., Cukic, B.: Adaptive Control Approach for Software Quality Improvement, vol. 20. World Scientific (2011)

Yin, R.K.: Case Study Research: Design and Methods. Sage Publications (2013)

Yin, R.K.: Qualitative Research from Start to Finish. Guilford Publications (2015)

Investigative Study of Unigram and Bigram Features for Short Message Spam Detection

Rasheed G. Jimoh(✉), Kayode S. Adewole, Tunbosun E. Aderemi, and Abdullateef O. Balogun

Department of Computer Science, University of Ilorin, Ilorin 1515, Nigeria
{jimoh_rasheed,adewole.ks,balogun.ao1}@unilorin.edu.ng

Abstract. Nowadays, it is very imperative to maintain a high standard of security for ensuring reliable and trusted communication channel across various organizations. Studies have shown that spammers have used mobile short message service (SMS) and microblogging as a channel of communication for disseminating unsolicited spam messages containing novel words and abbreviation. Over the years, a growing number of research works have been conducted in the area of short message spam detection where many of these approaches yielded promising results. However, investigating the effect of unigram and bigram features for short message spam detection remains an open research issue. This paper proposed unigram and bigram features for short message spam detection. The performance of the proposed features was evaluated using four classification algorithms: Support Vector Machine (SVM), Naïve Bayes, LibSVM and Random Forest. The results were evaluated on two datasets based on Tweets and mobile SMS data. Results obtained were compared for unigram and bigram features based on the individual classifiers. Experimental results showed that unigram features produced the best accuracy of 98.72% with Naïve Bayes algorithm compared to bigram features.

Keywords: SMS · Unigram · Bigram · Features extraction · Spam detection

1 Introduction

In recent times, the benefits of information technology (IT) as a tool for rapid communication with Short Message Service (SMS) and social media cannot be over-emphasized. SMS and social media applications have made the dissemination of information easier and faster. However, the usage of SMS and the huge increase in the number of spam messages sent by the spammer has been worrisome. This makes research on SMS spam detection becoming increasingly more important [1, 2]. The increasing number of unsolicited SMS text messages are a common occurrence that seriously annoys mobile phone subscribers leading to frequent change from one service to other [3, 4].

This kind of unsolicited messages has become widespread on social media. Nearly all smartphone owners visit social networking sites via their mobile phones [5]. A large number of users spend ample time in communicating with their acquaintances and friends on social media. After establishing the relationship, unsolicited messages in the form

J. H. Abawajy et al. (Eds.): EATI 2020, LNNS 254, pp. 70–81, 2021.
https://doi.org/10.1007/978-3-030-80216-5_6

of tweets, wall post or status updates are been sent. In most cases, spammers hijack these communication channels to distribute unsolicited spam messages [6]. Besides, a spammer can obtain mobile phone numbers with devious methods to engage in spamming activities [7].

Therefore, it is very crucial to provide an effective framework for short message spam detection that can address the evasion strategies posed by spammers. Spam detection and prevention is not a trivial research issue since spammers are devising new techniques to evade current detection system. A large number of studies have been conducted on short message spam detection [3, 7, 8]. However, existing studies failed to investigate unigram and bigram features across different communication media. Eliminating spam could significantly boost customer satisfaction ratings and experience primarily from authorized mobile advertisement agencies or individuals, where it requires an urgent need for more security measures to identify spammers sending spam messages.

This study investigates the performance of unigram and bigram features for short message spam detection. In summary, the main contributions of this study are highlighted as follows:

i. Conduct a comparative analysis of short message spam detection using two datasets that cut across different communication media.
ii. Propose a framework based on unigram and bigram features for short message spam detection using two datasets from different domains.
iii. Investigate the performance of four classification algorithms for short message spam detection based on the two datasets.

The remainder of this paper is structured as follows. A review of related works on short message spam detection is presented in Sect. 2. In Sect. 3, the methods considered for data collection, preprocessing, modelling and evaluation metrics were discussed. Section 4 presents the results of the analyses conducted based on unigram and bigram features across different classifiers and datasets. Section 5 concludes the paper and presents future direction.

2 Related Works

In this section, a brief review of work related to SMS spam detection. Several researchers have applied and proposed machine learning algorithms for short message services for spam detection. Shafi'I, et al. [3] proposed a spam filter method to distinguish between legitimate spam and non-spam messages. The proposed filter model is based on three components: the N-gram to extract features from short messages, information gain ration to select the most relevant features and classification. Adewole et al. [8] proposed a unified framework for spam messages and spam account detection tasks to identify a minimal number of features for spam account detection in Twitter. The research study focused on bio-inspired evolutionary search method.

Lota and Hossain [9] carried out a systematic literature review on SMS spam detection techniques based on research works from 2006 to 2016. The study presented various algorithms and methods, advantages and disadvantages of those algorithms and evaluation measures that have been conducted in the literature. The authors also presented the datasets used in those research works and compare their results.

Afzal and Mehmood [10] in their study, conducted a comparative analysis of selected classification algorithms on spam datasets from five major cities. The authors used four features from a tweet, which include the User Id, Hash-tags, Numbers and URLs. Performance of four classification algorithms was comparatively analysed on these datasets and Naïve Bayes multi-nominal (95.42%) was reported superior for the training data.

Fusilier et al. [11] proposed a method for character n-grams as features. The approach is evaluated on a standard corpus composed of 1600 hotel reviews, considered both positive and negative reviews. The analysis was carried between character n-grams and word n-grams and the result showed that character n-grams are a good feature for opinion spam detection. Kanaris et al. [12] explored different low-level pattern using character n-grams that avoid tokenization and other language-dependent approaches. Considering experiments on two benchmark corpora and different evaluation measures, the results show that character n-gram is a more dependable feature than word-token despite the increase in the dimensionality of the problem.

3 Methodology

To perform experiments on unigram and bigram features for short message spam detection, a framework based on unigram and bigram features is proposed which is aimed at providing better classification result. Figure 1 shows the proposed framework to generate an effective pattern to detect spam messages, which depicts all the component and the processes involved. The first stage involves data collection from Twitter and SMS corpus. These data were preprocessed and annotated to generate unigram and bigram features for analysis. The extracted unigram and bigram features from the datasets are subjected to classification algorithms to produce results. The results are presented based on the performance of the selected classification algorithms.

3.1 Data Collection

This study used two sets of data: social media tweets and SMS dataset to carry out the experiment on unigram and bigram features for short message spam detection. Tweets dataset is the first dataset, which contains 2597 tweets collected using Twitter API in raw JSON format. The SMS spam dataset which is publicly available online (http://www.comp.nus.eu.sg/~rpnlpir/downloads/corpora/smsCorpus/) contains 5574 instances (747 spam and 4827 ham).

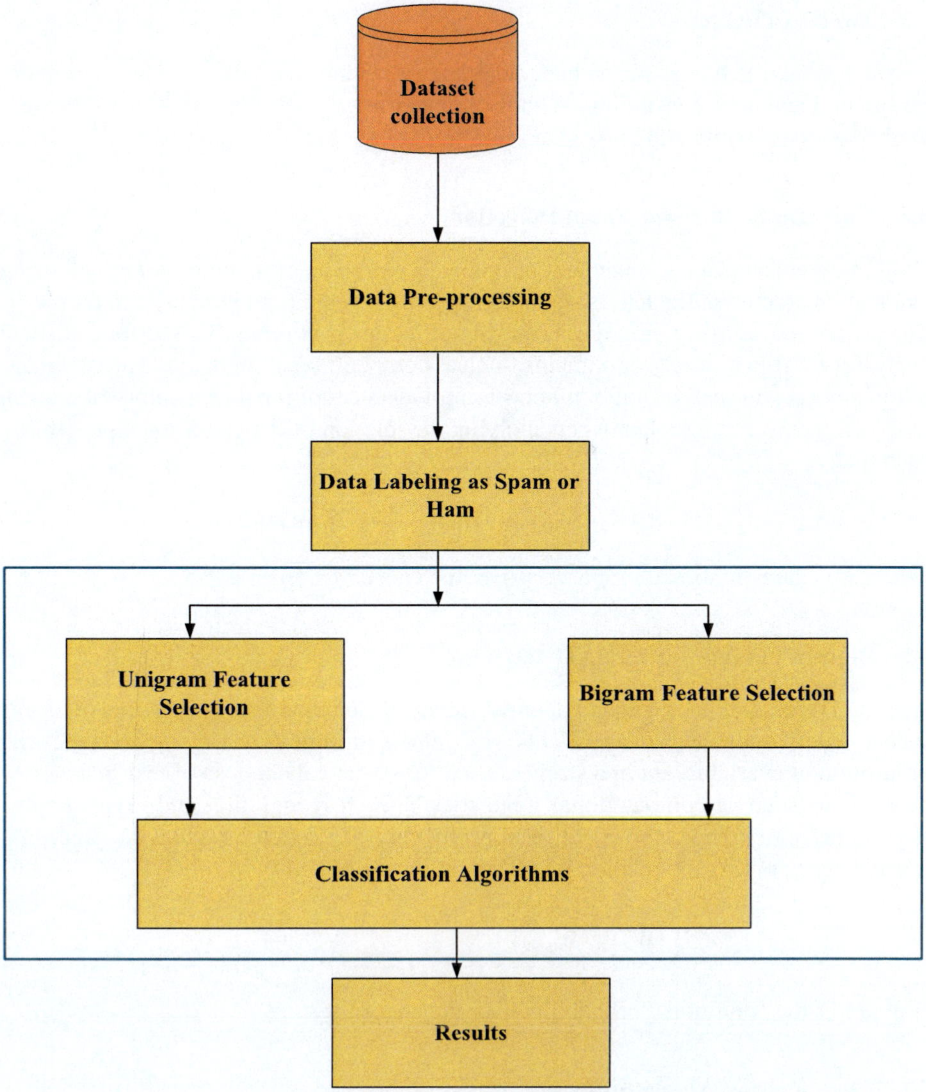

Fig. 1. The proposed framework for short message spam detection

3.2 Pre-processing of Tweets

Tweets were processed to remove characters which do not beer information in the context of spam detection [13]. Tweets of the size fewer than 20 characters were also removed to produce a better dataset for the analysis. The tweets in this category have a less useful context that can be used to conclude whether a message is spam or legitimate. After cleaning the tweets, a total of 1483 tweets were used for the analysis.

3.3 Tweets Labeling

A tweet is labelled as spam or ham (legitimate) by the researchers, which are more resourceful and time-consuming. With the number of tweets collected, 421 tweets were marked as spam while 1062 were ham tweets.

3.4 Unigram Feature for Spam Detection

Unigram represents an arrangement of one adjacent element based on tokens of string elements characterized as letters, syllables, or words. A unigram is an n-gram for $n = 1$. The proportion of every unigram in a string of tokens is commonly used for statistical analysis of text in a variety of domains, such as computational linguistics, speech recognition and cryptography. Unigram helps to provide a conditional probability of a token based on the preceding token when applying the relation of the conditional probability such that.

$$P\ (W_i \mid W_0 \ldots W_{i-1}) \quad = \quad P\ (W_i) \tag{1}$$

Where P is the conditional probability over the chosen feature W.

3.5 Bigram Feature for Spam Detection

A bigram represents an arrangement of two adjacent elements based on a string of tokens such as letters, syllables, or words. This arrangement represents an n-gram for $n = 2$. The proportion of every bigram in a string is used for statistical analysis of text in a variety of domains such as computational linguistics, speech recognition and cryptography. Bigram helps to provide the conditional probability of a token based on the preceding token when applying the relation of the conditional probability as follows.

$$P\left(W_{n\,|} \mid W_{n-1}\right) \quad = \quad \frac{P(W_{n-1},\ W_n)}{P\ (W_{n-1})} \tag{2}$$

Where P is the conditional probability over the chosen feature W.

3.6 Classification Algorithms

The proposed framework for short message spam detection is based on four selected classification algorithms. These algorithms are Naïve Bayes, Support Vector Machine (SVM), LibSVM and Random Forest. To measure the performance of the different algorithms, accuracy and kappa statistics were used. Accuracy is the number of the correctly classified instance as compared to the incorrectly specified. Kappa statistics is a chance corrected measure of agreement between the classification and the true classes. It is calculated by taking the chance of expected agreement away from the observed agreement and divide by the maximum possible agreement.

3.6.1 Naïve Bayes (NB)

Naïve Bayes (NB) is a probabilistic classifier based on Bayes theorem with strong independence assumption among features of a dataset. Specifically, NB assumes that the occurrence of an attribute in a dataset is unrelated to the occurrence of any other attribute [14, 15]. NB is a relatively simple and fast classifier that is commonly used on high-dimensional datasets across different research domains. NB performs best when there is an assumption of independence amongst attributes of a dataset [16]. NB is a conditional probability model which is defined as:

$$p\ (c_k \mid x) = \frac{p(c_k)\ p(x \mid c_k)}{p(x)} \tag{3}$$

Where $p(c_k \mid x)$ is the posterior probability of class (c) given predictor (x)

3.6.2 Support Vector Machine (SVM)

Support Vector Machine is a supervised machine learning algorithm which can be used for classification or regression challenges. The objective of the SVM is to find a hyperplane in N-dimensional space (N is the number of features) that distinctly classifies the data points [17]. SVM separates the classes of data points based on the distance between the data point of the classes. Maximizing the margin distance provides some reinforcement so that future data point can be classified with more confidence [18].

3.6.3 Random Forest (RF)

Random Forest (RF) is an ensemble learning method that operates by constructing a multitude of decision trees at training time and outputting the class that is the mode of the class classification or means prediction [19]. Each decision tree is constructed by using a random subset of the training data. The training algorithm for RF applies the general techniques of bootstrap aggregating (bagging) to tree learners. In other words,

Given a training set $x = x_1, \ldots, x_n$ with responses $y = y_1, \ldots. y_n$, RF-based on bagging repeatedly (N times) select a random sample with replacement of the training set and fits decision trees to the dataset [20].

3.6.4 LibSVM Machine

LibSVM machine is an integrated software for support vector classification, (C-SVC), regression (epsilon-SVR) and distribution estimation (one-class SVM). It supports the multi-class classification [21]. LIBSVM typically involves training a dataset to obtain a model and using the model to predict information of a testing data set [22].

3.7 Performance Evaluation Metrics

- Accuracy: This refers to the closeness of a measured value to a standard or known value.

$$\text{Accuracy} = \frac{\text{TP} + \text{TN}}{\text{TP} + \text{FP} + \text{FN} + \text{TN}}$$

Where TP = True positive; FP = False positive; TN = True negative; FN = False negative

- Kappa metrics: This is a chance corrected measure of agreement between two sets of categorized data. Kappa result ranges between 0 to 1. The higher value of kappa the stronger the agreement. If kappa = 1, then there is perfect agreement. If kappa = 0, then there is no agreement.

4 Results

The results of unigram and bigram features are presented based on the performance of the selected classifiers. These results are presented in this section. Results have been produced using 10-fold cross-validation (Fig. 2).

Table 1. Classification results using unigram features on Tweets datasets

Classifier	Accuracy	Kappa
NB	**98.72%**	**96.48%**
SVM	97.88%	94.21%
LibSVM	96.22%	89.21%
RF	96.6%	90.55%

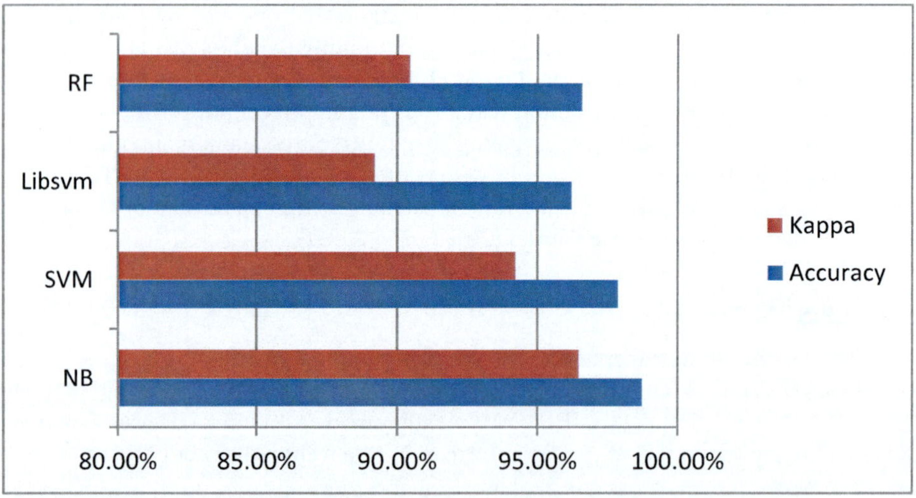

Fig. 2. Unigram results on Tweets datasets

Table 1 presents the results of the classifiers based on unigram features on Tweets dataset. According to the results, the highest accuracy is reported by Naïve Bayes classifier, which is 98.72% and Kappa statistics of 96.48%. This result show that using unigram feature Naïve Bayes produced promising result in detecting spam tweets. The implication of this is that spam tweets can be easily filtered based on the model with considerable reduction in error rate. SVM showed an accuracy of 97.88% with Kappa statistics of 94.71%. Random Forest produced an accuracy of 96.6% with Kappa statistics of 90.55%. LibSVM produced an accuracy of 96.22% and Kappa of 89.21%. All the selected algorithms produced promising results in terms of accuracy and Kappa statistics tested for unigram on tweets dataset for short message spam detection problem investigated in this study. However, Naïve Bayes has demonstrated superiority for the problem domain and particularly based on the unigram features investigated in this study (Fig. 3).

Table 2. Classification results using unigram features on SMS dataset

Classifier	Accuracy	Kappa
NB	98.13%	91.71%
SVM	**98.26%**	**92.38%**
LibSVM	93.7%	66.24%
RF	96.97%	85.63%

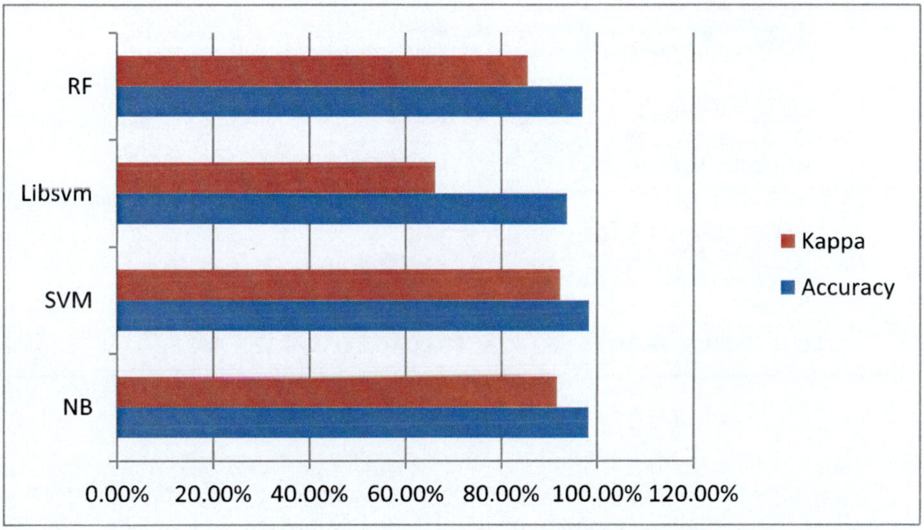

Fig. 3. Unigram results on SMS dataset

According to the results in Table 2, the highest accuracy is reported by SVM on unigram SMS dataset, which is 98.26% and Kappa statistics of 92.38%. The implication of this result is that the model based on SVM can detect SMS spam with reduction in error rate when compared with other models in this category. Naïve Bayes showed an accuracy of 98.13% with Kappa statistics of 91.71%. Random Forest produced an accuracy of 96.97% with Kappa statistics of 85.63%. LibSVM showed an accuracy of 93.7% and

Kappa of 66.24%. LibSVM classifier performed poorly based on Kappa statistics result for unigram SMS dataset. From the results obtained, it is clear that on SMS dataset, SVM classifier achieved better results compared to the other three classifiers (Fig. 4).

Table 3. Classification results using bigram features on tweets datasets

Classifier	Accuracy	Kappa
NB	94.71%	84.72%
SVM	**95.31%**	**86.76%**
LibSVM	82.01%	32.92%
RF	94.94%	85.55%

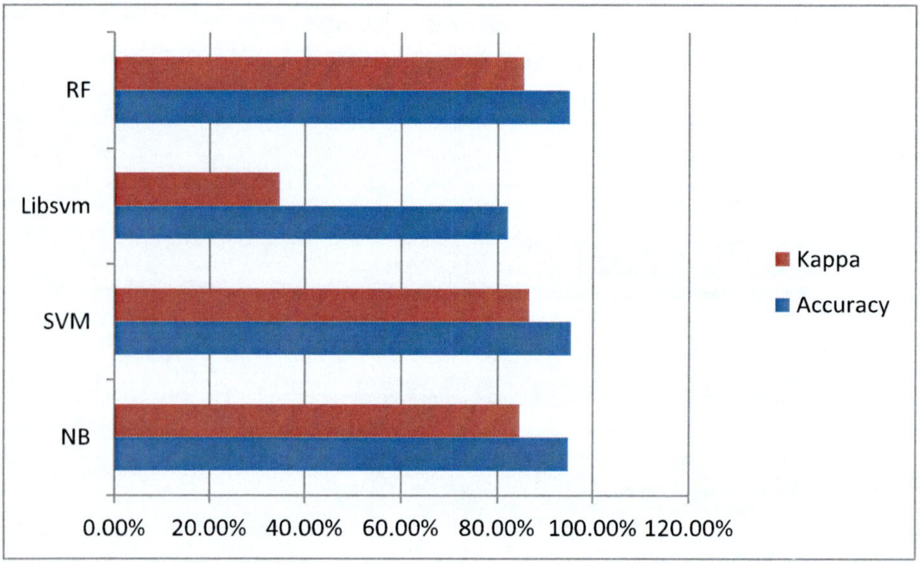

Fig. 4. Bigram results on Tweets datasets

The results of the classification using bigram features on tweets dataset revealed that SVM has the highest accuracy of 95.31% and Kappa statistics of 86.76%. RF also produced an accuracy of 94.94% with Kappa statistics of 85.55%. NB showed an accuracy of 94.71% with Kappa statistics of 84.72%. The worst result is obtained with LibSVM, which produced an accuracy of 82.01% and Kappa statistics of 32.92%. SVM classifier leads other algorithm using bigram features on tweets dataset (Fig. 5).

Table 4. Classification results using bigram features on SMS dataset

Classifier	Accuracy	Kappa
NB	97.34%	87.62%
SVM	97.24%	88.48%
LibSVM	86.61%	30.23%
RF	**97.56%**	**88.87%**

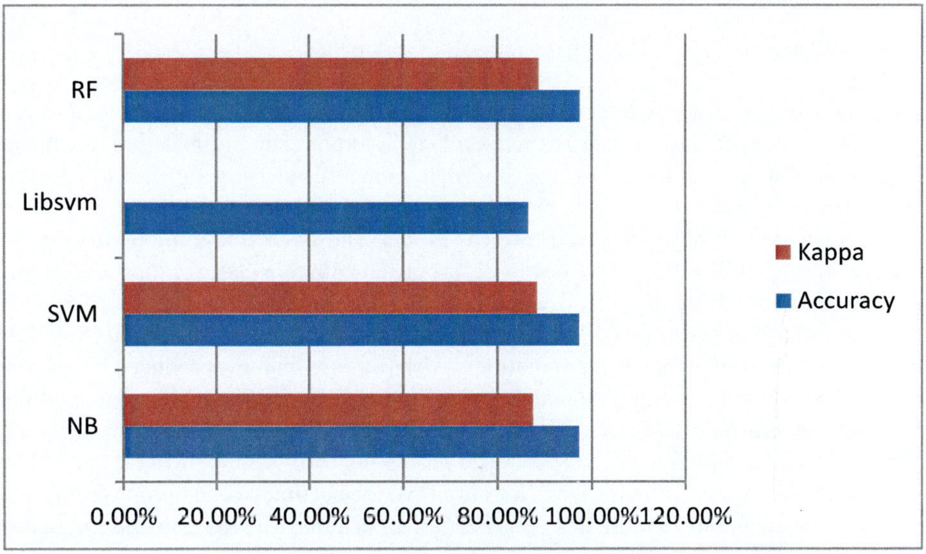

Fig. 5. Bigram results on SMS dataset

Random Forest classifier produced the best result based on bigram features using SMS dataset as shown in Table 4. Naïve Bayes and SVM also generated promising results that are very close to the results of Random Forest classifier as shown in Table 4. NB showed an accuracy of 97.34% with Kappa statistics of 87.62%. SVM reported accuracy of 97.24% and Kappa statistics of 88.48%. Similarly, LibSVM performed poorly based on bigram features on SMS dataset. Table 5 shows a summary of the performance of the classification algorithms on both Tweets and SMS datasets using unigram and bigram features. The best result for each task has been highlighted in bold. These results revealed that Naïve Bayes, SVM and Random Forest are candidate algorithms for selection, specifically, to detect short message spam.

Table 5. Performance comparison of the classifiers

Classifier	Unigram on tweets datasets	Unigram on SMS datasets	Bigram on tweets datasets	Bigram on SMS datasets
NB	**98.72%**	98.13%	94.71%	97.34%
SVM	97.88%	**98.26%**	**95.31%**	97.42%
LibSVM	96.22%	93.7%	82.01%	86.61%
RF	96.6%	96.97%	94.94%	**97.56%**

5 Conclusion

The growth of spam messages across communication media has been alarming in recent years. This paper presents an investigative study to ascertain if unigram features are better than bigram features for short message spam detection. This investigation is critical and it will help researchers in developing classification models to address the problem of short spam message distribution targeted towards mobile and micro-blogging platforms. To achieve the goal of this study, a framework is proposed with four classification algorithms investigated to ascertain the performance of both unigram and bigram features for short message spam detection. From the experimental; results, the use of unigram features produced better results than bigram features. Also, bigram features produced promising results based on the classifiers selected for the analysis. Conclusively, the results of the investigation showed that Naïve Bayes, SVM and Random Forest classifiers are good candidates for this problem as they produced promising results on both Tweets and SMS datasets. In the future, the authors wish to improve the accuracy of the proposed framework by introducing metaheuristics optimization algorithms for selecting discriminative features from the corpora. Furthermore, additional classification algorithms with varying characteristics will be investigated in future work.

References

1. Saxena, N., Chaudhari, N.S.: EasySMS: a protocol for end-to-end secure transmission of SMS. IEEE Trans. Inf. Forensics Secur. **9**, 1157–1168 (2014)
2. Adegbola, I., Jimoh, R., Longe, O.: An integrated system for detection and identification of spambot with action session and length frequency. Comput. Inf. Syst. Dev. Inform. Allied Res. J. **4** (2013)
3. Shafi'I, M.A., et al.: A review on mobile SMS spam filtering techniques. IEEE Access **5**, 15650–15666 (2017)
4. Jimoh, R., Coco, K.O., Abdel, M.: Design of mobile short message service (SMS) across a computer network for organisational communication. Int. J. Comput. Appl. Technol. Res. **2**, 6 (2013)
5. https://www.statista.com/topics/2478/mobile-social-networks/
6. Adewole, K.S., Han, T., Wu, W., Song, H., Sangaiah, A.K.: Twitter spam account detection based on clustering and classification methods. J. Supercomput. **76**(7), 4802–4837 (2018). https://doi.org/10.1007/s11227-018-2641-x

7. Almeida, T.A., Hidalgo, J.M.G., Yamakami, A.: Contributions to the study of SMS spam filtering: new collection and results. In: Proceedings of the 11th ACM Symposium on Document Engineering, pp. 259–262. ACM (2011)
8. Adewole, K.S., Anuar, N.B., Kamsin, A., Sangaiah, A.K.: SMSAD: a framework for spam message and spam account detection. Multimedia Tools Appl. **78**(4), 3925–3960 (2017). https://doi.org/10.1007/s11042-017-5018-x
9. Lota, L.N., Hossain, B.M.: A systematic literature review on SMS spam detection techniques (2017)
10. Afzal, H., Mehmood, K.: Spam filtering of bi-lingual tweets using machine learning. In: 2016 18th International Conference on Advanced Communication Technology (ICACT), pp. 710–714. IEEE (2016)
11. Fusilier, D.H., Montes-y-Gómez, M., Rosso, P., Cabrera, R.G.: Detection of opinion spam with character n-grams. In: Gelbukh, A. (ed.) CICLing 2015. LNCS, vol. 9042, pp. 285–294. Springer, Cham (2015). https://doi.org/10.1007/978-3-319-18117-2_21
12. Kanaris, I., Kanaris, K., Houvardas, I., Stamatatos, E.: Words versus character n-grams for anti-spam filtering. Int. J. Artif. Intell. Tools **16**, 1047–1067 (2007)
13. Aro, T.O., Dada, F., Balogun, A.O., Oluwasogo, S.A.: Stop words removal on textual data classification. Int. J. Inf. Process. Commun. **7**, 1–9 (2019)
14. Saritas, M.M., Yasar, A.: Performance analysis of ANN and Naive Bayes classification algorithm for data classification. Int. J. Intell. Syst. Appl. Eng. **7**, 88–91 (2019)
15. Chen, S., Webb, G.I., Liu, L., Ma, X.: A novel selective naïve Bayes algorithm. Knowledge-Based Syst. **192**, 105361 (2020)
16. Manino, E., Tran-Thanh, L., Jennings, N.R.: On the efficiency of data collection for multiple Naïve Bayes classifiers. Artif. Intell. **275**, 356–378 (2019)
17. Suthaharan, S.: Support vector machine. In: Machine Learning Models and Algorithms for Big Data Classification, Integrated Series in Information Systems, vol. 36, pp. 207–235. Springer, Boston (2016). https://doi.org/10.1007/978-1-4899-7641-3_9
18. Anitha, P., Neelima, G., Kumar, Y.S.: Prediction of cardiovascular disease using support vector machine. J. Innov. Electr. Commun. Eng. **9**, 28–33 (2019)
19. Biau, G., Scornet, E.: A random forest guided tour. TEST **25**(2), 197–227 (2016). https://doi.org/10.1007/s11749-016-0481-7
20. Schonlau, M., Zou, R.Y.: The random forest algorithm for statistical learning. Stand. Genomic Sci. **20**, 3–29 (2020)
21. Chang, C.-C., Lin, C.-J.: LIBSVM: a library for support vector machines. ACM Trans. Intell. Syst. Technol. **2**, 1–27 (2011)
22. Horvath, D., Brown, J., Marcou, G., Varnek, A.: An evolutionary optimizer of libsvm models. Challenges **5**, 450–472 (2014)

Application of K-Nearest Neighbor Algorithm for Prediction of Television Advertisement Rating

Rizqi Prima Hariadhy, Edi Sutoyo(✉), and Oktariani Nurul Pratiwi

Department of Information Systems, Telkom University, Bandung, West Java, Indonesia
rizqiphd@student.telkomuniversity.ac.id, {edisutoyo, onurulp}@telkomuniversity.ac.id

Abstract. Television is the most effective marketing media and is the choice of advertisers today. One parameter that is a measure of the success of a television is the rating of the programs on the television. This parameter is also the consideration of advertisers in choosing a television station. For companies engaged in the advertising industry generally will provide offers to prospective customers using historical data related to ratings that have been obtained. However, the company has not been able to provide rating predictions in the future with a rational method. One approach to overcoming this problem is to apply one of the techniques in data mining, the K-nearest neighbor algorithm. The results showed that the K-nearest neighbor was able to classify television advertising ratings with an accuracy of 91.18%. This achievement shows that the K-nearest neighbor algorithm remains powerful used for the classification of television advertising ratings. The results of this study can provide one of the options for advertising companies to provide offers related to the prediction of television ratings to advertisers as one of the considerations in choosing a television station.

Keywords: Data mining · K-nearest neighbor · Prediction · Classification · Television advertisement rating

1 Introduction

1.1 Background Problem

Marketing is one of the most important things in the development of a brand or product. According to Phillip and Duncan [1], marketing is everything about identifying and fulfilling human needs. Marketing is one of the main needs of running the business. Nowadays many channel options can be used as marketing media. For example, there are social media, magazines, billboards, and many more. One of them is to use advertisements on television stations.

Television is one of the mass communication media that has an important role in disseminating information, entertainment, and influence the community [2]. Television

J. H. Abawajy et al. (Eds.): EATI 2020, LNNS 254, pp. 82–91, 2021.
https://doi.org/10.1007/978-3-030-80216-5_7

is also a media that has more value than print media. The advantages of television compared to print media is because television is also a media that is more attractive and more actual than printed media [3]. Television as an advertising medium has many advantages because it can combine the vision, sound and motion, and wide range. Therefore, television can be used as a promotional media by the owner of the brand or product [4]. Through television, media delivering messages more freely with a variety of creativity, so it gets good attention from the audience.

The marketing method using television media is one of the methods that are desirable by the brand owners. One of the parameters used to measure the success of an advertisement on television is television rating. Television Rating is a unit that measures the audience's loyalty value in watching a television show [5]. Companies engaged in advertising that collaborates with television media generally offer advertisers using historical data related to television ratings based on channel, time, and event. The size of the rating affects the selling value of the ads in each time slot. High advertising costs are expected to be covered by the income obtained by the sale of its products. Therefore, the placement of what television programs, the selection of airtime, the duration of a particular time slot has a very significant role for an advertisement on television.

The rating value is a value in the form of a predicate used as a measure of the rating of an advertisement on a television program. This value is quite influential in determining the policy to be taken for a duration of the contract on the ongoing program, as well as making a contract on a new program that will be aired by the television station concerned. The obstacle faced by advertising companies is that they cannot provide rational predictions related to television ratings in the future. Therefore, if the rating value can be predicted, it will certainly be quite helpful in determining the policy for advertising on programs on the television station. From these problems, we need a rational way to predict television broadcast ratings so that advertisers can determine the channel, time, and program for the right advertising.

Data mining is a solution that can be used to predict the value of a variable in the future. The technique used is to analyze historical data. Data mining is a scientific field that unites techniques from machine learning, pattern recognition, statistics, databases, and visualization to deal with problems and information retrieval from large database storage [6, 7]. Data mining can be used to extract information from large data so that information can be obtained that can be used in predicting an event in the future.

Data mining has proven to be an alternative solution for predicting existing real-world problems [8–15]. One data mining algorithm that has been proven to have high performance is K-Nearest Neighbor (KNN) [16–19]. This algorithm is one of the algorithms used to classify objects based on learning data that is closest to the object. The basic concept of the K-nearest neighbor algorithm is to find the closest distance between the data that has been obtained with a number of k closest neighbors in the test data. K-nearest neighbors work by comparing test data and training data. K-nearest neighbors look for training data patterns that are the closest to the test data. The advantages of the K-nearest neighbor method in predicting are fast training time, resilient to noise data, and effective if the training data is large. Therefore, in this study, the KNN algorithm is used to predict television advertisement ratings. With this method, it is expected to be able to obtain prediction results that will be taken into consideration in policy-making for companies and advertisers.

The rest of this paper is organized as follows. In Sect. 2, we present related work including K-Nearest Neighbor (KNN) and then in Sect. 3, the research methodology is presented. We elaborate the result and discussion in Sect. 4. We finally present some conclusions in Sect. 5.

1.2 Research Question

Based on the problem that has been shown in the background problem section, the questions of this research are

1. How to implement the K-Nearest Neighbor (KNN) algorithm to predict television advertisement performance?
2. How to measure the performance of the K-Nearest Neighbor (KNN) algorithm for television advertisement performance prediction?

2 Related Work

2.1 Data Mining

Data mining is a search for information by looking for certain patterns or rules of a very large amount of data [20]. Data mining also has definitions as a series of processes to explore the added value of knowledge that has not been known manually from a data set. Data Mining is also referred to as KDD (Knowledge in the database). KDD is a series of activities that include collecting, using historical data to find regularity, pattern, or relationship in a large collection [21].

2.2 Information Gain

Information Gain is a simple feature selection method by ranking attributes and is widely used in text categorization applications, microarray data analysis, and image data analysis [22, 23]. Information Gain can help reduce noise caused by irrelevant features. Information Gain detects features that have the most information by class. Determination of the best attributes is done by calculating the value of entropy first. Entropy is a measure of class uncertainty using the probability of certain events or attributes [24]. The formula for calculating entropy is shown in Eq. (1).

$$S(c) = -\sum_{i=1}^{m} p(c_i)\log p(c_i) \tag{1}$$

After getting the entropy value, the Information Gain calculation can be done using Eq. (2).

$$IG(c,t) = S(c) + \sum_{j \in value(t)}^{n} \frac{|c_j|}{c} S(c_j) \tag{2}$$

2.3 K-Nearest Neighbor (KNN)

The K-nearest neighbor's algorithm is a method of classifying the object based on the learning data that is closest to the object [25]. The data is projected into a lot of dimensional space, where each dimension represents the features of data. K-Nearest neighbor's algorithm is one of the methods in the classification of documents where the method K-nearest neighbor classification based on the training/learning data that is closest to the object based on the K-neighbor. The training data will be a training sample, which will be projected into a lot of dimensional space, where each space represents the features of the training data. A point of the space will be marked class c if class c is the most widely encountered classification on the closest neighboring fruit from that point.
The steps of the K-Nearest neighbor algorithm:

1. Specify the K parameter (number of closest neighbors),
2. Calculating the object distance to the given training data,
3. Sort the results from step no. 2 becomes ascending order,
4. Collecting classes (classification of nearest-neighbor based on the value of k) and,
5. By using the most majority nearest neighbor category then it can be predicted as an object class.

To identify the distance between two points is the data train (x) and the point in data testing (y) used formula Euclidean distance which is shown in Eq. (3).

$$d = \sqrt{(x_2 - x_1)^2 + (y_2 - y_1)^2} \tag{3}$$

2.4 Confusion Matrix

According to Han and Kamber [21], a confusion matrix is a useful tool for analyzing how well classifiers recognize tuples from different classes. Confusion Matrix is created to map the performance of algorithms in tabular form.

Confusion Matrix		*Modeled Values: x_m*	
		True	*False*
Actual Values: x	*True*	*TP*	*FN (Type II error)*
	False	*FP (Type I error)*	*TN*

Fig. 1. Confusion matrix

As shown in Fig. 1, the results of the confusion matrix consist of 4 categories, namely true positive (TP), false positive (FP), false negative (FN), and true negative (TN). True positive represents data in the positive class, which means that it has been successfully predicted correctly by an algorithm. False-positive represents data that should be in positive class but the prediction results show in the negative class. False-negative represents data that should be in the negative class but the prediction results show in the positive class. True negative represents data that is in the negative class and the prediction results show are in the negative class.

2.5 Related Research

Recently, several researchers have researched TV rating predictions. These studies that have been carried out can be displayed in Table 1 below.

Table 1. Related research

Author	Algorithm	Research finding
Akarsu and Diri [26]	Naïve Bayes, Sequential minimal optimization, Random forest, J48	Random forest outperformed compared to other algorithms because it reaches an accuracy of 83.46%
Nugroho et al. [27]	Naïve Bayes	The results of accuracy, precision, recall reached 55.80%, 32.41%, and 46.70%, respectively
Cheng et al. [28]	Back-Propagation Neural Network	This study measures 4 TV programs with the best results of MAE, and MAPE achieved is 0.1775, and 7.59%, respectively
Zhang et al. [29]	Bayesian Network	The average accuracy rate, as a result of the final evaluation, reaches an accuracy level of 89%

3 Research Methodology

In carrying out the process of developing a tv rating prediction model using the k-nearest neighbor algorithm, the steps needed for problem-solving are divided into three stages, namely the stages of identifying problems and solutions, the stages of data preparation and processing, and the stages of data analysis. The systematic details of the research can be seen in Fig. 2.

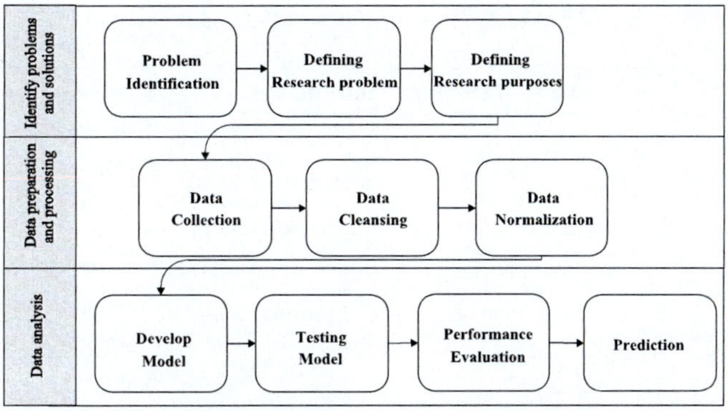

Fig. 2. Research methodology used in this study [21]

Based on the information obtained from Fig. 2, three stages in the research systematic will be explained below:

Identifying Problems and Solutions. At the stage of identifying problems and solutions, problem identification is carried out by digging information at one of the companies engaged in the advertising agency sector in Indonesia. Besides, at this stage, a literature study is also conducted by gathering information related to marketing with television media and television advertisement performance, then identifying solutions that can be used to solve problems and selecting algorithms and tools that will be used.

Data Preparation and Processing. In the data preparation and processing stages, data collection is carried out followed by data cleaning. The results of data cleaning will later be used in the data analysis stage.

Data Analysis. In the data analysis stage, testing is done to calculate the accuracy of the model. At this stage, we will get results that can be drawn as conclusions in this study and also suggestions that can be used for further research.

4 Result and Discussion

4.1 Dataset

Implementation is carried out by obtaining a model that can be used by users to be able to predict television advertisement ratings based on data from October 2019 to December 2019 using the K-Nearest Neighbor algorithm. The expected results of the system in the form of accuracy values that have been processed using the K-Nearest Neighbor algorithm. The dataset has 9 attributes and 22218 records, as shown in Table 2 below.

After the transformation data is performed, the next step is feature selection using the information gain method. Table 3 below is the result of the information gain process.

The results of the feature selection process will then be selected as many as 5 features with the highest information gain value. Based on the feature selection stage, the 5 features with the highest information gain are program, start time (hour), channel, product, and ads type. Table 4 below is the result of the dataset after the information gain process.

Table 2. The attributes of the dataset

No	Attribute	Data type	Information
1	No	Integer	Number of records
2	Day	String	Airtime
3	Channel	String	Television channel
4	Program	String	Television program
5	Product	String	Ad product name
6	Ads type	String	Ad type
7	Start time	Date time	Ad start time
8	Duration	Date time	Duration of ads
9	Cost	Float	Cost
10	TVR	Float	TV Rating or ad performance

Table 3. Result of information gain process

No	Attribute	Information gain
1	Program	0.374296
2	Start time	0.186507
3	Channel	0.109847
4	Product	0.024941
5	Ads type	0.018095
6	Day	0.004199
7	Cost	0.000940
8	Duration	0.000081

Table 4. Data sample results from data transformation

Channel	Program	Product	Ads type	Start time	TVR
0	0	0	0	13	1
0	0	0	1	13	1
1	1	1	2	18	1
1	2	2	3	20	1
⋮	⋮	⋮	⋮	⋮	⋮
2	3	0	1	15	1

4.2 Experimental Results

At this stage, the implementation of data that has been generated in the previous process will be carried out using the K-Nearest Neighbor algorithm. The process is done by dividing the dataset into 80% training data and 20% testing data, conducting data normalization, implementing the K-Nearest Neighbors classification function, and evaluating accuracy.

This test is done by using data testing with the aim to get optimal of the *k* value and also aims to determine the success rate of the model to make predictions that can be known with the accuracy value obtained. Table 5 below showed some test results of the model by using a value *k* with a range of *k* odd number between 3 to 15.

Table 5. Test result using *k* value with a range 3 to 15

Number of *k*	Accuracy	Precision	Recall
3	0.8981	0.6076	0.4865
5	0.9046	0.6073	0.4872
7	0.9082	0.4398	0.4443
9	0.9107	0.4411	0.4434
11	**0.9118**	**0.4417**	**0.4440**
13	0.9082	0.4416	0.4274
15	0.9075	0.4356	0.4370

Fluctuations of the accuracy of the K-Nearest neighbor algorithm for television rating predictions can be seen in Table 5 above. Based on the results of the research conducted on the dataset above, the K-Nearest neighbor classification algorithm with a value of $k = 11$ succeeded in occupying the highest accuracy rating of 91.18% and with a value of $k = 3$ occupying the lowest accuracy rate of 89.81%. Then it can be concluded that the most optimal *k* value to use is 11.

5 Conclusion

The marketing method using television media is one of the effective marketing methods. One of the parameters used to measure the success of an advertisement on television is television rating. Television rating is a unit that measures the value of audience loyalty in watching a television program. This parameter is also used for television station owners in promoting their media and for advertisers in making decisions to determine the channel, time, and program to be chosen as the media for placing advertisements in it. As a company engaged in the field of advertising in collaboration with television media offer advertisers using historical data related to television ratings based on channels, time, and events in a conventional manner. However, the obstacle faced by advertising companies is that they cannot provide rational predictions related to television ratings in the future. Therefore, if the rating value can be predicted, it will certainly be quite helpful in determining the policy for advertising on programs on the television station. From these problems, we need a rational way to predict television broadcast ratings so that advertisers can determine the channel, time, and program for the right advertising. K-Nearest Neighbor (KNN) which is one of the algorithms in data mining, can provide alternative solutions. KNN can reach an accuracy of 91.18% for the prediction of television advertising ratings. So, it can be concluded that the KNN algorithm is suitably applied to the case of the prediction of television advertising ratings.

References

1. Phillips, C.F., Duncan, D.: Marketing Principles and Methods. R.D Irwin, California (1968)
2. Syahputra, I.: Rezim Media: Pergulatan Demokrasi, Jurnalisme, dan Infotainment [Media Regime: The Struggle for Democracy, Journalism and Infotainment]. Gramedia Pustaka Utama (2013)
3. Abdullah, A., Puspitasari, L.: Media Televisi Di Era Internet. ProTVF. **2** 101 (2018). https://doi.org/10.24198/ptvf.v2i1.19880
4. Iskandar, M.S.: Pembentukan Persepsi Visual Pada Iklan Televisi. Visualita **3**, 14–33 (2011). https://doi.org/10.33375/vslt.v3i1.1095
5. Djamal, H., Fachruddin, A.: Dasar-dasar Penyiaran Sejarah, Organisasi, Operasional, Dan Regulasi. Kencana Prenada Media, Jakarta (2011)
6. Larose, D.T.: Discovering Knowledge in Data: An Introduction to Data Mining. Wiley-Interscience, New York (2005)
7. Han, J., Kamber, M., Pei, J.: Data Mining: Concepts and Techniques (2012). https://doi.org/10.1016/B978-0-12-381479-1.00001-0
8. Chiroma, H., et al.: An intelligent modeling of oil consumption. In: El-Alfy, E.-S., Thampi, S.M., Takagi, H., Piramuthu, S., Hanne, T. (eds.) Advances in Intelligent Informatics. AISC, vol. 320, pp. 557–568. Springer, Cham (2015). https://doi.org/10.1007/978-3-319-11218-3_50
9. Slavia, A.P., Sutoyo, E., Witarsyah, D.: Hotspots forecasting using autoregressive integrated moving average (ARIMA) for detecting forest fires. In: 2019 IEEE International Conference on Internet of Things and Intelligence System (IoTaIS), pp. 92–97 (2019)
10. Saedudin, R.R., et al.: A relative tolerance relation of rough set (RTRS) for potential fish yields in Indonesia. J. Coast. Res. **82**, 84–92 (2018). https://doi.org/10.2112/si82-011.1
11. Sutoyo, E., Saedudin, R.R., Yanto, I.T.R., Apriani, A.: Application of adaptive neuro-fuzzy inference system and chicken swarm optimization for classifying river water quality. In: 2017 5th International Conference on Electrical, Electronics and Information Engineering (ICEEIE), pp. 118–122 (2017)
12. Yanto, I.T.R., Sutoyo, E., Apriani, A., Verdiansyah, O.: Fuzzy soft set for rock igneous classification. In: Proceeding - 2018 International Symposium on Advanced Intelligent Informatics: Revolutionize Intelligent Informatics Spectrum for Humanity (SAIN 2018), pp. 199–203. Institute of Electrical and Electronics Engineers Inc. (2019). https://doi.org/10.1109/SAIN.2018.8673383.
13. Márquez-Vera, C., Romero Morales, C., Ventura Soto, S.: Predicting school failure and dropout by using data mining techniques. Rev. Iberoam. Tecnol. del Aprendiz. **8**, 7–14 (2013). https://doi.org/10.1109/RITA.2013.2244695
14. Ibáñez, I., Silander, J.A., Allen, J.M., Treanor, S.A., Wilson, A.: Identifying hotspots for plant invasions and forecasting focal points of further spread. J. Appl. Ecol. **46**, 1219–1228 (2009). https://doi.org/10.1111/j.1365-2664.2009.01736.x

15. Yanto, I.T.R., Sutoyo, E., Rahman, A., Hidayat, R., Ramli, A.A., Fudzee, M.F.M.: Classification of student academic performance using fuzzy soft set. In: 2020 International Conference on Smart Technology and Applications (ICoSTA). Institute of Electrical and Electronics Engineers Inc. (2020). https://doi.org/10.1109/ICoSTA48221.2020.1570606632.
16. Imandoust, S.B., Bolandraftar, M.: Application of K-nearest neighbor (KNN) approach for predicting economic events: theoretical background. Int. J. Eng. Res. Appl. **3**, 605–610 (2013)
17. Liao, Y., Vemuri, V.R.: Use of k-nearest neighbor classifier for intrusion detection (2002). https://doi.org/10.1016/S0167-4048(02)00514-X
18. Hu, Y., Lu, Y., Wang, S., Zhang, M., Qu, X., Niu, B.: Application of machine learning approaches for the design and study of anticancer drugs. Curr. Drug Targets **20**, 488–500 (2018). https://doi.org/10.2174/1389450119666180809122244
19. Mittal, K., Aggarwal, G., Mahajan, P.: Performance study of K-nearest neighbor classifier and K-means clustering for predicting the diagnostic accuracy. Int. J. Inf. Technol. **11**(3), 535–540 (2018). https://doi.org/10.1007/s41870-018-0233-x
20. Shasha, D.E., Bonnet, P.: Database systems. Dr. Dobb's J. **29**, 16–22 (2004). https://doi.org/10.4324/9781351228428-6
21. Han, J., Kamber, M.: Data Mining, Southeast Asia Edition: Concepts and Techniques, Morgan Kaufmann, San Francesco (2006)
22. Lee, C., Lee, G.G.: Information gain and divergence-based feature selection for machine learning-based text categorization. Inf. Process. Manag. **42**, 155–165 (2006)
23. Lei, S.: A feature selection method based on information gain and genetic algorithm. In: 2012 International Conference on Computer Science and Electronics Engineering, pp. 355–358 (2012)
24. Shaltout, N.A., El-Hefnawi, M., Rafea, A., Moustafa, A., El-Hefnawi, M.: Information gain as a feature selection method for the efficient classification of influenza based on viral hosts. In: Proceedings of the World Congress on Engineering. pp. 625–631 (2014)
25. Cover, T., Hart, P.: Nearest neighbor pattern classification. IEEE Trans. Inf. Theory **13**, 21–27 (1967)
26. Akarsu, C., Diri, B.: Turkish TV rating prediction with Twitter. In: 2016 24th Signal Processing and Communication Application Conference (SIU), pp. 345–348 (2016)
27. Nugroho, Y.S.: Others: Prediksi Rating Film Menggunakan Metode Naïve Bayes. J. Tek. Elektro. **8**, 60–63 (2016)
28. Cheng, Y.-H., Wu, C.-M., Ku, T., Chen, G.-D.: A predicting model of TV audience rating based on the Facebook. In: 2013 International Conference on Social Computing, pp. 1034–1037 (2013)
29. Zhang, J., Bai, B., Su, Y.: Study of predicting TV audience rating based on the Bayesian network. Sci. Technol. Eng. **19,** 63811159 (2007)

Check for updates

Predictive Decision Support Analytic Model for Intelligent Obstetric Risks Management

Udoinyang G. Inyang[1(✉)], Imoh J. Eyoh[1], Chukwudi O. Nwokoro[1], and Francis B. Osang[2]

[1] Department Computer Science, Faculty of Science, University of Uyo, Uyo, Nigeria
udoinyanginyang@uniuyo.edu.ng
[2] Department of Computer Science, National Open University of Nigeria, Abuja, Nigeria
fosang@noun.edu.ng

Abstract. Maternal complications (MC) frequently occur during pregnancies and significantly influence pregnancy outcome (PO). Screening test results, frequency and database query, and other conventional approaches used in assessing the impact of MC on PO produce unreliable results. The unavailability of screening tests for some MC also delay into critical stage of pregnancy before manifestation. This paper proposes an intelligent predictive model for mining pieces of knowledge for PO prediction using thirteen attributes selected from a feature space of forty two (42) based on eigenvalues. Five data splitting approaches consisting of cross validation and percentage splitting were used as the train and test datasets. Out of the five machine learning algorithms, Random Forest (RF) had 92.6% and 0.19, as accuracy and root mean squared error values respectively, followed by support vector machine. Although a single dataset used for both training and testing performed best with average accuracy of 85.5%, a 10-fold cross validation dataset which had an accuracy of 84.5% was chosen for better generalization and coverage capabilities. The area under the curve for all the four classes of PO (stillbirth, miscarriage, term and preterm) were above 95.5% while sensitivity and specificity were high for all classes. RF improves the performances of the prediction by statistical methods and further confirms the effectiveness of data mining approaches in the classification and management of obstetric risk.

Keywords: Pregnancy outcome · Random forest · ROC curve · Predictive analytics · Maternal complications

1 Introduction

Health systems are saddled with a vital and unending obligation of promoting, restoring and maintaining people's health throughout their lifetime. They play a pivotal role in the provision of health care services (HCS) in every country. Sustaining and strengthening health-care services delivery is the core goal of the health component of the Millennium Development Goals, which emphasizes the provision of mediations to minimize maternal morbidity and mortality; child mortality and the challenge of managing terminal diseases [1]. Inefficient and sub-standard maternal health care system delivery in most

J. H. Abawajy et al. (Eds.): EATI 2020, LNNS 254, pp. 92–108, 2021.
https://doi.org/10.1007/978-3-030-80216-5_8

developing economies account for high maternal deaths, especially while they are pregnant, during childbirth or few weeks after delivery [2]. In most instances, a significant proportion of women in low income countries lack access to maternal health care infrastructures, therefore record low utilization of crucial health services including emergency obstetric care (EOC). EOC is a proactive panacea for maternal mortality and morbidity. It facilitates easy access and timely life-salvaging HCS for maternal and neonatal health issues through trained and professional health service providers [3].

The provision of these basic services by expert birth assistants or well-timed referral for proper and broad investigations and care is capable of drastically decreasing maternal and neonatal mortalities and disabilities. EOC involves diagnosis, identification, referrals and management of obstetric complications, on the basis that all pregnant women have same likelihood of risks [3, 4]. Some of the pregnancy complications include ruptured membranes, blood strained mucus discharge, with palpable contractions and Human Immunodeficiency Virus (HIV) among others [5, 6]. The newborns may suffer from dehydration, infections, jaundice, neonatal jaundice, congenital malformation, congenital malaria, and chromosomal abnormality, hemorrhagic diseases, etc. [7–9]. These complications require proper and timely attention since they are life threating emergency situations which require proper management to reduce morbidity and mortality. World health organization (WHO) in 2017 affirmed that globally, over 340,000 maternal deaths, 2.7 million still births and 3.1 million neonatal deaths resulting from pregnancy complications occur annually [10, 11] observed that in some part of sub–Saharan Africa, women are exposed to life time risks and often to the point of death during pregnancy, child birth, or postnatal period.

Arising from these threats, complication risks prediction by obstetricians and other health personnel is very vital for early and timely interventions, treatment or management. Pregnancy outcome (POs) are largely determined by the complications and risk factors associated with the pregnancy. For example, placental dysfunction is connected to a number of key complications of pregnancy associated with adverse maternal and infant outcome, such as preeclampsia, fetal growth restriction, preterm birth and stillbirth. Health researches focus on utilizing information resulting from complications which is still largely based upon clinical grounds, ultrasonic and/or biochemical assessments [12], rather than intelligent mining knowledge for informed decision making. Like other emergencies, pregnancy complications manifest in varying degrees of symptoms (indicators), and also changes continuously, thereby creating confusion and making it difficult for stakeholders and responders, to create robust procedures [13] for short and long term maternal health care goals. Hence, the need to comprehend the nature, causes, impact and treatment plan required to prevent or control their occurrences. The reliance on screening test results, frequency and database query indicators and approaches for the prediction of complications are often unreliable largely because few symptoms (if not a single symptom) are considered while considering the appropriate course of action. The combination and association of clinical signs, test results, genetic make-up of the patient, health care facilities and other available features would produce more robust and accurate results. In addition, there are no basic screening tests available for some pregnancy complications (such as preeclampsia) and their detection depends on frequent maternal visits and records of vital signs. In this case complications might be detected at a critical stage of the pregnancy [14] thereby making it very arduous to manage.

A significant progress has been recorded in the development of statistical predictive models with better results than clinical tests, but always come at a later stage of pregnancy. The improvement of maternal HCS largely depends on an understanding of the associated risks, and will greatly impact on the future of maternal health care while attempting to curb maternal morbidity. This paper, aims at developing a predictive model capable of learning from pregnancy complications and risk factors data, for classifying and predicting pregnancy outcomes at an early stage, to improve maternal morbidity. This will be achieved by comparing the performances of machine learning algorithms and varying the dataset formation strategy for improved prediction results.

2 Related Works

Complications among mothers within child bearing age and newly born babies occur frequently. As, such, some works have focused on using statistical methods such as descriptive, retrospective and theoretical studies to create a lot of awareness on obstetric and neonatal complications. Reference [15] motivated by increasing frequency of maternal death, examined severe acute obstetric complications in rural Bangladesh by using five common obstetric complications; hemorrhage, puerperal sepsis, eclampsia, obstructed labour and induced abortion. Reference [16] carried out a study on barriers to emergency obstetric care service in perinatal deaths in rural Gambia. The study contributed to improvements on the response to EOC towards maternal and neonatal survival in poor rural settings in the Gambia. Reference [17] investigated essential basic emergencies of obstetric and newborn care using education, training to service delivery and quality of care as a case study. As transportation remains a vital tool to reducing traveling time for arrival at the HCS point [18–20], considered geographic access as a problem of emergency in obstetric and neonatal care and developed a geographic information system to bridge the gap between patients and HCS points. References [21, 22] stressed the need to reduce emergency obstetric referrals especially for patients with pre-eclampsia and eclampsia, and the need for constant supply of the right drugs which include magnesium sulphate. Despite the large access to literature describing the need for improving women's access to maternity care during emergencies, effective implementation still remains a huge problem. Reference [23] suggested that there should be more awareness in terms of referral interventions for patients.

Over the years machine learning approaches have been applied to investigate maternal and neonatal complications. Support Vector machine (SVM) has been known to outperform many machine learning algorithms in many applications, in terms of prediction accuracy and computational cost [24]. The work reported in [25] employed SVM-based decision support system for preterm birth risks prediction. The model predicted when the birth is likely to occur and the possible outcome of babies; status. The results of the empirical evaluations showcased SVM as the most performing in terms of intelligent and comprehensive inference mechanism regarding decision support for pregnant women who are at risk. The result yielded a true positive rate (TPR) of 83.9%, a false positive rate (FPR) of 0.27, and receiver operating characteristic (ROC) area of 0.79.

Moreso, [26–29] deployed decision support tools that would provide assistance to practitioners in ensuring safety of vaginal births after cesarean delivery for women of child bearing age and in the general management of PO. References [30, 31] have demonstrated the effectiveness of decision support systems in handling associations between two or more obstetric and neonatal emergencies. A comparative analysis of machine learning (ML) tools and statistical models (STs) was reported in [32] for the prediction of postpartum hemorrhage (PH) risks during labour with the aim of minimizing maternal morbidity and mortality. The experiments on data from 12 sites showed that all the ML and ST models adopted in the study produced satisfactory results, although the extreme gradient boosting model (XGboost) had the best ability to discriminate among PH followed by random forests (RF) and lasso regression model. Reference [33] compared linear and non-linear kernel based SVM and logistic regression models for forecasting the occurrence of preterm births. Private data were obtained from hospitals in India which included the following variables; age, number of times pregnant, obesity, diabetes mellitus and hypertension among other features were subjected to 10-fold cross validation test mode for each run. The results indicated that SVM provided a better prediction accuracy of 86% compared to the logistic regression model. Reference [34] applied eight (8) ML algorithms – artificial neural network (ANN), SVM, kNN, RF, classification and regression tree, logistic regression, C4.5 and radial basis function network were used to predict fetal pre-birth wellbeing using antepartum cardiotocography data. The authors adopted 10-FCV on the data with result revealing that RF outperformed other classifiers with an overall accuracy of 99.18%. The effectiveness of ML methods in mining of electronic health data in the domain of atrial fibrillation (AF) induced risks prediction was reported in [35]. Out of a total of 2,252,219 women used for the study, 1,225,533 developed AF during a selected 6-month interval. Two hundred (200) widely used electronic health record features, (age and sex inclusive), and random oversampling approach implemented with a single-layer, fully connected ANN yielded the optimal prediction of six-month incident AF, with an area under the receiver operating characteristic curve (AUC) of 80.0% and an F1 score of 11.0%. The ANN model performed only marginally better than the basic logistic regression model consisting of known clinical risk factors for AF, which had 79.4% and 79.0% as AUC and F1 value respectively. The results confirmed the effectiveness of ML in the prediction of AF in patients. The performance of Fuzzy approach, SVM, RF and Naïve Bayes (NB) for the prediction of cardiotocograph-based labour stage classification from patients with uterine contraction pressure during ante-partum and intra-partum period, the proposed algorithm tend to be efficient and effective in terms of visual estimation to incorporate automated decision support system, which will help to reduce high risk of hospitalized patients. Reference [36] proposed a hybrid system consisting of bijective soft set and back propagation ANN for the prediction of neonatal jaundice. The neonatal jaundice dataset comprising 808 instances with 16 attributes collected from January to December, 2007 in neonatal intensive care unit in Cairo, Egypt, was used for the experiment. The proposed system was compared with bijective soft set, back propagation ANN, MLP, decision table and NB and found to provide the best accuracy of 99.1%. In Reference [37] results of the investigation impact computational intelligence on the precision of cardiovascular medicine. The method was applied to neonatal coarctation classification and prediction by analyzing genome-wide DNA methylation of newborn blood DNA using 24 isolated, non-syndromic cases and 16 controls of the Illumina HumanMethylation450 BeadChip arrays using six artificial intelligence (AI) platforms including deep learning was used for detection. Deep learning achieved the optimal performance with an AUC and sensitivity of 95% and 98% specificity at 95% confidence interval. The related works considered were based on a single dataset test mode. The significance of this work is the assessment of each classifier on varying dataset test modes.

3 Materials and Methods

3.1 Classification Algorithms

Classification algorithms assigns objects to targeted clusters. It is a main approach in data mining (DM) for assigning categories of objects to targeted clusters. The reason for classification is the prediction of types of data-points based on the provided classes. Although there are many types of algorithms available in data mining for solving medical problems, RF, support vector machine (SVM), decision tree (DT), naïve Bayes (NB) and multi-layer perceptron (MLP) are widely used and have been considered for predicting pregnancy outcomes.

RF is an algorithm that creates and uses ensembles of classification trees which are grown on either bootstrap, bagging or voting scheme [38]. Features are selected at random to enhance their performances for efficient and effective results. The final class of individual objects is determined by voting over all trees in the ensemble. RF is a suitable and commonly adopted tool for medical and non-medical fields [39–43]. RF is robust and maintains accuracy irrespective of missing values in the dataset. It can handle a large number of input attributes consisting of both qualitative and quantitative attributes and ranks features before classification [38].

SVM is one of the widely used ML algorithms. It was introduced by [44]. It is vigorously used for ranking a function, regression and classification problems. It makes use of kernels to handle non-linear classification and maps its inputs into high-dimensional feature spaces because its peculiarity is being used for most medical research and it does not require a previous knowledge. Thus, SVM is a good classifier and particularly suitable for handling high dimensionality in data [45–48].

DT is an unsupervised ML tool capable of clustering predictions and modeling outcome explicitly in classes as well as visualized tree like representations. It has the high processing ability to handle large or small dataset in normal and numerical scales since they do not rely on domain knowledge. Reference [40] concluded that DTs are suitable for exploratory knowledge mining and statistical analysis. It algorithm supports only categorized values as the target domain [49–51].

NB is a widely used classifier that adopts bayesian reasoning—a logical methodology for updating the probability of any hypotheses given new instances. Hence, it plays a pivotal role in scientific researches. NB is a better choice for handling weighted data, incomplete data, small datasets, categorical variables and heterogeneous data types [52] NB uses class membership probabilities—the likelihood that a given data object is a member of a particular class. An NB classifier assumes that the presence (or otherwise) of members of a class is unconnected to the presence (or otherwise) of other variables when information on the class label is provided [53]. NB requires short computational time for model building and removes insignificant features from the dataset. It is used for comparative medical predications with other classifiers [54–56].

MLP is a feedforward Neural Network, with a hidden layer of nodes in addition to the input and output layers, and operates in human brain-like manner. It is suitable for non-linear modelling problems and can learn complex relationships between dependent and independent variables without assistance. Due to its performance and robustness it is applied to medical domain, education and speech technology [57–60].

3.2 Proposed Predictive Model

This work proposed and implemented a predictive model consisting of multiples classifiers for comparative analysis of their performances, voting and predictive analytics. The system model (Fig. 1) has knowledge base (KB), pre-processing engine, classification engine, decision support engine and user interface as main components. The user interface provides a means of communication and exchange of information between the environment and the system.

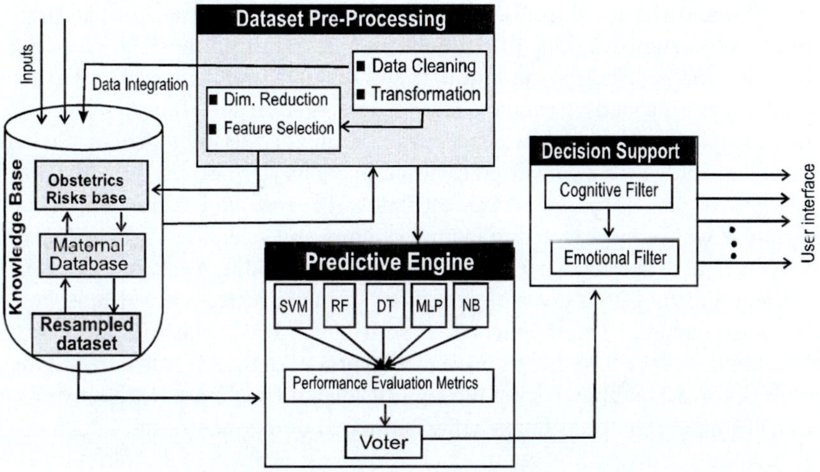

Fig. 1. Proposed system framework

The main end users of the system are the various stakeholders in obstetric management including health personnel, patients and relatives of patients. The KB is populated with information and data obtained through various locations and techniques of data acquisition (ah-hoc manual methods, automatic data generation and systematic approaches). Although, the suitability of any approach varies with the nature of data and source, data-driven knowledge acquisition systems is strongly recommended [61] to reduce computational resources and efforts.

3.3 Knowledge Base and Data Pre-processing Engine

The KB consists of semantically related network of maternal data objects linked together in a relational form. The major components include obstetric risk-base, maternal database and resampled datasets. The obstetric risk comprises complications, symptoms or clinical signs, diseases and their associated impact and risk on patients. It also holds information about the source and susceptibility of patients and other various maternal related hazards. Constituents of the maternal database are records generated from obstetric activities in health facilities. For example profile of patients during pregnancy, antenatal, postnatal or neonatal stages including obstetric risks outside pregnancy. In this work, maternal database was populated from data acquired manually from secondary health facilities in Uyo, Nigeria. A total of one thousand six hundred and thirty two (1632) records were obtained from archives of retrospective records of pregnant mothers taken while they enrolled for antenatal care, with an input feature space of forty two (42) attributes while the target attribute is PO. Some of the attributes include; maternal age, number of children delivered, previous medical history, abortion, miscarriage, prematurity, previous illness, number of attendances to antenatal care, antenatal registration, and mode of delivery, amongst other features.

Data pre-processing engine spans data cleaning, transformation, integration, dimension reduction and feature selection stages. Data cleaning and transformation produce uniform and consistent versions of the data for integration in the KB. The dimension reduction and feature selection, and grouping of the data into their respective areas preceded data integration step. Attribute cleaning, aggregation and elimination of attributes with only a single domain value was performed. Categorization of attributes where discrete textual attributes were converted to numeric codes was also performed. Composite attributes were broken down into atomic attributes. The resultant dataset which had thirty five (35) attributes were subjected to feature ranking and selection via Principal component analysis (PCA) in Waikato Environment for Knowledge Analysis (WEKA). PCA is an efficient approach that gives explanations of the variance a set of inter-correlated variables with a smaller set of independent attributes. It is a method that extracts the eigenvalues and eigenvectors from covariance matrix of default features [62]. This paper adopted PCA for the computation of the significance of each input feature by producing a new set of uncorrelated attributes called principal components, and which are organized sequentially with the first component explaining as much of the variation as it can. Every principal component was represented linearly by combining variables of the original dataset, in which the coefficients depicts the relative contribution of the variables to the variability of the target feature [63]. The feature selection involved the identification of the target variable in the dataset and selection of significant input features (features whose eigenvalues are equal to or greater than unity) from the resultant feature space, needed for the knowledge mining. Eigenvalues of the principal components were the basis for input rank analysis and selection. Attributes with eigenvalues score greater than or equal to unity were thirteen (13) and together accounted for 67.13% variation of the target feature. The resultant rank and attribute description given in Table 1, shows that average maternal blood pressure topped the list with eigenvalue of 3.86 (11.7% proportion of variance), followed by maternal weight (eigenvalue = 2.77, proportion = 8.39%).

Table 1. Significant attribute description, importance and rank

Rank	Attribute	Description	Eigenvalue	Proportion (%)	Cumulative (%)
1	Maternal BP	Average maternal blood pressure taken during antenatal visits	3.86	11.69	11.69
2	Maternal Weight	Average maternal weight during pregnancy and antenatal care	2.77	8.39	20.29
3	Hemoglobin Level	Average number of red blood cells count during antenatal visits	2.37	7.18	27.47
4	PCV level	Average Packed Cell Volume count during antenatal care	1.92	5.82	33.29
5	Pulse Rate	Average number of heart beats per minute taken during antenatal period	1.54	4.67	37.67
6	Mode of Delivery	Delivery method virginal delivery = 1; caesarean section = 2	1.42	4.30	42.26
7	Malaria Frequency	Number of times maternal malaria diagnosis	1.39	4.21	46.47
8	Hepatitis C	Indicates history of hepatitis C disease; presence = 1, absence = 2	1.26	3.82	50.29
9	Diabetes Status	Maternal Diabetic status non-diabetic = 0 type 1 = 1; type 2 = 2, others = 3	1.18	3.60	53.89
10	Herbal Ingestion	Use of herbal medicinal products during pregnancy	1.15	3.48%	57.37
11	Respiratory disorder	Maternal respiratory disease status; presence = 1, absence = 2	1.12	3.39%	60.76
12	Age	Maternal age during pregnancy	1.06	3.20%	63.96

(*continued*)

Table 1. (*continued*)

Rank	Attribute	Description	Eigenvalue	Proportion (%)	Cumulative (%)
13	Ascorbic acid Level	Average amount of ascorbic acid in the body during pregnancy	1.05	3.17%	67.13
14	Pregnancy outcome	Maternal delivery outcome miscarriage = 0; pre-term = 1; full-term = 2, stillbirth = 3	–	–	–

The five topmost attributes accounted for a cumulative variance of 37.67% while the last five ranked (9th–13th) significant features together earn a cumulative proportional effect of 16.84%. The thirteenth ranked attribute, Ascorbic Acid level accounted for 3.17% variation with eigenvalue score of 1.05. The target feature, PO has four distinct values as follows; miscarriage(1), pre-term(2), full-term(3) and stillbirth (4).

3.4 Predictive Analytic Workflow

The intelligent mining of pieces of knowledge is necessary to guide informed decision making in the domain of obstetric management in general and PO prediction in particular spanned five (5) major steps; 1) data set formation/splitting 2) model building 3) prediction and classification 4) evaluation, and 5) deployment. Five methods of data formation were performed based on dataset splitting, 100%–100% dataset approach—where the entire dataset serves as training and testing dataset (without splitting), 80% -20% split—80% for training and 20% for testing, 70%-30% split where 70% of the dataset is used for training and 30% for testing. The other two are k-fold cross validation (k-FCV). The k-FCV was structured to iterate for values k alternating between 5 and 10. In each case the dataset was split into k-folds and implemented in k iterations where the *kth* fold was used for testing. These datasets were used for model building with RF, MLP, SVM-SMO and DT-C4.5 classifiers for comparative analysis based on accuracy and model building time. The best performing model in conjunction with the best dataset formation strategy will be used for prediction of PO and the results evaluated with receivers operating characteristics (ROC) curve derivatives (i.e. sensitivity, specificity, recall, precision, area under the curve etc.) and other performance measures like kapa statistic (KS), mean absolute error (MAE), root mean squared error (RMSE) and relative absolute error (RAE) and the confusion matrix [64]. Once the analytic phases have been concluded and the model has been calibrated, results are integrated into the decision support system objective and subjective filtering by obstetrician and other stakeholders.

4 Experimental Results

4.1 Model Selection and Evaluation

The resultant attributes from the feature selection stage were used to rank the five classifiers—SVM optimized by sequential minimal optimization (SMO), RF, NB, DT and

MLP, based on some performance metrics regarding PO prediction. The dataset were resampled as follows; k-FCV variants (5-FCV and 10-FCV), percentage split datasets consist of 80-20 (20% resampled for testing/validation), 70-30 dataset (where 30% was reserved for testing/validation) and 100-100 (where no splitting was done by using the same dataset for model training and testing). The experiment spanned five phases of performance evaluation of each model, in the different dataset formations and implemented in WEKA using default parameters of classifiers. The results of the models' performances on the various datasets are summarized in Table 2 and in Fig. 2

Table 2. Accuracy of classifiers based on test dataset formation

Classifier	Accuracy (%)					
	5-FCV	10-FCV	70-30	80-20	100-100	Average
SVM-SMO	87.20	87.80	79.40	79.60	86.90	84.18
NB	79.90	80.60	76.10	76.80	76.90	78.06
MLP	86.60	87.30	79.50	81.10	85.10	83.92
RF	88.90	89.90	77.70	81.90	92.60	86.2
DT-C4.5	83.70	77.10	86.00	81.40	86.00	82.84

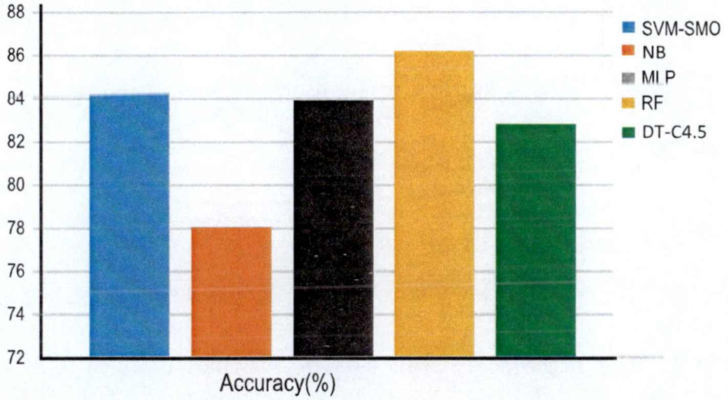

Fig. 2. Average accuracy of classifiers based on test dataset formation

As shown in Table 2, the best dataset formation strategy is the 100-100 dataset used as train and test dataset, it gave an average accuracy of 85.5% in all the classifiers followed by 5-FCV (85.26%) while 70-30 splitting earned the least average score of 79.74%. The overall top most accuracy score is 92.6% (RF classifier) and the least is 76.1% (NB). In the 100-100 data method the least score (76.9%) belongs to NB while the performances of MLP, DT-C4.5 and SVM-SMO are in the range 85.1%–86.9%. Figure 2 gives a visual of the average accuracy of classifiers across datasets; RF (86.2%)

is the best performing classifier followed by MLP (83.9%) while NB (78.06%) gave the least accuracy. RF stands out in terms of overall accuracy in all dataset formations with yielded an accuracy of 92.6% and 86.2% in 100-100 split and 10-FCV datasets respectively. However, the 100-100 test mode was only used as a control dataset since it is known for poor generalization capabilities and prone to overfitting when presented with new instances. That informed the choice of 10-FCV dataset for predicting PO. The resultant models generated from 10-FCV dataset were evaluated with KS, MAE, RMSE and RAE. The summary of results is presented in Table 3 and Fig. 3.

Table 3. Summary of classifiers' performances on PO prediction using 10-FCV

Classifier	Evaluation metric					
	KS	MAE	RMSE	RAE	Coverage of cases (%)	Time (in sec.)
SVM-SMO	0.38	0.27	0.35	1.40	97.90	0.0100
NB	0.33	0.16	0.32	0.80	90.44	0.0400
MLP	0.40	0.12	0.26	0.64	94.40	0.0001
RF	0.78	0.09	0.19	0.44	100	0.2100
DT-C4.5	0.44	0.14	0.2631	0.71	96.70	0.0001

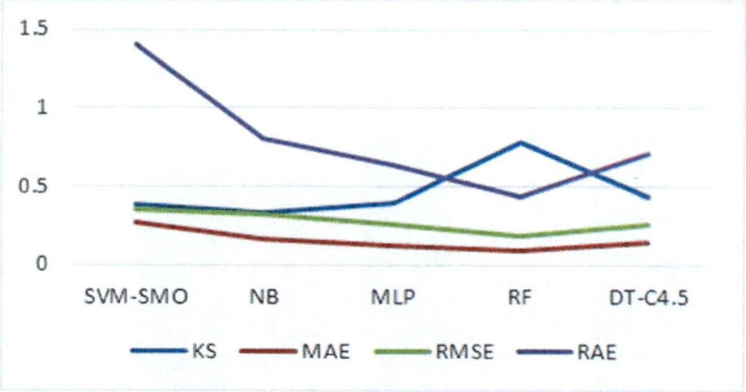

Fig. 3. Performance measurement scores for classifiers based on 10-FCV

The results in Table 3 show that MLP and DT-C4.5 require insignificant time (in seconds) compared to the other classifiers. RF algorithm requires the longest model building time (0.2100 s) followed by NB (0.0400) and SVM-SMO (0.0100). In terms of coverage of cases at 0.95 level in the dataset, RF recorded 100% coverage while SVM-SMO, DT-C.45 and MLP have 97.90, 96.70% and 94.40% respectively. KS, a commonly used to measure of inter-rater reliability, quantifies the degree of agreement between classifications with similar classes [65]. SVM-SMO and NB depicts fair agreement (0.2–0.4) while there is moderate agreement of classification schemes in MLP (0.40) and DT-C4.5 (0.40). A substantial agreement is depicted by kappa's coefficient earned by RF

(0.78). This confirms the suitability of RF in the classification of PO. The error resulting from the classification and predictions are also reported in Table 3 with RF having the lowest values in all the three metrics—MAE (0.09), RAE (0.44) and RMSE (0.19). The performance of RF classifier confirms its suitability for prediction and classification in the domain of health care and also in tandem with results reported in [34, 66], although accuracy reported in [34] is 6.58% higher. The class accuracy of RF model is satisfactory.

4.2 RF Model Evaluation

Specificity and sensitivity are key and effective indicators that quantify the inherent strength of predictive and prescriptive tasks for discretize and continuous results [67, 53]. The components of the ROC curve—specificity, sensitivity, area under the curve (AUC), and partial area under the curve (PRC), are useful in medical researches. The sensitivity (True Positive rate) measure in the domain of PO prediction, gives the degree to which the classifier correctly predicts all outcomes of the maternal delivery while specificity (true negative rate) quantifies the correctness to predicting those pregnancies not resulting to a particular PO [68, 54]. The details of classification by RF model are depicted in the ROC curve (Fig. 4), Tables 3 and 4.

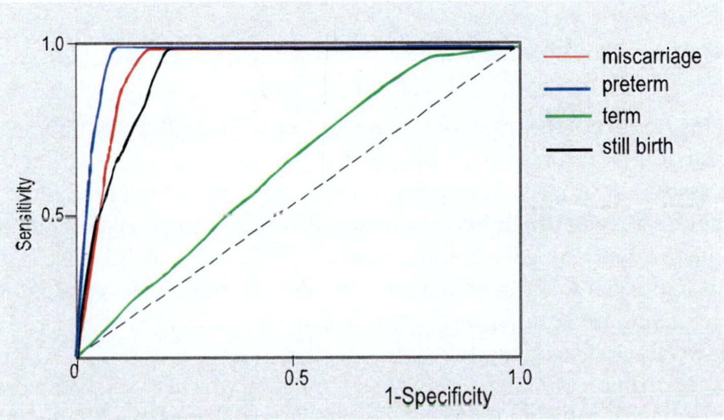

Fig. 4. ROC curve for pregnancy outcome prediction

As shown in Fig. 3 and Table 3, AUC for each class is above 95.5% depicting a high sensitivity, specificity, accuracy and precision values of predictions. The results show that out of 1632 instances, 114 (6.9%) were stillbirth and 1255 (76.9%) were full term births. Preterm and miscarriages represented 39.8% and 12.20% respectively. However, out of 114 still-birth cases, 31 (27.20%) were correctly classified while 71.90% and 0.010% were misclassified as term-birth and miscarriage cases respectively. The sensitivity was highest in the prediction of term-births with 99.4% (1247) followed by the prediction of miscarriage instances which recorded 82.23% sensitivity value. The classification of still-births and miscarriages recorded the lowest FPR (which implies high specificity) while FPR of 37.70% is recorded when term-births are classified (Table 5).

Table 4. Details of RF predictions by class

	TPR	FPR	Precision	Recall	F-Measure	AUC	PRC	Class
	0.272	0.004	0.838	0.272	0.411	0.956	0.640	Stillbirth
	0.994	0.377	0.898	0.994	0.943	0.962	0.988	Term
	0.569	0.001	0.949	0.569	0.712	0.993	0.832	Preterm
	0.823	0.003	0.976	0.823	0.893	0.996	0.971	Miscarriage
Weighted Avg.	0.906	0.290	0.905	0.906	0.891	0.967	0.956	

Table 5. Confusion matrix of PO classification

	Still birth	Term	Preterm	Miscarriage
Still birth	31	82	0	1
Term	5	1247	1	2
Preterm	0	27	37	1
Miscarriage	1	33	1	163

5 Conclusion

Complications frequently occur during pregnancies and significantly influence PO. The reliance on laboratory test results, frequency and traditional record querying approaches adopted to assess the impact of complications on PO produce unreliable results and delay. In addition, the unavailability of screening tests for some pregnancy complications causes the manifestation of such complications to delay into critical stage of pregnancy. This paper proposes an intelligent predictive model for mining pieces of knowledge for PO prediction using thirteen attributes selected from a feature space of forty two (42) based on eigenvalues. The methodology for attribute selection was PCA which produced eigenvalues for attribute ranking. Five data splitting approaches consisting cross validation and percentage-split were the test and train datasets formation methods. Out of the five machine learning algorithms, RF had 92.6% and 0.19 as accuracy and RMSE values respectively, followed by SVM. Although a 100-100 dataset used for both training and testing performed best with average accuracy of 85.50%, 10-FCV with an accuracy of 84.50% was chosen due to its better generalization and coverage capabilities. The AUC for all four classes of PO (stillbirth, miscarriage, term and preterm) was above 95.50% while sensitivity and specificity were high for all classes. Class based assessments show that out of 114 still-birth cases, 31 (27.20%) were correctly classified while 71.90% and 0.010% were misclassified as term-birth and miscarriage cases respectively. These results proves that among the machine learning algorithms tested, RF classifier has the potential to significantly improve the conventional classification methods for use in medical researches, supports multi-class problems. It further confirms the effectiveness of data mining approaches in the classification and management of obstetric risks. Furthermore, the impact of the dataset formation strategies on classification results has been demonstrated. The results show that, the performance of predictive analytics is dependent on the data formation methodology and the choice must support generalization and coverage. A comparative analysis and hybridization of more supervised learning algorithms and the determination of the optimum parameters for the classifiers is for future work.

References

1. Austin, A., Ana L., Salam, R., Lassi, Z., Das, J., Bhutta, Z.: Approaches to improve the quality of maternal and newborn health care: an overview of the evidence. Reprod. Health **11**(2), S1 (2014)
2. Nuamah, G.B., et al.: Access and utilization of maternal healthcare in a rural district in the forest belt of Ghana. BMC Pregnancy Childbirth **19**(1), 6 (2019)
3. Bhandari, T.R., Dangal, G.: Emergency obstetric care: strategy for reducing maternal mortality in developing Countries (2014)
4. Adeyi, O., Morrow, R.: Concepts and methods for assessing the quality of essential obstetric care. Int. J. Health Plann. Manage. **11**(2), 119–134 (1996)
5. Amenu, G., Mulaw, Z., Seyoum, T., Bayu, H.: Knowledge about danger signs of obstetric complications and associated factors among postnatal mothers of Mechekel District Health Centers, East Gojjam Zone, Northwest Ethiopia 2014 (2016)
6. Filippi, V., et al.: Effects of severe obstetric complications on women's health and infant mortality in Benin. Tropical Med. Int. Health **15**(6), 733–742 (2010)
7. Hossain, M., Begum, M., Ahmed, S., Absar, M.: Causes, management and immediate complications of management of neonatal jaundice? A hospital-based study. J. Enam Med. Coll. **5**(2), 104–1095 (2015)
8. Grgić, G., Brkičević, E., Ljuca, D., Ostrvica, E., Tulumović, A.: Frequency of neonatal complications after premature delivery. J. Health Sci. **3**(1), 65–69 (2013)
9. Ward, R. M., & Beachy, J. C.: Neonatal complications following preterm birth. BJOG Int. J. Obstet. Gynaecol. **110**, 8–16 (2003)
10. Banajeh, S.: Learning from low income countries: Investing in traditional birth attendants may help reduce mortality in poor countries. BMJ **330**(7489), 478–479 (2005)
11. Khan, M., Hashim.: Boundary layer flow and heat transfer to Carreau fluid over a nonlinear stretching sheet. AIP Adv. **5**(10), 107203 (2015)
12. Gaccioli, F., Lager, S., Sovio, U., Charnock-Jones, D.S., Smith, G.C.: The pregnancy outcome prediction (POP) study: investigating the relationship between serial prenatal ultrasonography, biomarkers, placental phenotype and adverse pregnancy outcomes. Placenta **59**, S17–S25 (2017)
13. Inyang, U.G., Akinyokun, O.C.: A hybrid knowledge discovery system for oil spillage risks pattern classification. Artif. Intell. Res. **3**(4), 77–86 (2014)
14. Leemaqz, S.Y., et al.: Maternal marijuana use has independent effects on risk for spontaneous preterm birth but not other common late pregnancy complications. Reprod. Toxicol. **62**, 77–86 (2016)
15. Sikder, S.S., et al.: Accounts of severe acute obstetric complications in rural Bangladesh. BMC Pregnancy Childbirth **11**(1), 76 (2011)
16. Jammeh, A., Sundby, J., Vangen, S.: Barriers to emergency obstetric care services in perinatal deaths in rural gambia: a qualitative in-depth interview study. ISRN Obstet. Gynecol. 2011 (2011)

17. Otolorin, E., Gomez, P., Currie, S., Thapa, K., Dao, B.: Essential basic and emergency obstetric and newborn care: from education and training to service delivery and quality of care. Int. J. Gynecol. Obstetr. **130**, S46–S53 (2015)
18. Chen, Y.N., Schmitz, M.M., Serbanescu, F., Dynes, M.M., Maro, G., Kramer, M.R.: Geographic access modeling of emergency obstetric and neonatal care in Kigoma Region, Tanzania: transportation schemes and programmatic implications. Global Health Sci. Pract. **5**(3), 430–445 (2017)
19. Keyes, E.B., Parker, C., Zissette, S., Bailey, P.E., Augusto, O.: Geographic access to emergency obstetric services: a model incorporating patient bypassing using data from Mozambique. BMJ Global Health **4**(Suppl 5), e000772 (2019)
20. Ntambue, A.M., Malonga, F.K., Cowgill, K.D., Dramaix-Wilmet, M., Donnen, P.: Emergency obstetric and neonatal care availability, use, and quality: a cross-sectional study in the city of Lubumbashi, Democratic Republic of the Congo, 2011. BMC Pregnancy Childbirth **17**(1), 40 (2017)
21. Singh, A., Nandi, L.: Obstetric emergencies: role of obstetric drill for a better maternal outcome. J. Obstet. Gynecol. India **62**(3), 291–296 (2012)
22. Akaba, G.O., Ekele, B.A.: Maternal and fetal outcomes of emergency obstetric referrals to a Nigerian teaching hospital. Trop. Doct. **48**(2), 132–135 (2018)
23. Hussein, J., Kanguru, L., Astin, M., Munjanja, S.: The effectiveness of emergency obstetric referral interventions in developing country settings: a systematic review. PLoS Med. **9**(7), e1001264 (2012)
24. Kim, S., Yu, Z., Kil, R.M., Lee, M.: Deep learning of support vector machines with class probability output networks. Neural Netw. **64**, 19–28 (2015)
25. Moreira, M.W.L., Rodrigues, J.J.P.C., Marcondes, G.A.B., Venancio Neto, A.J., Kumar, N., de la Torre Diez, I.: A preterm birth risk prediction system for mobile health applications based on the support vector machine algorithm. In: 2018 IEEE International Conference on Communications (ICC), pp. 1–5. IEEE (2018)
26. Kuppermann, M., et al.: Effect of a patient-centered decision support tool on rates of trial of labor after previous cesarean delivery: the PROCEED randomized clinical trial. JAMA **323**(21), 2151–2159 (2020)
27. Vinks, A.A., et al.: Electronic health record–embedded decision support platform for morphine precision dosing in neonates. Clin. Pharmacol. Ther. **107**(1), 186–194 (2020)
28. López-Martínez, F., Núñez-Valdez, E.R., García-Díaz, V., Bursac, Z.: A case study for a big data and machine learning platform to improve medical decision support in population health management. Algorithms **13**(4), 102 (2020)
29. Løhre, E.T., Thronæs, M., Brunelli, C., Kaasa, S., Klepstad, P.: An in-hospital clinical care pathway with integrated decision support for cancer pain management reduced pain intensity and needs for hospital stay. Support. Care Cancer **28**(2), 671–682 (2019). https://doi.org/10.1007/s00520-019-04836-8
30. Pick, R.A.: Benefits of decision support systems. In: Handbook on Decision Support Systems, vol. 1, pp. 719–730. Springer, Heidelberg (2008). https://doi.org/10.1007/978-3-540-48713-5_32
31. Ekong, V., Inyang, U.G., Onibere, E.A.: Intelligent decision support system for depression diagnosis based on neuro-fuzzy-CBR hybrid. Mod. Appl. Sci. **6**(7), 79 (2012)
32. Venkatesh, K.K., et al.: Machine learning and statistical models to predict postpartum hemorrhage. Obstet. Gynecol. **135**(4), 935–944 (2020)
33. Prema, N.S., Pushpalatha, M.P.: Machine learning approach for preterm birth prediction based on maternal chronic conditions. In: Sridhar, V., Padma, M.C., Rao, K.A.R. (eds.) Emerging Research in Electronics, Computer Science and Technology. LNEE, vol. 545, pp. 581–588. Springer, Singapore (2019). https://doi.org/10.1007/978-981-13-5802-9_52

34. Sahin, H., Abdulhamit, S.: Classification of "the cardiotocogram data for anticipation of fetal risks using machine learning techniques." Appl. Soft Comput. **33**, 231–238 (2015)
35. Tiwari, P., Colborn, K.L., Smith, D.E., Xing, F., Ghosh, D., Rosenberg, M.A.: Assessment of a machine learning model applied to harmonized electronic health record data for the prediction of incident atrial fibrillation. JAMA Network Open. **3**(1), e1919396 (2020)
36. Azar, A.T., Hannah Inbarani, H., Udhaya Kumar, S., Own, H.S.: Hybrid system based on bijective soft and neural network for Egyptian neonatal jaundice diagnosis. Int. J. Intell. Eng. Inform. **4**(1), 71–90 (2016)
37. Bahado-Singh, R.O., et al.: Precision cardiovascular medicine: artificial intelligence and epigenetics for the pathogenesis and prediction of coarctation in neonates. J. Maternal-Fetal Neonatal Med. **4**, 1–8 (2020)
38. Nedjar, I., El Habib Daho, M., Settouti, N., Mahmoudi, S., Chikh, M.A.: Random Forest based classification of medical x-ray images using a genetic algorithm for feature selection. J. Mech. Med. Biol. 15(02), 1540025 (2015)
39. Moreira, M.W., Rodrigues, J.J., Oliveira, A.M., Saleem, K., Neto, A.J.V.: Predicting hypertensive disorders in high-risk pregnancy using the random forest approach. In: 2017 IEEE International Conference on Communications (ICC), pp. 1–5. IEEE (2017)
40. Senthilkumar, D., Paulraj, S.: Prediction of low birth weight infants and its risk factors using data mining techniques. In: Proceedings of the 2015 International Conference on Industrial Engineering and Operations Management, pp. 186–194 (2015)
41. Horning, N.: Random Forests: an algorithm for image classification and generation of continuous fields data sets. In: Proceedings of the International Conference on Geoinformatics for Spatial Infrastructure Development in Earth and Allied Sciences, Osaka, Japan, vol. 911 (2010)
42. Ricordeau, J., Lacaille, J.: Application of random forests to engine health monitoring. In: ICAS, pp. 1–10 (2010)
43. Mei, J., He, D., Harley, R., Habetler, T., Qu, G.: A random forest method for real-time price forecasting in New York electricity market. In: 2014 IEEE PES General Meeting| Conference & Exposition, pp. 1–5. IEEE (2014)
44. Cortes, C., Vapnik, V.: Support-vector networks. Mach. Learn. **20**(3), 273–297 (1995)
45. Ahadi, B., et al.: Using support vector machines in predicting and classifying factors affecting preterm delivery (2016)
46. Spilka, J., Frecon, J., Leonarduzzi, R., Pustelnik, N., Abry, P., Doret, M.: Sparse support vector machine for intrapartum fetal heart rate classification. IEEE J. Biomed. Health Inform. **21**(3), 664–671 (2016)
47. Shirwaikar, R.D., Acharya, D.U., Makkithaya, K., Surulivelrajan, M., Lewis, L.E.S.: Machine learning techniques for neonatal apnea prediction. J. Artif. Intell. **9**, 1–3 (2016)
48. Santoso, N., Wulandari, S.P.: Hybrid support vector machine to preterm birth prediction. IJEIS (Indonesian J. Electron. Instrum. Syst.) **8**(2), 191–200 (2018)
49. Lakshmi, B.N., Indumathi, T.S., Ravi, N.: A study on C. 5 decision tree classification algorithm for risk predictions during pregnancy. Procedia Technol. **24**, 1542–1549 (2016)
50. Kavitha, D., Balasubramanian, T.: Predicting the mode of delivery and the risk factors associated with cesarean delivery using decision tree model. Int. J. Eng. Sci. Res. Technol. **2277–9655**, 1–9 (2018)
51. Kamat, A., Veenal, O., Manalee, D.: Implementation of classification algorithms to predict mode of delivery. Int. J. Comput. Sci. Inf. Technol. **6**(5), 4531–4534 (2015)
52. Trovato, G., Chrupała, G., Takanishi, A.: Application of the naive bayes classifier for representation and use of heterogeneous and incomplete knowledge in social robotics. Robotics **5**(1), 6 (2016)
53. Jadhav, S.D., Channe, H.P.: Comparative study of K-NN, naive Bayes and decision tree classification techniques. Int. J. Sci. Res. (IJSR) **5**(1), 1842–1845 (2016)

54. Ide, M.A., Mathias, D., Anireh, V.: An optimized data management model for maternal mortality in Bayelsa state. Int. J. Sci. Eng. Res. **10**(7), 1–10 (2019)
55. Tesfaye, B., Atique, S., Azim, T., Kebede, M.M.: Predicting skilled delivery service use in Ethiopia: dual application of logistic regression and machine learning algorithms. BMC Med. Inform. Decis. Making **19**(1), 1–10 (2019)
56. Aleksandrowicz, L., Shestopaloff, A.Y., Alam, D., Tollman, S., Samarikhalaji, A., Jha, P.: Naive Bayes classifiers for verbal autopsies: comparison to physician-based classification for 21,000 child and adult deaths. BMC Med. **13**(1) (2015)
57. Amin, M., Habib, A.: Comparison of different classification techniques using WEKA for hematological data. Am. J. Eng. Res. **4**(3), 55–61 (2015)
58. Etemadi, M., Chung, P., Heller, J.A., Liu, J.A., Rand, L., Roy, S.: Towards birthalert—a clinical device intended for early preterm birth detection. IEEE Trans. Biomed. Eng. **60**(12), 3484–3493 (2013)
59. Hu, Y., Wang, J., Li, X., Ren, D., Driskell, L., Zhu, J.: Exploring geological and socio-demographic factors associated with under-five mortality in the Wenchuan earthquake using neural network model. Int. J. Environ. Health Res. **22**(2), 184–196 (2012)
60. Raghavendra, B.K., Srivatsa, S.K.: Evaluation of logistic regression and neural network model with sensitivity analysis on medical datasets. Int. J. Comput. Sci. Secur. (IJCSS) **5**(5), 503 (2011)
61. Ali, M., et al.: A data-driven knowledge acquisition system: an end-to-end knowledge engineering process for generating production rules. IEEE Access **6**, 15587–15607 (2018)
62. Gajbhiye, S., Sharma, S., Awasthi, M.: Application of principal components analysis for interpretation and grouping of water quality parameters. Int. J. Hybrid Inf. Technol. **8**(4), 89–96 (2015). https://doi.org/10.14257/ijhit.2015.8.4.11
63. Inyang, U.G., Akpan, E.E., Akinyokun, O.C.: A hybrid machine learning approach for flood risk assessment and classification. Int. J. Comput. Intell. Appl. **19**(2), 2050012 (2020). https://doi.org/10.1142/S1469026820500121
64. bin Othman, M.F., Yau, T.M.S.: Comparison of different classification techniques using WEKA for breast cancer. In: Ibrahim, F., Osman, N.A.A., Usman, J., Kadri, N.A. (eds.) 3rd Kuala Lumpur International Conference on Biomedical Engineering 2006. IP, vol. 15, pp. 520–523. Springer, Heidelberg (2007). https://doi.org/10.1007/978-3-540-68017-8_131
65. Yang, Z., Ming, Z.: Weighted kappa statistic for clustered matched-pair ordinal data. Comput. Stat. Data Anal. **82**, 1–18 (2015)
66. Alsadah, A., Moretti, F.: 568: Validation of a scoring system for prediction of morbidly adherent placenta in high risk population. Am. J. Obstet. Gynecol. **222**(1), S364–S365 (2020)
67. Kumar, R., Indrayan, A.: Receiver operating characteristic (ROC) curve for medical researchers. Indian Pediatr. **48**(4), 277–287 (2011)
68. Lalkhen, A.G., McCluskey, A.: Clinical tests: sensitivity and specificity. Continuing Educ. Anaesth. Crit. Care Pain **8**(6), 221–223 (2008)

An Evaluation of the Frameworks for Predicting COVID-19 in Nigeria Using Time Series Data Analytics Model

Collins N. Udanor[1](✉), Agozie H. Eneh[1], and Stella-Maris I. Orim[2]

[1] Department of Computer Science, University of Nigeria, Nsukka, Nsukka, Nigeria
collins.udanor@unn.edu.ng
[2] School of Computing, Electronics and Mathematics, Coventry University, Coventry, UK

Abstract. When the novel coronavirus, COVID-19 broke out in Wuhan China, not many in this generation knew how much impact it was going to have on our daily lives. The pandemic has resulted in the loss of thousands of lives, the collapse of businesses, indefinite closure of schools, and other activities resulting to the need for a 'new normal'. Virologists, epidemiologists, and all other scientists are working round the clock to bring a stop to this ravaging disease. The virus which exhibits severe acute respiratory syndrome was first detected in Wuhan China towards the end of 2019, with a fatality rate of 2 to 3% . By the first week of May 2020, the infection rate stood at over 3.4 million confirmed cases, with more than 238,000 deaths across 215 countries. Many virologists and epidemiologists have come to trust computer-based approaches in finding solutions to diseases of this kind. This study seeks to evaluate the appropriateness and effectiveness of relevant models e.g. SIR (Susceptible, Infective, Recovered) model in predicting the spread of nCov-19 using live time-series data while making Nigeria a case study. Results from the prediction suggest that, like many other countries, Nigeria has entered the exponential state of the pandemics. Although the pandemic was well managed at the onset, the results also depict that full relaxation of the lockdown will raise the moderate transmission rate (Ro) value of 1.22.

Keywords: Coronavirus · COVID-19 · Epidemiology · Nigeria · Pandemic · World Health Organization · Susceptible infectious recovery model · Time series

1 Introduction

The history of Pandemics stretches back to several years Before Christ (B.C.), much to the extent that we can recall the biblical accounts of the Egyptian plague and similar stories. However, and also, notable in literature are some pre-historic epidemics, such as the Circa of 3000 B.C., which claimed several lives of people from all age groups in China about 5000 years ago. There were also the plagues of Athens of 430 B.C., the Antonine plague of 165 to 180 A.D., the plague of Cyprian from 250 to 271 A.D., the Justinian plague of 541 to 542, the Black Death of 1346 to 1353, and so on to American plagues of the 16th Century, 17th Century great plague of London, 18th Century great

J. H. Abawajy et al. (Eds.): EATI 2020, LNNS 254, pp. 109–123, 2021.
https://doi.org/10.1007/978-3-030-80216-5_9

plague of Marseille, Russian plague, and Philadelphia yellow fever epidemic, which precisely occurred in 1793. The 19th Century also saw notable flu pandemic from 1889 to 1890 [1]. The twentieth century began with American polio epidemic of 1916, and then the very remarkable Spanish flu of 1918 with a staggering number of about 500 million cases and over 50 million deaths, having some indigenous communities pushed to brink of extinction. Even though the Spanish flu was first reported by the Spanish press, the exact origin of this flu is still being debated as a result of the world war at that time. There was also the Asian flu of 1957 to 1958 and the AIDS pandemic and epidemic from 1981 to present day. The rise of the twenty first Century has seen the H1N1 Swine flu pandemic of 2009 to 2010, the West African Ebola of 2014 to present day, as cases are still being recorded, the Zika Virus epidemic of 2015 to present day, the Lassa fever pandemic that was first described in 1950s but identified in 1969 and continues to spring up till present, usually during the dry season. Lassa fever, caused by a virus of the Arenavira family, is normally transmitted directly or indirectly through the urine and excreta of infected rats and rodents, it is an epidemic seen largely in Nigeria and some other West African countries. And now, the new COVID 19 that is currently ravaging the entire world, recorded in late 2019 and believed to have started in Wuhan, China, and having neither vaccine nor treatment as at the time of this report.

Arguably, Corona Virus, codenamed COVID-19, has profoundly informed the current world view of Pandemic. Debuting towards the end of 2019 and at an era of reasonable technological breakthroughs and innovation, it is a matter of common sense that appropriate technologies will inadvertently be sought and deployed for the purposes of diagnosis, treatment, prevention, prediction, and analysis of this Pandemy and any other that may be incidental or active in this era and beyond.

The world's response, usually administered by the World Health Organisation (WHO), to any onslaught of epidemic or pandemic is treatment and then vaccines to prevent further escalation of the pandemic. Notable amongst the set of human efforts in the combat of pandemics and epidemics is the effort of the frontline health personnel and responsible governments, in addition to collective efforts of people in various communities. What successful periods of termination or receding of pandemics have in common is the synergy and coordination of all the efforts to tackle the pandemics. In the present times, technology has very significant roles to play across all the efforts, both locally and globally, towards addressing the challenges posed by pandemics. To this extent and more, medical and other associated health professionals are finding computer based technologies indispensable for effective and efficient delivery of services. Generally, technology provides support in all the efforts to save lives through the dissemination of health messages, awareness campaigns, increased access to healthcare, and disease surveillance [2].

Over the years it has become necessary to track the spread of diseases and control their advancement among the people. The field of epidemiology models the behaviour of diseases, predicts its spread using mathematical and statistical models, and suggests how they can be controlled, either by vaccines or by quarantine. Differential equation has been an important tool in modelling epidemiology.

Invariably, Computer Science presents the opportunity and the enablement for the use of algorithms and programs to manipulate, store, communicate, and analyse data or digital information. The advances in computational power have enabled rapid and accurate data analysis and prediction, which are most probably beyond the imagination of early inventors and developers. With the ubiquity of computer devices, for virtually all works of life, and availability of reliable means of information communication and the Internet, Computer Science indispensably lends itself to all areas of health care services delivery and support.

In the perspective of disease detection and prevention, there are various automated devices for monitoring; for instance, body temperature and all the other vital signs, there are scanners, and systems for medical imaging and diagnostics. Interestingly, these computer-based devices effectively use mainly electromagnetic wave theory for disease detection by usually avoiding the risky and costly evasive approaches. There are remarkable advances in the areas of computer aided analytical and predictive models, and of bio-surveillance and contact tracing to monitor, control, and prevent the escalation of pandemics [2, 3].

Predicting pandemics poses a lot of challenges to epidemiologists and data scientists given the novellity of this particular disease and its other unknown characteristics. As a result of these, popular automated means of predcition are still very much rudimentary in providing accurate and relaible trends analysis in the spread of the disease for the purposes of both effeicient and effective management and containment of the pandemic.

This paper presents a study of computer-based epidemiology modelling algorithms based on differential equations. Particular attention is given to the Susceptible, Infectious and Recovery (SIR) model, comparing its operation and suitablility to predicting the spread, control and management of COVID-19. Realtime COVID-19 dataset is applied using time series to predict Nigeria's nCOV-19 cases in particular, while comparing the result with some selceted countries with similar demography.

The rest of the paper is organized as follows; Sect. 2 examines existing works by investigating the computer based approaches and models employed in tracking a pandemic of this sort. Section 3 discusses the data collection and curation, as well as the methods employed in the study. Section 4 presents the results of the prediction and compares the result with that of 4 other countries. A critical evaluation of the SIR model is also presented. And Sect. 5 concludes the study, while highlighting the direction of the future work.

2 Review of Related Works

Tracking pandemics helps provide early warning signals to both the populace and policy makers. Such information enables people to begin to change their style of living, and also enables government and health agencies to plan.

Tanok [4] is of the opinion that global travels for work and leisure contribute to the spread of infection to all places globally in a matter of days. These travels also include participation in professional conferences. Infectious diseases are transmitted from person to person as people move from one location to another oblivious that they have been infected, especially a disease that is asymptomatic like COVID-19. Since February 27 when the first covid-19 case was registered in Nigeria through an Italian national who came in through the Lagos international airport [5], Nigeria has been fighting to keep the number of cases low. In less than one month the contacts of the index case have reached 216 in 5 states including the Federal Capital Territory, Abuja. As at the time of writing, exactly 3 months after the index case was reported, the total confirmed cases stood at 9,302 and counting [6] across all but 2 of the 37 states. Nigeria was listed as one of the countries that had a moderate risk of importing the disease from China, with variable capacity and high vulnerability [7]. It was predicted that many African countries may be ill-prepared in terms of resources, surveillance, and capacity building in fighting the pandemic. But the WHO thinks otherwise. According to the WHO, Africa is better prepared now to combat the pandemic based on previous experiences in dealing with epidemics like Ebola, Cholera, etc, in addition to the creation of an African Centre for Disease Control and Prevention, and development of a funding consortia [8]. The Nigerian Centre for Disease Control (NCDC) in anticipation of the outbreak of covid-19 had set up a number of test centres in different regions of the country, as well as established the Surveillance and Outbreak Response Management System (SORMAS) for case-based reporting for epidemic prone diseases in some states of the federation [9].

Tracking COVID-19 requires computer-based approaches stemming from Computer Science algorithms derived from mathematical and statistical theorems. Besides algorithms AI systems have evolved into machines that have the ability to detect diseases [10]. Systems based on artificial intelligence can learn from historical data, patterns, and other inputs to track and predict the progression of a pandemic. Amobi [11] applied Supervised Machine Learning and Empirical Bayesian Kriging (EBK) techniques to discover patterns and correlates of the COVID-19 pandemic in Sub-Saharan Africa. Their model predicted seven variables which significantly influence the spread of the disease. Tracking different strains of influenza during the 2009 pandemic to determine their linage was achieved in [12] by converting the statistical data gathered from the OSN into a flu score, after which a combination of algorithms like decision trees and support vector machine (SVM) were employed to model the system. The results obtained from the decision tree method was applied to the Hidden Markov Model (HMM) to predict the influenza A virus in a host through a web application.

McCall [13] in tracking the spread of COVID-19 employed a web platform, Healthmap, which represented pandemics according to location, time, and agent through which the infectious disease spreads. Their model predicted that the main entry points of COVID-19 into Sub-Saharan Africa would be through South Africa, Nigeria and Ethiopia, which according to the author are high population centres. Many flights move from China the origin of this disease to these parts of Africa. The use of tweets from the Twitter microblogging site [12] was of great assistance in tracking the Flu pandemic of 2009 by using textual markers or n-gram to extract relevant symptom information such as fever, temperature, headache, etc. from users as they update their status. Tracking disease spread through online social networks (OSN) provides real time information that is cheaper than the traditional methods of disease surveillance. Researchers in [14] provide a review of pandemic tracking using OSN, extracting user keywords that match the pandemic terms on Twitter and comparing them with traditional surveillance data for correlation. Oyelola et al. [15], using Bayesian method attempted estimating early transmissibility reproduction number of COVID 19 in Nigeria at different time intervals.

They were of the opinion that the transmission rate in Nigeria has been lower than anticipated accounting it to the preparedness of the government. Zifeng et al. [16] applied the SEIR model and AI to predict the trend of the pandemic in China in which they predicted that the pandemic would peak in February ending and decline by April ending. Hytham et al. [17] presented a study that used time series models to compare day level forecasting models of countries affected by COVID-19. They observed that countries that did not mandate lockdown, closing of schools, and social distancing experienced exponential growth of the pandemic. Authors in [18] used mathematical model to calculate the average reproduction number for COVID-19 and herd immunity in India. They also analysed the public health capacity available to combat the pandemic.

3 Material and Methods

A number of disease modelling techniques exist for tracking and predicting pandemics in the field of epidemiology. Examples are the SIR, SIS, SIER models, etc. We shall limit our study to the SIR model. The Susceptible, Infected and Recovered (SIR) model was applied in predicting the COVID 19 pandemic in Nigeria using time series daily data. The results were compared with that of four other countries. The model was evaluated by comparing the results of the prediction from other countries with actual records to determine the level of accuracy of the model.

3.1 The SIR Model

The SIR model is a model generally used for modelling epidemics. This is a simple system of differential equations that qualitatively behaves in a reasonable way [19, 23]. In the model, people are grouped into three class statuses [21], such as:

- Susceptible, S(t)
- Infected, I(t)
- Recovered, R(t)

Everyone is born susceptible to be infected by an epidemic or can move from being infected to infecting others, i.e. "Infective". The model assumes that anyone who was infected and recovered can no longer be infected by the disease again, so they are removed. The model also assumes that once an infected patient recovers they cannot transmit the disease to anyone else.

The model is given by Eq. (1);

$$S(t) + I(t) + R(t) = N \tag{1}$$

Where N is the total number of persons in the population, such as a country or region. At the beginning when there is no infection at time to, we have the classes:

$$S(0) = S0; I(0) = I0; R(0) = R0$$

The transition of state from one class to another at time, t is shown in Fig. 1

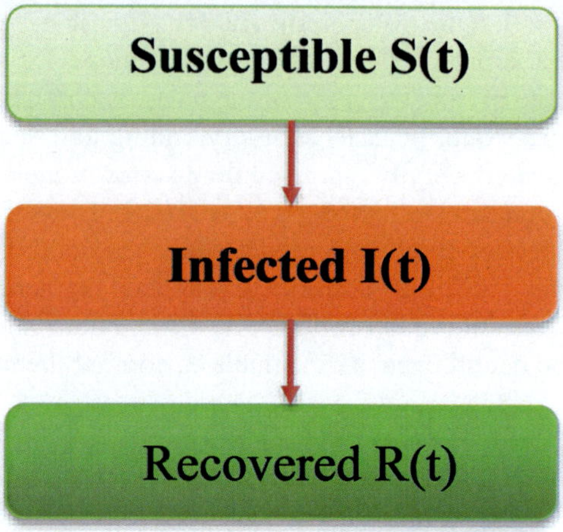

Fig. 1. The transition in the SIR model

3.2 Determining the Rate of Change

To determine the rate of change with respect to time for each of the classes above, we refer to calculus by using differential equations;

$$dS/dt = -\beta SI/N \tag{2}$$

What does this depend on? People must come in contact for infection to occur. The more the connection with people, the more the interaction and the higher will be the probability of people getting infected.

From Eq. 2, 'β' is the average contact per an infected person per time, it is a positive proportionality constant that multiplies SI, the probability of disease transmission between susceptible person and an infected person. dS/dt is negative because when people who are susceptible become infected we minus them from that class because they have left that pool. The larger the interaction, the larger the multiplication will be.

Rate of infection = $\beta \times$ (proportion of susceptible) $\times$ (proportion of infected) = $\beta s(t)i(t)$, or βsi β is also called the transmission rate and is the average number of people each infectious person spreads the disease to each day. It can be calculated by multiplying the transmission risk with the average number of contacts per day.

$$dI/dt = \beta SI/N - \gamma I \tag{3}$$

The βSI that was lost in Eq. 2 is added to Eq. 3, since they have joined the class of infected people. – γI is the loss of infected persons either by death or recovery.

$$dR/dt = \gamma I \tag{4}$$

From Eq. 4, we again regain the people we lost to the infected class, that is bI.

For the initial people who were infected, we see Eq. (5),

$$(dI)/dt|t = 0 = \beta SI\_0 - \gamma I \tag{5}$$

We ask the question, is $\beta S0 - \gamma < 0$?, by moving γ to the right, we ask the question; Is $(\beta S\_0)/\gamma < 1$?, we assign this to R0

$$R0 = \beta/\gamma \tag{6}$$

R_0 or R-zero (Eq. 6), is known as the average number of previously unexposed individuals infected by a single, freshly introduced disease carrier [20]. It is called the reproduction number. This ratio is derived as the expected number of new infections or secondary infections from a single infection in a population where all subjects are susceptible. If R0 is less than 1 then the disease will die off. But if it is larger, that is, if this rate of change is greater than zero, then the number of people who will be infected will go up and we will have an epidemic, as the disease's spread will grow exponentially. This is a standard metric in epidemiology. This exponential growth usually follows a sigmoid curve. But if the rate of change is less than one, they leave the class of those who are infected faster. If we lower the coefficient of R0, that is a, by washing of hands, and social distancing, the rate of transmission will be lowered. This will increase the probability of R0 being less than 1. The presence of vaccine can lower S0, but lowering γ is difficult. When COVID-19 first manifested R0 was predicted to be between 1.5 and 4, with at least 2, between December 2019 and January 2020. If R0 is 2, the first person (the index case), who contracts it will spread it to 2 other persons, who each will spread to 2 other persons until it grows exponentially. Experiences show that the exponential curve flattens after sometime due to decrease in contact either by isolation or social distancing. If R0 is 1.5 for example, the SIR model predicts that 58% of the population will be infected, if R0 is 4, 98% will be infected except a vaccine is introduced [24, 25].

3.2.1 Exponential Growth Model

The basic exponential growth model is described (Balram et al. 2020) as;

$$N(t) = \alpha^{*}\exp(r^{*}t) \tag{7}$$

Where N(t) = number of cases at any time 't.', α = constant representing average incubation period, we assume it is 6.4 days, r = growth rate, t = time (day).

The mathematical model for calculating the number of incidence cases based on reproduction number is defined as:

$$I(t) = (Ro)t/SI \tag{8}$$

I(t) = Number of incidence cases at time t, Ro = Reproduction number, SI = Serial interval, i.e. t0 − ti, t = Prediction time.

3.3 Times Series Analysis of COVID-19

In logistic or linear regression, there are two variables, dependent and independent. Time series is used when one of the components is a time variable, and the data is changing. Time series is a series of data points indexed in a timely order. You can use the time series data to create a model. With this model you can create a prediction.

Whenever there is a time component we use time series. You will be expected to predict for the future using the time variable.

Time series or discrete-time stochastic process analysis helps us to understand the past and predict the future using historical random data gathered over a series of time intervals. Most simulations are derived from time series analysis [21]. The main components of a time series are trends and seasonal variations which can be modelled deterministically using mathematical equations. Time series analysis enables policy makers, business managers, etc. to make informed decisions. Time series analysis also aims at finding close relationships (correlation) among data entities, and using appropriate statistical methods explains these correlations and features in the data. When a good model has been formulated, it can be used to predict future trends. Time series analysis is also useful in epidemiology research. Subsequent sections will show how we used time series data on the daily progression of the COVID-19 pandemic from the John Hopkin's University (JHU) repository to monitor and predict the epidemic.

3.4 Using COVID19.Analytic Package

This is an R package developed and maintained by Marcelo Ponce [22]. The package was released on 3rd May, 2020 and is available on the CRAN Repository as well as on Github. The package uses live data from the John Hopkin's University (JHU) (https://github.com/CSSEGISandData/COVID-19) to perform several analytics such as time series, regression, etc. It also uses the SIR model for the prediction of the pandemic.

In this section, we provide a description of the experiment conducted using the COVID19.analytic package in R. R is a powerful software tool used for statistical computation of all types, as well as Data Science and Machine Learning programming and analytics. R rivals Python in machine learning and Data Science applications. One of the major strengths of R is its numerous inbuilt packages, libraries and functions. R also comes with a good number of datasets, such as the popular Iris dataset, car dataset, breast cancer dataset, etc. R can be run from the console or from the R Studio IDE. Most importantly, R is open source and is free, with a good active support community. The R software or studio is available at the CRAN Project website.

The COVID-19 daily global live data was streamed from the JHU's portal using R codes and stored as a .csv file. The dataset contained daily reports of COVID-19 confirmed cases, deaths, and recoveries from 3,522 combined distinct geographical locations made up of countries/regions, cities and provinces (see Fig. 2). Similar datasets were also downloaded from the NCDC, and the European Disease Control and protection (ECDC) websites for comparison and analysis. We isolated Nigeria's data and performed time series analysis on it, after which we used the SIR model to predict the cases of infection using time value (to = day 64), and it converged on (time tfinal = day 90). We also generated a linear regression model in which a correlation was sought for the number of infected cases and number of days. We also performed similar prediction for US, China, Italy, and South Africa and compared the results with Nigeria's prediction result.

```
##########################################################################################
  Number of Countries/Regions reported:  187
  Number of Cities/Provinces reported:  83
  Unique number of distinct geographical locations combined: 266
------------------------------------------------------------------------------------------
  Worldwide ts-confirmed  Totals: 3343777
------------------------------------------------------------------------------------------
   Country.Region Province.State  Totals GlobalPerc LastDayChange   t-2   t-3   t-7  t-14  t-30
1              US                1103461      33.00         34037 29515 27327 32796 32491 30380
2           Spain                 213435       6.38             0   518  2144  2915   887  7947
3           Italy                 207428       6.20          1965  1872  2086  2357  3491  4668
4  United Kingdom                 177454       5.31          6201  6032  4076  4913  5525  4244
5          France                 165764       4.96             0   671 -2512  1656   -17  2116
6         Germany                 164077       4.91          1068  1470  1627  1514  1945  6922
7          Turkey                 122392       3.66          2188  2615  2936  2861  3783  2456
8          Russia                 114431       3.42          7933  7099  5841  5966  4785   771
9            Iran                  95646       2.86          1006   983  1073  1134  1374  2875
10         Brazil                  92202       2.76          5015  7502  6450  5281  2976  1208
------------------------------------------------------------------------------------------
  Global Perc. Average:  0.38 (sd: 2.19)
  Global Perc. Average in top  10 :  7.35 (sd: 9.11)
```

Fig. 2. A screen shot of an output showing top 10 countries with the highest number of cases.

4 Results and Discussion

In this section we present the results of the prediction, evaluate the model and discuss our findings.

Applying R time series function with live data from the JHU and NCDC's portals we produced the result in Table 1, which shows that the total confirmed cases in Nigeria as at 30th May was 9,302, total deaths, 261, number of persons who recovered, 2,697, and the active cases were 6,344. The respective percentages were also shown in the table.

Table 1. Nigeria's COVID 19 status as at May 30, 2020.

Confirmed cases	Confirmed perc.	Deaths	Deaths perc.	Recovered	Recovered perc.	Active	Active perc.
9302	0.16	261	2.81	2697	28.99	6344	68.20

Table 2 shows the result of the prediction produced by the SIR model. The SIR model using beta value of 0.55, and gamma value of 0.45, and Eq. 6 produces Ro value of 1.22 with these values the model predicted that on the 24th of June, 2020, which is day 90, the number of infected cases would have peaked to a maximum value of 386,654.4 translating to 0.2% of the confirmed cases, using a population value of 195,899,949.

Table 2. SIR model result

Time interval:	t0	tfinal	Fatality rate	beta	gamma	Ro	Fatality (assuming 25% fatality rate)
t0-tfinal	(day 64) 2020-03-26	(day 90) 2020-06-24	0.02	0.550531	0.449468	1.224848	7,733.09

Table 3. Linear regression models

Coefficients:	Estimate	Std. Error	t value	Pr(>|t|)	Adjusted R-squared	P-value	Standard Error
(Intercept)	-1.972490	0.148788	-13.26	<2e-16	0.9418	< 2.2e-16	0.84
	0.090406	0.001986	45.52	<2e-16			

Table 3 shows the result of the regression model which showed that there is a correlation between the number of cases and the number of days represented as the x.var. They were found to be statistically significant with p-value less than 0.001. An accuracy of 94.18% was achieved by the model (Figs. 3 and 4).

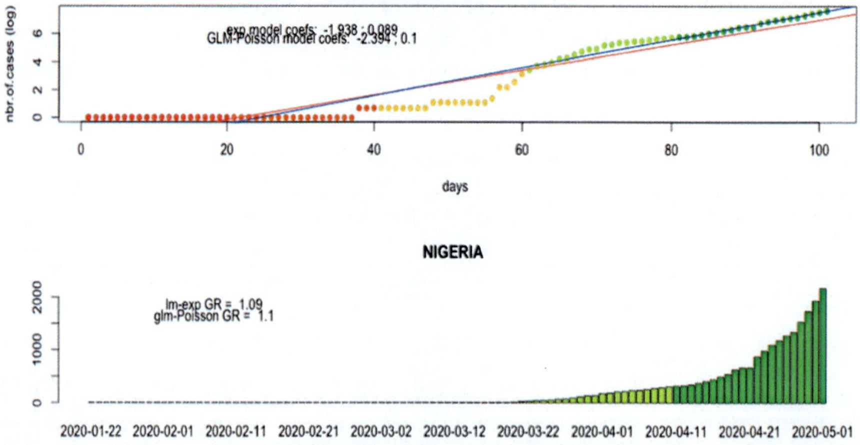

Fig. 3. COVID-19 growth rate in Nigeria

In Fig. 5 when we compare the growth of the epidemics in Nigeria with the rest of the world, we see the same pattern, only that the values are less than that of the rest of the world. Fig. 6 depicts a graphical representation of the predicted cases for Nigeria using the SIR model.

The covid19 analytics package in R is still at its experimental stage, the results predicted here may still be improved in future. From the plots in Fig. 6, the y-axis

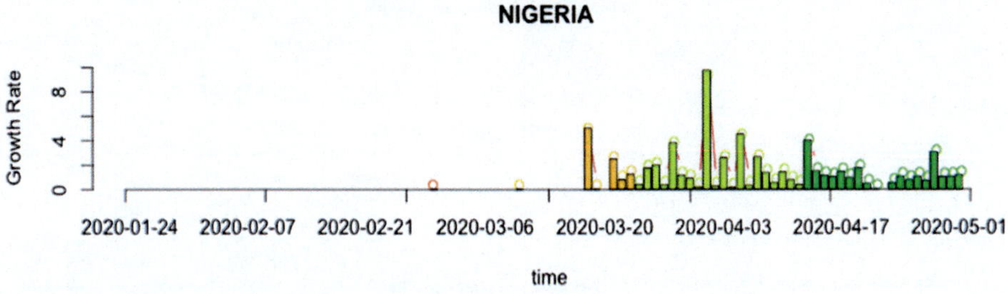

Fig. 4. Rate of growth of confirmed cases in Nigeria

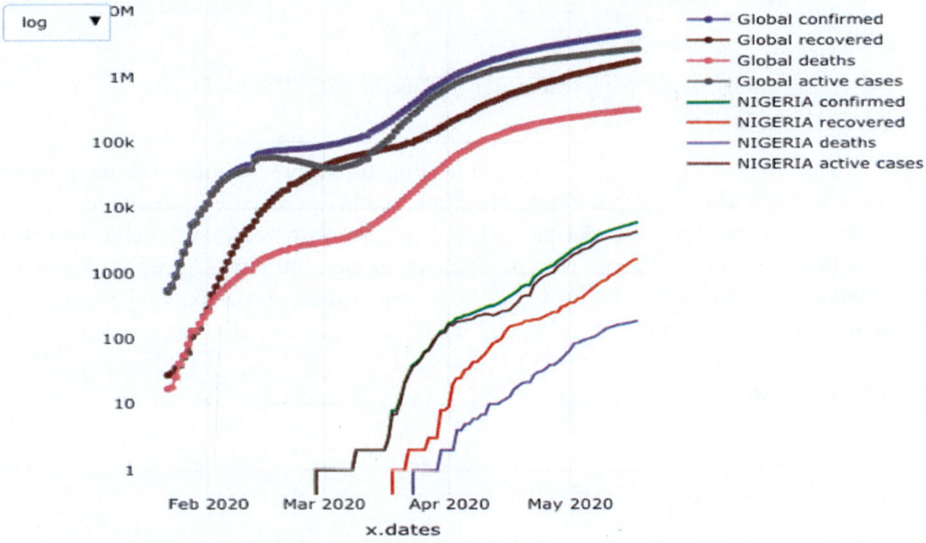

Fig. 5. Log plot of global cases and Nigeria's cases

represents the number of people who are infected, while the x-axis represents the number of days. The first pair of plots above is based on the first 25 days of data. The next plot to it is the same plot, only that it is log scale in which the y-axis is logarithm. It is also called log-linear plot or semi-log plot. The dotted line in the plot represents the real data, while the straight line represents the model. Looking at the data and the model, we see that there is a good fit.

4.1 Comparing Nigeria's Prediction with Other Countries' Predicted Cases

The lower plots show the number of subjects in log scale. The blue line represents the number of persons who are susceptible based on the entire population, the red line, the number of people infected, and the green line shows the number of persons who have recovered from the virus. From the plot, the green line and the red line cannot be differentiated, meaning that they are both moving at the same rate and at the same level, there is no peak yet. This can be compared with the plots for countries like South Africa

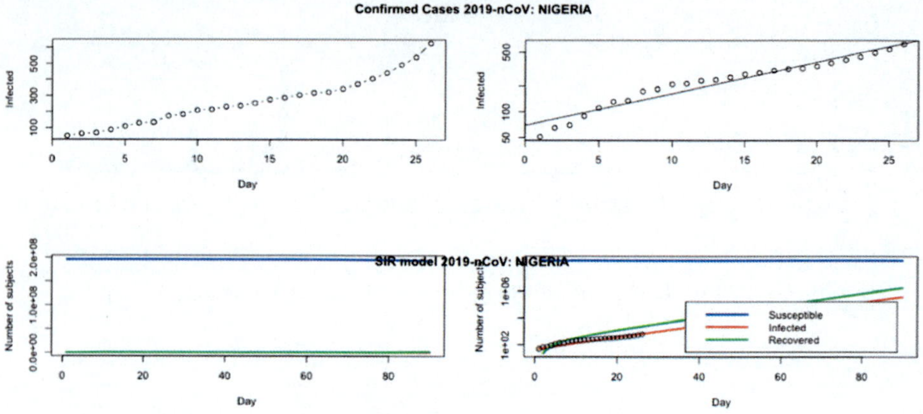

Fig. 6. SIR model prediction of Nigeria's COVID-19 cases

(Fig. 7), Italy (Fig. 8), US (Fig. 9), and China (Fig. 10) where the infection line peaked on day 67, 43, 62, and 67, respectively. The Nigeria's infection case is predicted to peak on the day 90, being the 24th of June, 2020. It is also observed that the blue line still remains flat, showing that there is still high risk of susceptibility. A large number of the population is still at the risk of infection if urgent measures are not taken, such as observing lockdown, social distancing and handwashing protocols.

Comparing with

i. **South Africa**

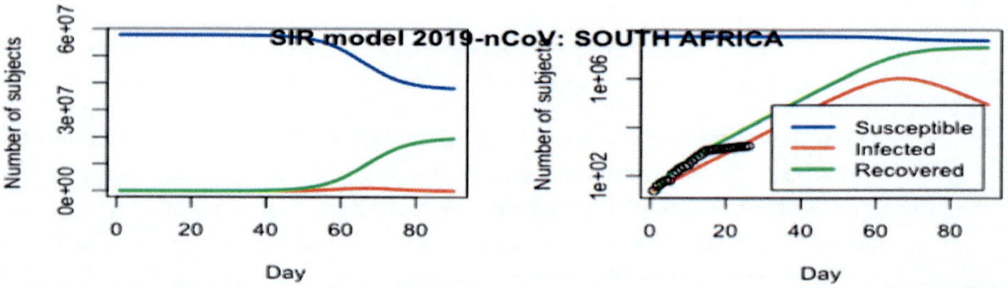

Fig. 7. SIR model prediction of South Africa's COVID-19 cases

ii. **Italy**
iii. **US**
iv. **China**

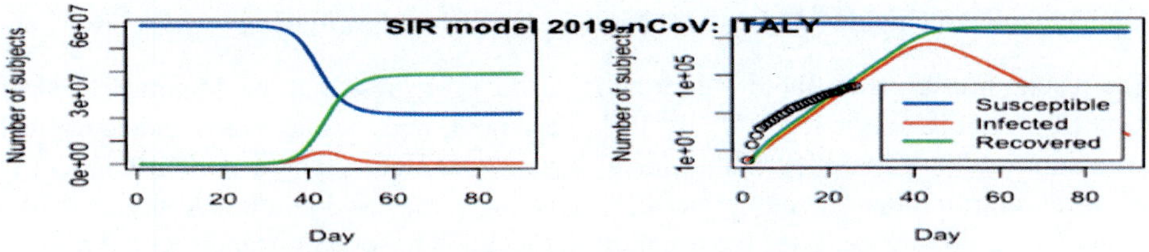

Fig. 8. SIR model prediction of Italy's COVID-19 cases

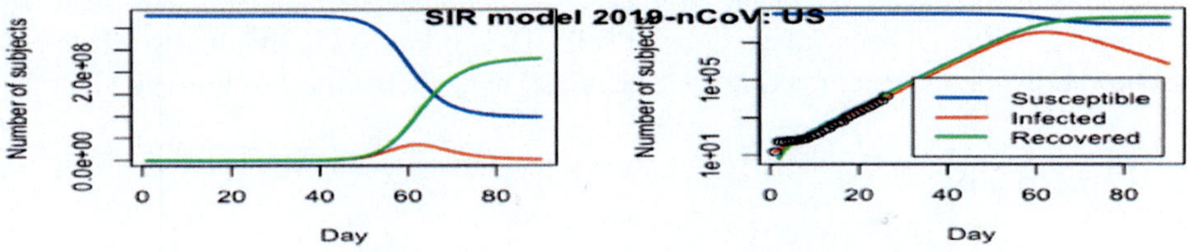

Fig. 9. SIR model prediction of US's COVID-19 cases

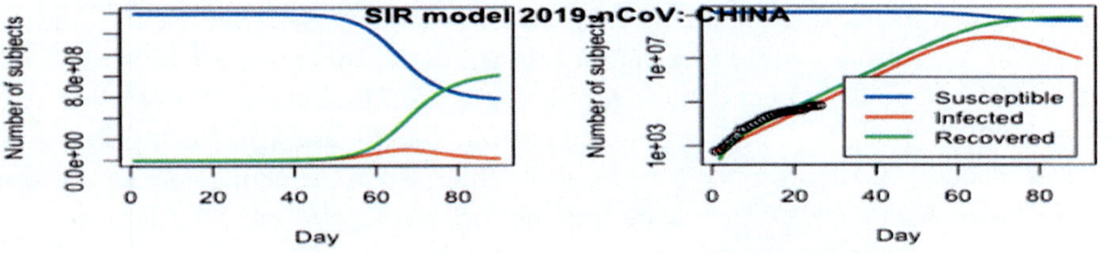

Fig. 10. SIR model prediction of China's COVID-19 cases

Table 4. A comparison of actual data with prediction by Covid19 analytics tool in R.

S/N	Country	Predicted Peak	Ro value	Actual Peak	Predicted Date	Actual Date	Accuracy(%)
1	China	91,806,058.71 (6.59%)	1.52	15,152	30/03/20	13/02/20	0.017
2	US	34,404,705.62 (10.48%)	1.73	38,509	26/04/20	26/04/20	0.112
3	Italy	4,872,496.45 (8.07%)	1.60	46,925	04/04/20	21/03/20	0.963
4	South Africa	1,062,208.92 (1.84%)	1.22	15,478	20/05/20	24/05/20	1.46
5	Nigeria	386,654.4 (0.2%)	1.22	?	24/06/2020	?	?

4.2 Modelling Covid19 Using SIR Model: A Critical Evaluation

The results presented in Table 4 and the result of [18] show that the accuracy level of the SIR model in predicting COVID 19 is very poor, and therefore demands a critical evaluation on the suitability of the model for such a disease. The use of SIR model is on the assumption that the model predicts what the spread of an outbreak will look like if there were no interventions like quarantine, lockdown, social distancing, no vaccine, etc. The error margins found in the predictions are so huge. This has been neglected because there are no means of assigning quantitative values to the interventions which governments introduced to keep the numbers down. Another issue is the error in assuming that anyone who had been infected by COVID-19 is immuned to the disease. This has not been proved, therefore making the SIR model not suitable for the current pandemic.

5 Conclusion

Predicting the nCov-19 still poses some challenges to epidemiologists and data scientists because not much information has been gathered that will of a certainty give the full characteristics of the disease. Experiments are still at empirical stages in determining drugs and treatments that could fight the virus. A number of efforts geared towards developing vaccines are still at the initial testing stages. In similar circumstance, data scientists, virologists and epidemiologists are still testing to see what models may be best suitable in predicting, tracking and controlling the disease. Our finding shows that the SIR model is not appropriate for modelling this current pandemic. Our future work will focus on evaluating other models like the SIS, SIER, etc, and even formulation a new model if none of the existing models is suitable.

References

1. Al-garadi, M.A., Khan, M.S., Varathan, K.D., Mujtaba, G., Al-Kabsi, A.M.: Using online social networks to track a pandemic: a systematic review. J. Biomed. Inform. **62**, 1–11 (2016)
2. Onovo, A., et al.: Using Supervised Machine Learning and Empirical Bayesian Kriging to Reveal Correlates and Patterns of COVID-19 Disease Outbreak in Sub-Saharan Africa: Exploratory Data Analysis. https://www.medrxiv.org/content/medrxiv/early/2020/05/02/2020.04.27.20082057.full.pdf. Accessed 27 May 2020
3. Rai, B., Shukla, A., Dwivedi, L.K.: COVID-19 in India: Predictions, Reproduction Number and Public Health Preparedness. https://www.medrxiv.org/content/10.1101/2020.04.09.20059261v1.full.pdf. Accessed 27 May 2020
4. Yates, C.: How to model a pandemic. https://theconversation.com/how-to-model-a-pandemic-134187. Accessed 8 May 2020
5. Smith, D., Moore, L.: The SIR Model for Spread of Disease - Background: Hong Kong Flu, Convergence, December 2004. https://www.maa.org/press/periodicals/loci/joma/the-sir-model-for-spread-of-disease-background-hong-kong-flu
6. Smith, D., Moore, L.: The SIR Model for Spread of Disease - The Differential Equation Model. https://www.maa.org/press/periodicals/loci/joma/the-sir-model-for-spread-of-disease-the-differential-equation-model. Accessed 25 May 2020
7. Ebenso, B., Otu, A.: Can Nigeria contain the COVID-19 outbreak using lessons from recent epidemics? Lancet Glob. Health **8**(6), e770 (2020)

8. Fu, Y.C., Wang, D.W., Chuang, J.H.: Representative contact diaries for modeling the spread of infectious diseases in Taiwan. PLoS ONE **7**(10), e45113 (2012). https://doi.org/10.1371/journal.pone.0045113
9. Gilbert, M., et al.: Preparedness and vulnerability of African countries against importations of COVID-19: a modelling study. The Lancet **395**(10227), 871–877 (2020)
10. Elmousalami, H.H., Hassanien, A.E.: Day Level Forecasting for Coronavirus Disease (COVID-19) Spread: Analysis, Modeling and Recommendations. https://arxiv.org/pdf/2003.07778.pdf
11. Aziz, J., Malik, A., Fatima, N.: 2020: The COVID-19 Law and Policy Challenge: Public Health vs. Individual Privacy in the Age of Cyber Surveillance. World Health Organization, Emergency Guidelines – Implementation of Contact Tracing for the Ebola Virus Disease, September 2015. https://apps.who.int/iris/bitstream/handle. Accessed 8 April 2020
12. Owen, J.: 20 of the Worst Epidemics and Pandemics in History (2020). https://www.livescience.com/worst-epidemics-and-pandemics-in-the-world.pdf. Accessed 20 May 2020
13. Collins, J., Abdelal, N.: Spread of Disease. https://calculate.org.au/wp-content/uploads/sites/15/2018/10/spread-of-disease.pdf
14. Kapata, N., et al.: Is Africa prepared for tackling the COVID-19 (SARS-CoV-2) epidemic. Lessons from past outbreaks, ongoing pan-African public health efforts, and implications for the future. Int. J. Infect. Dis. **93**, 233–236 (2020)
15. Prem, K., Liu, Y., Russell, T.W., Kucharski, A.J., Eggo, R.M., Davis, N.: The effect of control strategies to reduce social mixing on outcomes of the COID-19 epidemic in Wuhan: a modelling study. The Lancet Public Health **5**(5), e261–e270 (2020)
16. Lampos, V., Cristianini, N.: Tracking the flu pandemic by monitoring the social web. In: 2010 2nd International Workshop on Cognitive Information Processing, pp. 411–416. IEEE, June 2010
17. Ponce, M.: Load and analyze live data from the CoViD-19 pandemic (2020)
18. McCall, B.: COVID-19 and artificial intelligence: protecting health-care workers and curbing the spread. The Lancet Digit. Health **2**(4), e166–e167 (2020)
19. Mossong, J., et al.: Social contacts and mixing patterns relevant to the spread of infectious diseases. PLoS Med. **5**(3), e74 (2008). https://doi.org/10.1371/journal.pmed.0050074
20. Mustapha, J.O., Adedokun, K.A., Nasir, I.A.: Public health preparedness towards COVID-19 outbreak in Nigeria. Asian Pac. J. Trop. Med. **13**, 197–198 (2020). http://www.apjtm.org/text.asp?2020/13/5/197/279650
21. Jibril, M.L., Sharif, U.S.: Power of Artificial Intelligence to Diagnose and Prevent Further COVID-19 Outbreak: A Short Communication. https://arxiv.org/pdf/2004.12463.pdf. Accessed 27 May 2020
22. NCDC: Covid-19 Nigeria, National Centre for Disease Control. https://covid19.ncdc.gov.ng/. Accessed 27 May 2020
23. Adegboye, O.A., Adekunle, A.I., Gayawan, E.: Early transmission dynamics of novel coronavirus (COVID-19) in Nigeria. Int. J. Environ. Res. Public Health **17**, 3054 (2020). https://doi.org/10.3390/ijerph17093054. www.mdpi.com/journal/ijerph. Accessed 27 May 2020
24. Cowpertwait, P.S.P., Metcalfe, A.V.: Introductory Time Series with R. Use R, 1st edn. Springer, New York (2009). https://doi.org/10.1007/978-0-387-88698-5
25. Tárnok, A.: Machine learning, COVID-19 (2019-nCoV), and multi-OMICS. Cytometry **97**(3), 215 (2020)

Multi-objective Wrapper-Based Feature Selection Using Binary Cuckoo Optimisation Algorithm: A Comparison Between NSGAII and NSGAIII

Ali Muhammad Usman[1,2](✉), Umi Kalsom Yusof[1], Syibrah Naim[3], Ali Usman Abdullahi[2], Abubakar Mu'azu Ahmed[1], Osama Ahmad Alomari[4], and Mohammed Joda Usman[5]

[1] School of Computer Sciences, Universiti Sains Malaysia, 11800 Pulau Pinang, Malaysia
[2] Computer Science Department, Federal College of Education (Technical), Gombe, Nigeria
[3] Technology Department, Endicott College of International Studies (ECIS), Woosong University, Daejeon, Korea
[4] MLALP Research Group, University of Sharjah, Sharjah, United Arab Emirates
[5] Department of Mathematics, Bauchi State University Gadau, Bauchi 81007, Nigeria

Abstract. Feature selection (FS) is one of the data mining methods that is used to select the most relevant subsets of features from a stored data while eliminating the irrelevant ones and yet minimising the error rate of the selected subsets of features. Based on that, FS is considered as a two objectives minimisation problem. FS can be filter-based or wrapper-based reliant on its assessment conditions. In the wrapper-based approach, each subset of features are evaluated, and the one with least error rate is selected. However, this process is computationally tricky for a dataset with many features. Majority of the previous work treats FS as a one objective by joining both error rate and the number of selected features in a single fitness function. To create a balance between the error rate and the number of selected features, and consequently improve the computational cost. This study proposes the use of non-dominated sorting genetic algorithms, along with binary cuckoo optimisation algorithm, BCOA (BCOA-NSGA3). The results show that the proposed methods exhibit promising results and can successfully develops a set of non-dominated solutions with a reduced error rate and less computational cost compared with the single-objective and other related works, with BCOA-NSGA3 been the best of all on the majority of the datasets.

Keywords: Multi-objective feature selection · Wrapper-based · Binary cuckoo optimisation algorithm · NSGA-II · NSGA-III

1 Introduction

The era of high-performance technologies led the way to large and incremental advances in the complexity of data stored [30]. This enormous data seems to be an emerging issue to

J. H. Abawajy et al. (Eds.): EATI 2020, LNNS 254, pp. 124–136, 2021.
https://doi.org/10.1007/978-3-030-80216-5_10

manage successfully. Manual control of such dataset is nearly unreliable. Thus, machine learning is used to uncover new knowledge and trends out of this data pool.

The data collected are mostly associated with noise due to faulty of the device while gathering data like in medical imaging whereby a slight deficit may end of up affecting the entire process. And sometimes, due to the grammatical mistakes, wrong punctuations and spelling errors when writing text messages. Extracting knowledge from these kinds of data is indisputably tricky [7, 9].

Dimensionality reduction is used to eliminate the noise from the dataset which is otherwise considered unnecessary or redundant features [27]. Dimensionality reduction could be the extraction of features or selection of features (FS). New features are generated by upgrading the existing features. Whereas, in feature selection (FS), the nicest subset of features is chosen from the old ones. Both feature extraction and FS aim to improve classification performance, minimise computing time and storage, and yet boost the generalisation of the model [10, 19].

The main goal of FS is to reduce the number of features in the data set and therefore enhance its classification performance while upholding its requirement [9]. Enhancing the performance of classification can be measured on the basis of the error rate of the classifiers concerned [21].

FS could be either filter or wrapper contingent on the assessment criteria [27]. Filters are assessed without any classifiers, which ends up making the computational time faster. Conversely, they disregard the characteristic, relationship or dependence in between selected or scored features, which later affects the performance of the classification [7].

Wrappers, on the other hand, use a classifier to evaluate each subset of chosen features and therefore obtained a lesser classification error rate [11]. Because of the significant search space, however, seraching the promising subsets of features with the least computational time is a very challenging task, and the number of features required to search in the solutions is too many. For example, if a dataset has N set of features; there may be 2^M subsets to be produced and assessed, but for a sizeable value N is almost impracticable. Therefore, FS is regarded as non polynomial hard-problem [13].

FS has two main competing objectives, which are reducing both the error rate of the classifiers and the number of selected features. As a result, it is regarded as a two-objective minimisation problem [10]. Many of the latest works view FS as a single-objective optimisation problem [20, 22, 23, 27]. The superiority of a solution over the other is quickly resolved in the single-objective optimisation problem by comparing their error rates. Hence, only the least error rate of the best subsets of features is chosen in single-objective FS to the disadvantage of the selected features. There is no connection between the number of features chosen and the classification performance. Consequently, the chosen features could have some level of redundancy, but yet smaller subsets of features can achieve similar or even better classification performance. By contrast, the multi-objective FS treats both the error rate and the feature size concurrently [9]. The critical difference between them is that in multi-objective FS the objective functions create extra multi-dimensional space in addition to the standard decision variable space in the case of single-objective function [14]. This extra space is marked as the objective space [31]. If there are numerous-objectives to be improved simulteneously, many research teams use the notion of non-dominated or solutions of the Pareto optimal set [30, 31].

Deb et al. in [4] offer a general multi-objective optimization technique referred to as non-dominated genetic sorting algorithm (NSGAII). Theses days, academics are adventurous to use it and fix multi-objective FS difficulties. For instance, [15] used the NSGAII idea with PSO to develop a multi-objective PSO relying on the NSGAII. Then [25] and [26] have used the approach to solve filter-based multi-objective FS problems using BPSO. Likewise, in the work of [24, 28] and differential evolution in [11], GA was used for a multi-objective FS using PSO. NSGAII success is motivating for PSO, GA and DE. After all, their other evolutionary computing techniques, like COA, that is reported to have faster convergence and done much better than most other ECs. Still, their capability for wrapper-based multi-objective FS is not carefully examined. NSGAII also appears to lack a reference point; instead, it used the operators of crowding distance and mutation for its computation. Thus, the crowded contrast limits the NSGAII convergence. On that basis, a robust NSGAIII was developed by [3]. Lately [21] suggested the use of the ideas of NSGAIII together with cuckoo optimisation algorithm and success rates were documented, but it's restricted to the only filter-based method. Thus, according to NSGAIII and BCOA a multi-objective framework for FS was introduced, i.e., BCOA-NSGA3 to balance the FS' conflicting objectives, reduce computational time and improve classification performance.

2 Proposed Method

COA developed by Rajabioun in [18] is exclusively designed for continuous optimisation while the binary version (BCOA) by [16] is meant for the binary optimisation problems such as FS. It can be noted that Yang and Deb in [29] presented a cuckoo search algorithm (CSA) before the development of COA. Both COA and CSA use the same source of inspiration, i.e. the cuckoos. However, their mode of operation is different. CAO is evolutionary-based, whereas CSA is swarm-based, the detail of the difference between these cuckoo-inspired algorithms can be found in [1]. Besides, the recent work of Del Ser et al., in [5] sheds more light on how the majority of these bio-inspired algorithms are categories, including cuckoo-inspired algorithms.

1. COA

   - The optimisation problems variables should be interpreted in the form of a "Habitat" list, H. This means that the habitat array, $1 \times NUar$, in which the cuckoo's position is represented, is specified in Eq. 1, in $NUar$:

$$H = [x_1, x_2, \ldots, x_{M\ var}] \tag{1}$$

   - Each cuckoo produced around 5 to 20 eggs. For each iteration, these values are used for the top and the bottom limit.
   - They place their eggs in the egg-laying radius (ELR) in a given range from their habitat. The ELR depends on the variable limit, the number of eggs currently in the cuckoo, and the number of eggs.

$$ELR = \alpha \times \frac{NCC}{TNE} \times uar_{ki} - uar_{low} \tag{2}$$

An integer that manages the maximum ELR value is α. The number of current cuckoos is *NCC*, while total number of cuckoos is *TNE*. Whereas, Uar_{ki}, Uar_{low}, represents upper and lower limits respectively.
- The eggs are usually put at 10% with lower income and higher costs after cuckoo laid eggs.
- Cuckoo societies are grouped using the k-means clustering system. In the simulation, a $k = 5$ was used.

2. BCOA
BCOA often employed to resolve FS issues; whereas, the expression of the habitat is a *n-bit* binary string. Also, the habitat's position is a Boolean, whereby 1 implies that a feature is chosen and 0 instead. If X_G and X_{CP} respectively describe the entire target point and existing habitat of the cuckoo populace. Then, the following cuckoo X_{MH} is computed as shown in Eq. 3.

$$X_{MH} = X_{CP} + rand(X_G - X_{CP}) \tag{3}$$

A sigmoid function *Sig* in the Eq. 4 is applied to create new habitat *XNhabitat* desirable for solving discrete binary issues. The rationale for this is to save *XNhabitat* within the interval [0, 1]. After this, Eq. 5 can change the habitat values into binary values of 0 and 1. The term *rand* is a random number systematically generated in Eq. 5.

$$Sig = \frac{1}{(1 + e^{-X_{MH}})} \tag{4}$$

$$IF\ (Sig > rand)\ THEN\ X_{MH} = 1\ AND\ IF\ (Sig < rand)\ THEN\ X_{M\,kabitat} = 0 \tag{5}$$

3. Fitness Function
Equation 6 is used for calculating the error rate of the selected features based on the fitness function using K-nearest neighbour (KNN).

$$Error\ Rate = \frac{(FP + FN)}{(TP + TN + FP + FN)} \tag{6}$$

TP, *TN*, *FP*, and *FN* represent the true positive, true negative, false positive and false negative, correspondingly.
4. Multi-objective Optimisation
Two concepts of the most commonly used multi-objective optimisation; NSGAII [4] and NSGAIII [3] are being used in this study. NSGAII is mostly used to solve two-objective optimisation problems, whereas NSGAIII is meant to solve many-objective optimisation problems. The detailed of how these algorithms works can be found in [3, 4].
5. The Proposed BCOA-NSGA2 and BCOA-NSGA3
This study epitomises the first research that used BCOA along with the principles of non-dominated sorting in NSGAIII and NSGAII for multi-objective wrapper-based FS. The comprehensive pseudo-code for the BCOA-NSGA2 and BCOA-NSGA3 could be seen in Algorithm 1 and Algorithm 2.

3 Experimental Design

3.1 Experimental Settings

The initial and maximum population are set to five and twenty resepctively, and then all the methods are run for 40 separate times per each dataset. The parameters used for the BCOA-NSGA2 and BCOA-NSGA3 are based on the work of [17, 18] and [28]. The maximum iteration had been set to one hundred.

The datasets in the experiments were partitioned into two groups: a training group and a test group. In this study, 70% of the rows are classified as the training group. Whereas, the remaining 30% as the test group, as suggested in [6]. Besides, a less computational cost classifier KNN [2], with $K = 5$ was used throughout the experiment.

3.2 Datasets Description

The twelve (12) datasets used in this study can be found in the UCI machine learning repository in [8]. The detail of the dataset can be found in Table 1. All the datasets has continuous outcome with varying degree of complexity, with the feature size from 13 until 500.

4 Results and Discussion

The results of the BCOA-NSGA2 and BCOA-NSGA3 are displayed in Fig. 1. In the Figure, each graph represents a dataset used in the experiments. On top of every graph, the numbers inside the brackets represent the total number of feature together with the error rate realised by using the whole features. In each graph, the x-axis indicates the number of selected features, whereas the y-axis represents the error rate of the classifier. In every chart, the "−1" in BCOA-NSGA2-1 and BCOA-NSGA3-1 signifies the average Pareto front obtained from BCOA-NSGA2 and BCOA-NSGA3 in the forty independent runs. On the other hand, "−2" in BCOA-NSGA2-2 and BCOA-NSGA3-2 represents the set of non-dominated solutions obtained from BCOA-NSGA2 or BCOA-NSGA3 during the forty separete runs. "BCOA-2S" shows the 40 sets of solutions through the BCOA-2S in our previous work that is based on single-objective [22].

Algorithm 1 Proposed BCOA-NSGA2

```
Begin
   Divide the dataset into training set and test set;
   Initialise the habitat;
   Allow the cuckoos to lay their eggs in their matching ELR according to Equation 2 and Equation 4;
   Recognise the cuckoos in the nondominated solutions;
   Compute reference point of the cuckoos and generate initial population;
   Use nondominated population sorting mechanism;
   WHILE maximum iteration is not reached DO
      Evaluate the two fitness values of each cuckoo
                          *Feature size together with their error rate using Equation 6
      Identify the habitats (nonDomCOAList);
      Calculate the crowding distance in nonDomCOAList and sort them;
      FOR i = 1 TO PopulationSize DO
         Update the ith habitat;
         Select the best habitat_i from the highest ranked solutions in nonDomCOAList;
         Update the ELR;
         Update the ith habitat;
      END FOR
      Add, the original habitats and update the habitats in the unioun;
      identify the different levels of nondominated fronts Fit = (Fit_1, Fit_2, Fit_3, ...);
      Empty the cuckoos for the next iteration;
      i = 1 + 1;
      WHILE |H| < PopulationSize
         IF (|H| + |Fit_i| <= PopulationSize) THEN
            Add Fit_i to H;
            i = i + 1;
         END IF
         IF (|H| + |Fit_i| > PopulationSize) THEN
            compute the crowding distance for each habitat in Fit_i;
            Sort the habitats in Fit_i;
            Add the (PopulationSize - |H|) least crowded habitat to cuckoo;
         END IF
      END WHILE
   Calculate and save the error rate of the feature subsets (solutions);
   Calculate the error rate of the solutions or feature subsets in Fit_i;
   Return the feature subsets along with their error rates in Fit_i;
END
```

Algorithm 2 Proposed BCOA-NSGA3

Begin
    Divide the dataset into training set and test set;
    Initialise the habitat;
        * Initialise the habitat H as in Equation 1 on the training set
    Evaluate the two fitness values of each cuckoo
        *Feature size together with their error rate in BCOA-NSGA3 of Equation 6 on the training set
    Allow the cuckoos to lay their eggs in their matching ELR according to Equation 2 and Equation 4;
    Recognise the cuckoos in the non-dominated solutions;
    Compute reference point of the cuckoos and generate initial population;
    Use non-dominated population sorting mechanism;
    WHILE maximum iteration is not reached **DO**
        Apply tournament selection and crossover with two parents as probability;
        Again, apply non-dominated population sorting mechanism;
        Apply normalization on the population;
        Create reference set;
        Find out reference points and solution with associated member based on associate procedure;
        Apply the niche preservation;
        Keep the niche obtained solutions for the next generations;
    END WHILE
    Compute and store the error rate of the feature subsets (solutions);
    Return the feature subsets along with their error rates;
End

Table 1. Experimental datasets

S/N	dataset	Features	Classes	Instances	Data Type
1	Wine	13	3	178	Continuous
2	Australian	14	2	690	Continuous
3	Zoo	17	7	101	Continuous
4	Vehicle	18	4	846	Continuous
6	Ionosphere (Ionosp)	34	2	351	Continuous
7	LungCancer (Lung)	56	3	32	Continuous
8	Hillvalley	100	2	606	Continuous
9	Musk1 (Clean1)	166	2	476	Continuous
10	Madelon	500	2	4400	Continuous

On the results of some of the datasets, BCOA-2S possibly evolve similar subset of features in different runs, and they are displayed in the same point in the graph. Thus, there would be less than 40 different points shown in the graph.

BCOA-2S has a single solution in every forty independent execution for each dataset. Due to its favourable performance in resolving the single-objective problem, BCOA-2S was favoured. The proposed multi-objective wrapper-based FS, BCOA-NSGA2 and BCOA-NSGA3, has at every turn acquired a set of non-dominated solutions. The comparison of BCOA-NSGA2 and BCOA-NSGA3 results along with BCOA-2S resulting from the forty independent runs are shown in the graphs.

The findings in Table 2 indicates that it takes a minimum of 5.8559 s and 6.4692 s to finish the cycle by the BCOA-NSGA3 and BCOA-NSGA2, respectively, on those ten datasets. Therefore, BCOA-NSGA3 showed less computational time compared to BCOA-NSGA2.

4.1 Comparison Between BCOA-NSGA2 and BCOA-NSGA3 Approaches

Comparative analysis in between the two proposed multi-objective wrapper-based approaches is presented in Fig. 1. The graph shows that BCOA-NSGA3 outperformed its BCOA-NSGA2 counterpart in most of the datasets. Likewise, the result in Table 2 attests to BCOA-NSGA3 on all datasets computationally faster than BCOA-NSGA2. The key reason for this is that BCOA-NSGA3 has another system to preserve its variety and uses some mutation operators to prevent stagnation in the local optimum.

The results show that both BCOA-NSGA2-1 and BCOA-NSGA3-1 together with BCOA-NSGA2-2 and BCOA-NSGA3-2 have one or more sub-sets of features which can select features with a lower error rate than full-size features. Examining the effect of BCOA-NSGA2 on Hillavlley and LungCancer datasets, the feature size decreases by 2% of the whole features but yet realizes an impressive performance on the error rate than the whole feature size. The feature size has been reduced by 15% for the residual datasets with the exception of Madelon and Zoo data sets. By contrast, on most datasets, BCOA-NSGA3 also performed well. A clear indication can be seen in the Zoo datasets, where BCOA-NSGA3 selects almost 1.80% of the seventeen (17) total features. For other datasets such as Madelon and Musk1, the number of selected features seems to be more than 14% of their corresponding total features.

Analyzing the performance of both BCOA-NSGA2 and BCOA-NSGA3 with BCOA-2S, in most datasets it can be concluded that the methods proposed exceeded BCOA-2S. This means that the methods proposed have the capacity to achieve the best results.

Regarding the BCOA-NSGA2 and BCOA-NSGA3 methods, it is clear that BCOA-NSGA3 did best in almost all datasets than BCOA-NSGA2 with regard to the chosen features, the error rate and the processing costs. The graphs across each chart show that BCOA-NSGA3 is at the bottom, that is, closer to the lowest error rate than that of the BCOA-NSGA2.

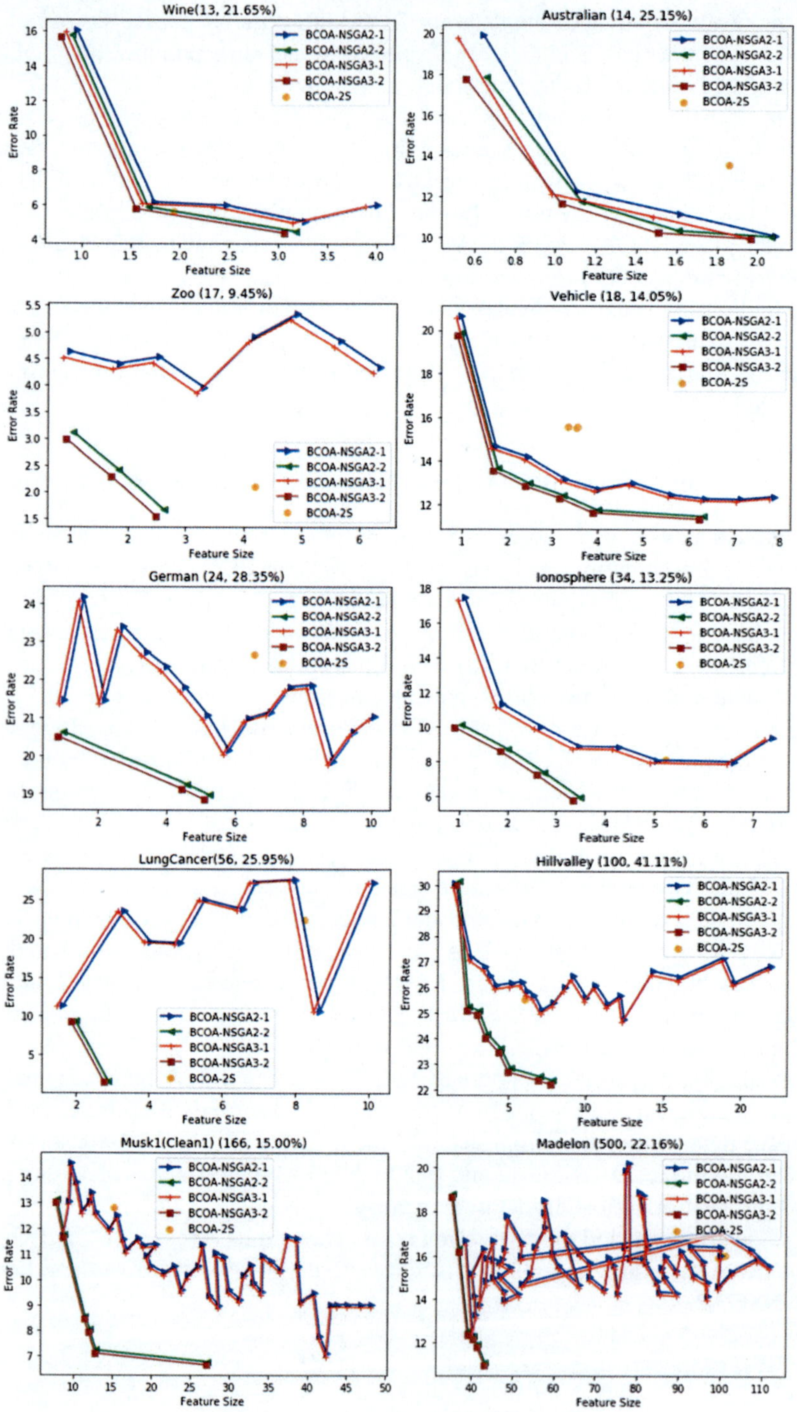

Fig. 1. Experimental results of BCOA-NSGA2 and BCOA-NSGA3

Table 2. Comparison based on computational time (in seconds)

Datasets	BCOA-NSGA2	BCOA-NSGA3	Bin-MOABC	Num-MOABC
Wine	1.568	1.439	-	-
Australian	2.310	2.159	-	-
Zoo	2.544	2.447	-	-
Vehicle	1.320	1.131	2.370	2.361
Germany	1.403	1.414	1.971	1.998
Ionosphere	1.348	1.208	-	-
LungCancer	4.153	3.984	-	-
HillValley	1.622	1.459	1.643	1.669
Musk1	1.689	1.583	1.680	1.734
Madelon	46.735	41.735	46.845	47.949
Average Time	**6.4692**	**5.8559**	**10.9018**	**11.1422**

4.2 Comparisons Between BCOA-NSGA2 and BCOA-NSGA3 with Existing Multi-objective Wrapper-Based Approaches

The results obtained were analysed with NSPSOFS and CMDPSOFS similar works on identical datasets in [24]. And Num-MOABC together with Bin-MOABC on the matched datasets in [12].

While the results of the NSPSOFS and CMDPSOFS methods can effectively develop a set of non-dominated solutions and acquire a Pareto front on a fewer datasets, e.g. on Zoo, Wine, and Australia datasets, the graphs tend to be almost identical to the current research. Nonetheless, the error rate achieved on the remaining seven datasets is in favour of BCOA-NSGA2 and BCOA-NSGA3 methods on both in term of the Pareto fronts achieved by the feature subsets (non-dominated solutions) and the error rate.

In most datasets, the proposed approaches have achieved a better error rate. The figures for each dataset can be seen from the respective graph. The plotline in the chart that is nearer to the bottom of the chart has a better result than the one at the top, both in terms of the error rate and the number of features selected.

The comparative results, therefore, indicated that the proposed BCOA-NSGA2 and BCOA-NSGA3 could successfully develop a set of non-dominated solutions with fewer features than full features and looked better than CMDPSOFS and NSPSOFS on most datasets.

Comparing Bin-MOABC to Num-MOABC in [12], one can see that the proposed methods have outpaced both Num-MOABC and Bin-MOABC in every way. For instance, in Hillvalay, Musk1 and Madelon datasets, an increase of around 5% signifies that our proposed approaches did much better on datasets with higher dimensional size.

In addition, almost 3% increase is recorded for the German and Vehicle datasets compared to the Bin-MOABC and Num-MOABC datasets. Conversely, a good set of non-dominated solutions was realized using the proposed approaches as opposed to the previous study in [12]. As a result, both BCOA-NSGA2 and BCOA-NSGA3 could successfully develop a set of non-dominated solutions with a better error rate. Num-MOABCB and Bin-MOABC achieved an average computation cost of 10.9018 and 11.1422 s, that is almost double the computational time of the proposed BCOA-NSGA3, as shown in Table 2. Therefore, the proposed method is computational faster compare to the existing approach.

5 Conclusions

Most of the previous FS wrapper research is singular-objective; that works by combining both the number of selected feature and classification performance into a single fitness function. But that could not resolve the issue better; there could be some redundancy between the number of features selected. There is, therefore, a need for a multi-objective FS that addresses both the number of selected features and the achievement of the classification at the same time. This study thus describes the initial work on wrapper-based multi-objective FS using BCOA in conjunction with NSGAIII.

The findings suggest that the proposed BCOA-NSGA3 achieved more significant subsets of features compared to BCOA-2S, and that is the best wrapper-best single-objective used for comparison. BCOA-NSGA2 has obtained better results than the two multi-objective wrapper-based approaches (MOABC, and Num-MOABC, together with NSPSOFS, CMDPSOFS) reported in the literature. Regardless of the features in the dataset, BCOA-NSGA3 achieves a superior result in classification with fewer features and shorter computational time than other methods.

The research findings also demonstrated that the BCOA-NSGA3 solution space could be more effectively searched for a range of non-dominated solutions as an option to a single, best solution. Observing the Pareto front that the proposed multi-objective wrapper-based FS has acquired, this can be seen; it can help users choose their suitable solution that suits their needs. Meanwhile, this study recognizes that, despite the potential benefits of habitat conservation strategies, the potential limitation of the rapid loss of diversity in BCOA-NSGA2 restricts its efficiency as a multi-objective FS. As well as a binary tournament collection, a crowding factor will effectively filter out individual crowded non-dominated solutions in the leader group. Various mutation operators in numerous cuckoo categories can successfully preserve habitat diversity and align their global and local search capabilities. Similar techniques account for BCOA-NSGA3's superior performance over BCOA-NSGA2 on all datasets, irrespective of the number of features.

References

1. Abdel-Basset, M., Hessin, A.-N., Abdel-Fatah, L.: A comprehensive study of cuckoo-inspired algorithms. Neural Comput. Appl. **29**, 345–361 (2018). https://doi.org/10.1007/s00521-016-2464-8. ISSN: 0941-0643
2. Alpaydin, E.: Introduction to Machine Learning. MIT Press, Cambridge (2014). ISBN: 0262325756

3. Deb, K., Jain, H.: An evolutionary many-objective optimization algorithm using reference-point-based nondominated sorting approach, part I: solving problems with box constraints. IEEE Trans. Evol. Comput. **18**(4), 577–601 (2014)
4. Deb, K., et al.: A fast and elitist multiobjective genetic algorithm: NSGA-II. IEEE Trans. Evol. Comput. **6**(2), 182–197 (2002). ISSN: 1089-778X
5. Del Ser, J., et al.: Bio-inspired computation: where we stand and what's next. Swarm Evol. Comput. **48**, 220–250 (2019)
6. Dobbin, K.K., Simon, R.M.: Optimally splitting cases for training and testing high dimensional classifiers. BMC Med. Genomics **4**(1), 1–8 (2011). ISSN: 1755-8794
7. Faris, H., et al.: An efficient binary Salp Swarm Algorithm with crossover scheme for feature selection problems. Knowl.-Based Syst. **154**, 43–67 (2018). ISSN: 0950-7051
8. Frank, A., Asuncion, A.: UCI machine learning repository. Sch. Inf. Comput. Sci. **213**, 21–22 (2010). University of California, Irvine, CA. http://archive.ics.uci.edu/ml
9. González, J., et al.: A new multi-objective wrapper method for feature selection–accuracy and stability analysis for BCI. Neurocomputing **333**, 407–418 (2019). ISSN: 0925-2312
10. Hammami, M., Bechikh, S., Hung, C.-C., Said, L.B.: A multi-objective hybrid filter-wrapper evolutionary approach for feature selection. Memetic Comput. **11**(2), 193–208 (2019). https://doi.org/10.1007/s12293-018-0269-2. ISSN: 1865-9284
11. Hancer, E., Xue, B., Zhang, M.: Differential evolution for filter feature selection based on information theory and feature ranking. Knowl.-Based Syst. **140**, 103–119 (2018). ISSN: 0950-7051
12. Hancer, E., et al.: Pareto front feature selection based on artificial bee colony optimization. Inf. Sci. **422**, 462–479 (2018). ISSN: 0020-0255
13. Jiménez, F., et al.: Multi-objective evolutionary feature selection for online sales forecasting. Neurocomputing **234**, 75–92 (2017). ISSN: 0925-2312
14. Khan, A., Baig, A.R.: Multi-objective feature subset selection using non-dominated sorting genetic algorithm. J. Appl. Res. Technol. **13**(1), 145–159 (2015). ISSN: 1665-6423
15. Li, X.: A non-dominated sorting particle swarm optimizer for multiobjective optimization. In: Cantú-Paz, E., et al. (eds.) GECCO 2003. LNCS, vol. 2723, pp. 37–48. Springer, Heidelberg (2003). https://doi.org/10.1007/3-540-45105-6_4
16. Mahmoudi, S., Rajabioun, R., Lotfi, S.: Binary cuckoo optimization algorithm. Nature 1–7 (2013)
17. Mahmoudi, S., Rajabioun, R., Lotfi, S.: Binary cuckoo optimization algorithm. In: 1st National Conference on New Approaches in Computer Engineering and Information Retrieval Young Researchers and Elite Club of the Islamic Azad University, Roudsar-Amlash Branch, pp. 1–7 (2013)
18. Rajabioun, R.: Cuckoo optimization algorithm. Appl. Soft Comput. **11**(8), 5508–5518 (2011). ISSN: 1568-4946
19. Tang, J., Alelyani, S., Liu, H.: Feature selection for classification: a review. In: Data Classification: Algorithms and Applications, pp. 37–64. CRC Press (2014)
20. Usman, A.M., Yusof, U.K., Naim, S.: Cuckoo inspired algorithms for feature selection in heart disease prediction. Int. J. Adv. Intell. Inform. **4**(2), 95–106 (2018). ISSN: 2548-3161
21. Usman, A.M., Yusof, U.K., Naim, S.: Filter-based multi-objective feature selection using NSGA III and cuckoo optimisation algorithm. IEEE Access **11**(11), 76333–76356 (2020). ISSN: 2169-3536
22. Usman, A.M., Yusof, U.K., Naim, S.: Two-steps wrapper-based feature selection in classification: a comparison between continuous and binary variants of cuckoo optimisation algorithm. In: Chiroma, H., Abdulhamid, S.M., Fournier-Viger, P., Garcia, N.M. (eds.) Machine Learning and Data Mining for Emerging Trend in Cyber Dynamics, pp. 145–175. Springer, Cham (2021). https://doi.org/10.1007/978-3-030-66288-2_6

23. Usman, A.M., et al.: Comparative evaluation of nature-based optimization algorithms for feature selection on some medical datasets. i-Manager's J. Image Process. **5**(4), 9 (2018)
24. Xue, B., Zhang, M., Browne, W.N.: Particle swarm optimization for feature selection in classification: a multi-objective approach. IEEE Trans. Cybern. **43**(6), 1656–1671 (2012). ISSN: 2168-2267
25. Xue, B., Zhang, M., Browne, W.N.: Particle swarm optimization for feature selection in classification: a multi-objective approach. IEEE Trans. Cybern. **43**(6), 1656–1671 (2013). ISSN: 2168-2267
26. Xue, B., et al.: A multi-objective particle swarm optimisation for filter-based feature selection in classification problems. Connection Sci. **24**(2), 91–116 (2012). ISSN: 0954-0091
27. Xue, B., et al.: A survey on evolutionary computation approaches to feature selection. IEEE Trans. Evol. Comput. **20**(4), 606–626 (2016). ISSN: 1089-778X
28. Xue, B., et al.: Multi-objective evolutionary algorithms for filter based feature selection in classification. Int. J. Artif. Intell. Tools **22**(4), 1–31 (2013). ISSN: 0218-2130
29. Yang, X.-S., Deb, S.: Cuckoo search via Lévy flights. In: World Congress on Nature and Biologically Inspired Computing, NaBIC 2009, pp. 210–214. IEEE (2009). ISBN: 1424450535
30. Zhang, Y., et al.: Binary differential evolution with self-learning for multi-objective feature selection. Inf. Sci. **507**, 67–85 (2019). ISSN: 0020-0255
31. Zhou, Z., et al.: Multi-objective based radiomic feature selection for lesion malignancy classification. IEEE J. Biomed. Health Inform. **10**(2), 1–11 (2019). ISSN: 2168-2194

A Logistic Predictive Model for Determining the Prevalent Mode of Financial Cybercrime in Sub-Saharan Africa

C. N. Udanor(✉), I. A. Ogbodo, O. A. Ezugwu, and C. H. Ugwuishiwu

Department of Computer Science, University of Nigeria, Nsukka, Nsukka, Enugu State, Nigeria
collins.udanor@unn.edu.ng

Abstract. Cyber fraud is considered as the utilization of the Internet to commit financial-related crimes such as email phishing to craftily retrieve personal and sensitive information from a target. Any crime perpetrated using a computer and/or the Internet is considered as cybercrime, of which cyber fraud is an aspect of. Cybercrime cost has been on the increase, estimated to cost the world over 6 trillion US Dollars by 2021. Its impact on all aspects of life, especially the economy, is very grave and worrying. This study focuses on the analysis of the prevalence of cyber fraud in Sub-Saharan Africa. Google online survey tool was employed to gather responses from users of digital infrastructures for financial transactions, through modes such as mobile banking, SMS USSD codes, ATMs, etc. The questionnaire sought to know if the users had at any time been victims or targets of financial cybercrime and the electronic platforms on which the positive cases of cyber fraud occurred. Machine learning techniques were applied in analyzing the user responses using R Studio version 3.6. A supervised machine learning logistic regression model was developed which showed that two variables representing two modes of financial cybercrime attacks were statistically significant having p values less than 0.001. The accuracy of the model was found to be **81.38%**, which was promising for the generalization of the model. The probability accuracy for ATM cybercrime is **90.37%,** while that of Phone mode is **87.76%**.

Keywords: Cybercrime · Cyber fraud · Logistic regression · Machine learning · Financial cybercrime victim · Cybercrime prevalence

1 Introduction

Financial losses due to cybercrime are huge and increasing daily. Clement [1] presented a ranking of countries most affected by cybercrime in the year 2017 in terms of billions of Dollars as shown in Fig. 1. China topped the list with a loss in the excess of 66.3 Billion Dollars, followed by Brazil with losses amounting to 22.5 billion U.S. dollars, etc. He stated that on the global scale, the average cybercrime victim lost 142 U.S. Dollars. Lewis [2] in the third report of the CSIS and Mcafee calculated an estimated loss of 600 US Dollars equivalent to 0.8 percentage of the global GDP to cybercrime. Morgan [3] predicted that cybercrime would cost the world about 6 trillion US Dollars

J. H. Abawajy et al. (Eds.): EATI 2020, LNNS 254, pp. 137–151, 2021.
https://doi.org/10.1007/978-3-030-80216-5_11

by 2021, coming from a figure of 3 trillion USD in 2015. These cybercrime losses are measured in terms of loss of money, theft of intellectual properties, loss of productive hours, damage to reputation, fraud, destruction of data, forensic investigation costs, etc. Cybercrime is the greatest threat to every company in the world today [3]. The gravity of cybercrime was according to [4], as a result of the increasing production of harmful software regarded as crime-ware. The US Internet Crime Complaint Centre, in 2009 got 336,655 cases of such, leading to the loss of 559.7 million US dollars. Kshetri [5] observed that cybercrime is gradually moving over to emerging economies where low hanging fruits are easy to reach. The author also observed that the cost of cybercrime in 2017 to the economies of Africa was 3.5 billion US Dollars. In the same year, annual losses to Nigeria, Kenya, and South Africa were estimated at $649 million, $210 Million, and $157 million, respectively. From the foregoing, it is obvious that the widespread of cybercrime today can no longer be ignored seeing its impact on the economic, social and political life of the society.

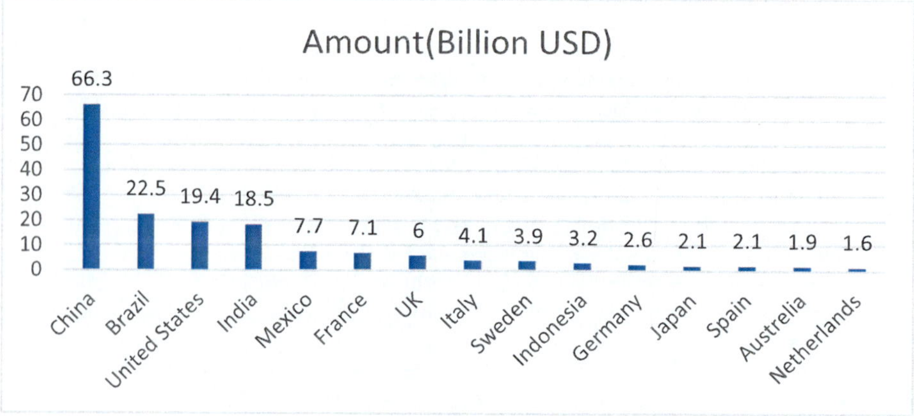

Fig. 1. Global cybercrime losses in 2017 [1]

Cybercrime has been perceived as both local computer crimes and internet crimes. Whereas the local computer crimes are the unlawful electronic actions that expose computer systems and their associated data to security risks, internet crimes are the unlawful actions perpetuated through a computer system or a computer network [6]. Furthermore, any behaviour considered as a crime and operated using the computer and/or the internet is considered as cybercrime. This implies that any known traditional crime committed but through the use of computers or the internet qualifies to be regarded as cybercrime [7]. Cybercrime can be of two categories: Crimes directly affecting the computer networks and devices e.g. harmful codes, computer viruses, malware, e.t.c. and computer network/device-facilitated crimes targeting internet users, individuals, institutions e.t.c., such as cyberstalking, theft of identity, phishing, cyber fraud, and cyber-information warfare. In the same vein, cyber fraud is regarded as the use of the internet or the computer to commit dishonest acts to illegally obtain money/swindling an unsuspecting individual or company of either property, money, or other valuables [8]. Cyber fraud could either be dependent, enabled, or assisted by cyber technology, although, cyber frauds are most times not dependent on cyber technology [9]. The term prevalence refers statistically to the number of events of a phenomenon in a given population at a particular point in time. The majority of cybercrimes are slightly different online versions of fraudulent activities that existed previously and were carried out offline. The internet, however, provided the grounds for cybercriminals to operate without being discovered since their frauds know no geographical boundary. Majority of the cybercrimes are perpetrated by youths, majorly through phishing [10, 11]. Some researchers identified the causes of cybercrime to include unemployment, desire for wealth, lack of strong cybercrime laws, incompetent security on personal computers, recognition/fame, easy to perpetrate, intellectual pursuit, frustration, revenge, display of wealth by corrupt politicians and 'yahoo yahoo' boys, laziness, lack of good moral upbringing from parents and guardians [10, 12]. Some opinion leaders are attributing cybercrime to the decay noticeable in the educational sector, pointing out that the damage to the sector is nearing complete collapse with more than 90% of cyber fraudsters in Nigeria alone originating from that sector.

This study analyzes the prevalence of cyber fraud in Sub-Saharan Africa, and establishes the most frequent means of cyber fraud occurrences. Google online survey tool was employed in gathering responses from users of digital infrastructures for financial transactions, through modes such as mobile banking, SMS USSD codes, ATMs, etc. The questionnaire sought to know if the users had at any time been victims or targets of financial cybercrime and the electronic platforms on which the positive cases of cyber fraud occurred. Logistic regression model was employed in providing analytic insights into the probability and methods of cyber frauds.

The rest of the paper is laid out as follows: Sect. 2 takes a deeper look into existing works to discover the prevalence of cybercrime among different ages and professions. Section 3 lays out the procedures for data collection and curation, as well as model formulation and development using machine learning. Section 4 presents the results and findings from the survey and experiments as well as their implications. And Sect. 5 concludes the findings and pre-empts the direction of further work.

2 Review of Related Works

In an attempt to find the connection between cyber fraud and poverty, Adesina [13] discovered that poverty is a key factor for the prevalence of cybercrime in Nigeria of which unemployment of capable and competent youths is a significant pointer. Cyber frauds in Nigeria are popularly known as "Yahoo Yahoo" because email addresses such as Yahoo mails or messengers of internet users were mainly targeted. The paper highlighted that cyber frauds perpetrators have resorted to what is known as "Yahoo Plus" that involves the use of fetish and diabolic means to get their victims to fall to their criminal schemes because people have become aware of their activities, and strict measures are currently adopted by firms and financial organizations that carry out e-commercial activities to minimize cyber fraudsters' activities.

Taiwo and Igbekoyi [11] studied the prevalence and consequence of cyber fraud in public institutions of learning in Ekiti State, Nigeria, as regards the economic and national development. They observed that the abusive use of technological tools by cybercriminals has led to insecurities in the economy of Nigeria, and has resulted in the tarnishing of the image of the country, as well as economic losses. The prevalence of fraudulent activities carried out on the internet and through internet technologies by youths in Onitsha North LGA, in the eastern part of Nigeria was investigated in [14]. The authors opined that cyber frauds in that area were motivated by the mindset that cybercrime is an innovative quick means of acquiring wealth with limited risks as compared to traditional crimes. The authors also identified the prevalence of cybercrime to be majorly dominant among youths within the age bracket of 24–29 years, and that advance fee fraud, hacking, identity theft, and cyber terrorism were among the most prevalent cyber frauds in that environment.

Igba et al. [15] highlighted the following as contributors to cybercrime among youths: upsurge in the use of technology, greed, phobia of unemployment among undergraduate youths, peer-group pressure, negative choice of role models, poor parenting, and poverty. The paper also reported that the sum of 7.3 billion Naira was lost in 2007 to cyber fraud in Nigeria, while Nigeria cyber fraudsters defrauded "netizens" of the sum of $28000 on fake lotteries to Spain. Further, cyber frauds among undergraduates have resulted in an increased number of dropouts, low moral standards, hooliganism, ritualism, spiritualism, reduced academic performance because undergraduate youths see people with little or no education amassing much wealth and being respected in the society, and so on.

Tade and Aliyu [16] in their research discovered that cyber frauds in higher institutions are highly organized, and associated with crimes that depend on "connections" for sustenance and survival. The authors observed that surprisingly, bank workers were part of the cybercrime networks, disclosing important information about their customers to fraudsters and helping them to receive money successfully either through Western Union, Money Gram, Credit Cards, domiciliary accounts or checks, so as to benefit from the fraudster's deals. The cyber fraudsters were reported to use the money gotten from their illegal activities to pay school fees, live 'big' on campus, and fend for a living. These frauds according to the authors continue to increase in Nigeria because of the inordinate nature of socialization in the country that promotes getting wealth by all means whatsoever possible and also due to failure on the part of parents to inculcate values and monitor their children. This is sad because they are equally involved in the rat race for money.

Banerjee et al. [17] observed that cybercrimes might occur when users wrongly type a URL that lands to a fake URL thereby, exposing them to risk of being duped or their personal information deceitfully collected. The authors also discovered that prevalent among this type of cybercrime is the phenomenon of cloned URLs called website poaching that mimics the website names of genuine websites mostly involving a character change in the original URL and largely affecting URLs that are in length below eight (8) characters. It also noted that some companies purchase the phony URLs in a bid to manage the situation but unfortunately end up encouraging the fraudsters that create fake URLs since they do that, expecting that the URLs would be purchased.

Amongst the numerous cyber frauds, the online shopping/auction scam and dating scams were the most dominant frauds [9]. The victims of cyber fraud were majorly business organizations, though, not often reported. The reported cyber frauds were predominantly done by individuals. Frauds related to the theft of identity were the dominant form of e-commerce frauds, mostly consumer-related [8].

To determine the prevalence of cybercrime, an investigation was carried out in [18] which discovered that a very high percentage of people are aware of cybercrimes perpetrated through the use of the internet and/ or the computer and that these cybercrimes are most times not reported by the victims.

Oluwadera et al. [19] carried out an evaluation study to discover the number of victims of loss due to financial cybercrime out of the 48.3 million people in Nigeria using the internet. It estimated that internet users in Nigeria lost up to ₦2, 146, 666, 345, 014.75 ($13,547,910,034.80) to cyber fraud.

A 2016 survey on the prevalence of cybercrime discovered that out of the questionnaire respondents, 88% agreed to the widespread of cybercrime in Nigeria, while 12% thought cybercrime is not common in Nigeria. It further showed that the same percentage of the respondents that affirmed the prevalence of cybercrime in Nigeria supports the notion that cybercrime tarnishes the global image of Nigeria while as expected, the other 12% disagree on the prevalence of cybercrime in Nigeria opposed the fact of cybercrime damaging Nigeria's image [20].

In his report, Abdulkabir [21] said that 55% of youths between 15–22 years, 30% of adults between 23–45 years, and more than 20% of elderly persons were recorded to have been caught in this heinous cybercrime business. The paper, therefore, argues that 'yahoo-yahoo' and the '419' syndrome are two words of the same meaning and the slight difference is that 'yahoo-yahoo' is mainly an online scam with the use of the internet, while 419 refers to the word borrowed from section 419 of the Nigeria criminal act and purposefully used to describe swindling or dishonesty.

According to Fadare [22], Nigeria experienced a great rise in the number of cellular telephone users and internet subscribers from the year 2000 to 2012. The author highlighted that there was an increase in the use of mobile phones from 30,000 (that is, in every 100 people, there are 0.02 promoters) to 112,777,785 (that is, in every 100 people, there are 67.68 approvers), and surge in internet payment from 0.06 to 32.88 approvers out of every 100 people.

3 Materials and Methods

This research employed an online survey tool using Google forms to gather responses from users of digital infrastructures for financial transactions through platforms such as mobile banking, SMS USSD codes, ATM, etc. The questionnaire sought to know if the users have at any time been victims or targets of financial cybercrime and if yes, the electronic platforms through which it occurred. Other information sought from the respondents included Age range, device type used for the transactions (smartphones, PCs, ATMs, etc.), the network used (GSM, institutional VPN, Wi-Fi, LAN), payment platforms used (mobile money, ATM, Digital wallet, e-banking, etc.), and the financial cybercrime mode they had experienced (ATM, SMS, Mobile app, Phone calls, etc.). Two Hundred and Thirty (230) insightful responses were received from professionals and students after one month.

The researchers applied machine learning techniques to analyze the user responses using R Studio version 3.6 from the CRAN project. The experiment is described in the following sections.

3.1 Machine Learning and Data Analytics

R is a powerful statistical tool that can be used to solve any statistical problem, as well as machine learning tasks due to its rich libraries and many inbuilt functions. R is an open-source tool with many contributors from the R Community. R can be run from the console as a standalone application or run though R Studio integrated development environment (IDE).

Machine Learning Processes. The following steps in Fig. 2 were taken in building the machine learning model to analyze the data and predict the likelihood of one being a victim of financial cybercrimes.

a. Data Collection
b. Data visualization
c. Data cleaning and pre-processing
d. Choosing a machine learning algorithm
e. Training the data
f. Testing the result
g. Using the model

a. *Data Collection.* As earlier stated this research applied an online Google form-based data collection technique from real users. A total of two hundred and thirty observations with eight variables were collected before this report. The form is still open for more respondents. The respondents cut across all works of life with ages ranging from sixteen years and above. R is capable of reading different file formats. When the responses were downloaded they were saved in Excel format as.xlsx file. Though R can read Excel format, we decided to save the file in.csv (comma-separated value) format, which is more conventional for machine learning tasks.
b. *Data Visualization.* The purpose of data visualization is to provide a bird's-eye view of the data distribution. This shows if the data is evenly distributed or skewed if there are outliers, etc. R has several powerful tools to plot different types of visualizations in 2D and 3D views. The initial responses are shown in the bar plot in Fig. 3.
c. *Data cleaning and pre-processing.* Looking at the plot in Fig. 3, it is obvious that there exists some noise in the data. A closer look at the dataset showed that there were some fields with NAs (not available). It means there are empty fields. These NAs have to be expunged, and R can identify and remove them effectively. Another thing we notice in the figure is that there are three levels of responses, No, Not sure, and Yes. But we are only interested in Yes and No responses. Taking a look at the structure of the dataset we had to make sure that the dataset is munged into categorical data types if they were not. This is because the responses, Yes and No are categorical types. Figure 4 shows the cleaned and pre-processed dataset with 199 observations and 8 variables, out of which the No responses are 113 and the Yes are 86.

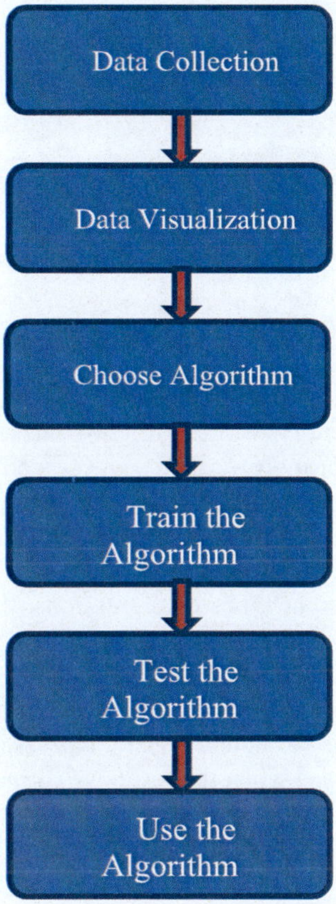

Fig. 2. Process flow of machine learning

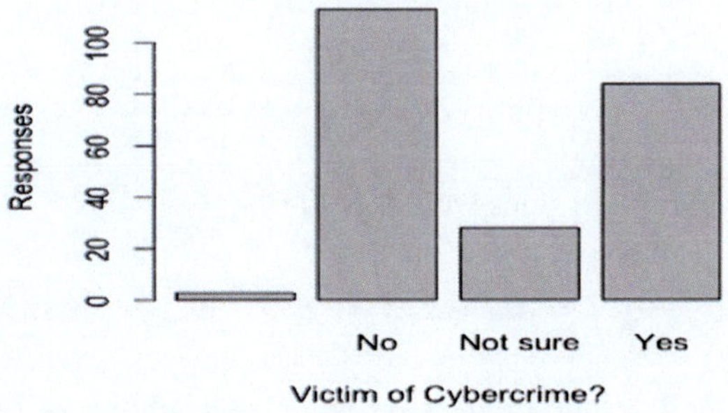

Fig. 3. Initial user responses

d. *Choosing a machine learning algorithm.* Several machine learning algorithms exist for classification and prediction within the contexts of supervised and unsupervised learning. It is usually a tough task selecting the right algorithm for a machine learning task. Most often the researcher may have to employ several of them and compare their performances on the given dataset before settling on one.

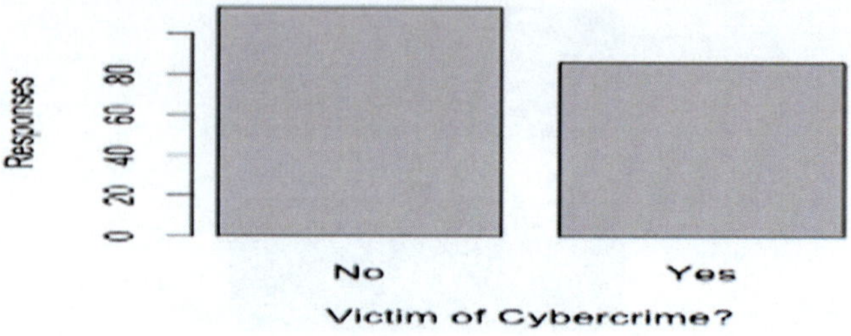

Fig. 4. Cleaned dataset

Using Regression algorithms: To predict the impact of the independent (response) variable on the dependent (explanatory) variable, regression is needed. Regression is a statistic relationship between two or more variables where a change in the independent variable is associated with a change in the dependent variable. The independent variable drives the dependent variable. If there is some relationship or correlation between the two variables, then we may be able to use the independent variable to predict the dependent variable. There are different types of regressions, the most popular being linear regression, and logistic regression.

Logistic Regression: unlike linear regression which answers the question, 'how much?' logistic regression answers the question, 'will it happen or not?', While linear regression predicts continuous variables, logistic regression classifies categorical variables. Logistic regression is used when the response variables have only two outcomes, see Fig. 4. Again, while linear regression uses linear function as in Eq. (1):

$$Y = B_0 + B_1X_1 + B_2X_2 + \ldots. + B_nX_n \tag{1}$$

Where B_0 is the intercept and B_1 to B_n are coefficients of the variables $X_1 \ldots X_n$, logistic regression uses a probability or (Sigmoid) function distributed along the X-axis, which is represented by Eq. (2) and shown in Fig. 5.

$$\text{Probability or logit (p)} = \frac{1}{1 + e^{-y}} \tag{2}$$

Taking the natural log of both sides would yield Eq. (3)

$$\ln\left(\frac{p}{1-p}\right) = B_0 + B_1X_1 + B_2X_2 + \ldots.. + B_nX_n \tag{3}$$

This research used the Logistic Regression model. The choice of logistic regression is predicated on the type of outcomes we are searching for. The simplicity and relative ease and quick implementation makes it a ready choice for machine learning users. Logistic regression is very efficient, as it uses less computational resources, and is highly interpretable. Furthermore, it does not need scaling of input feature. The performance of more complex algorithms can be measured using logistic regression as a good baseline method [19].

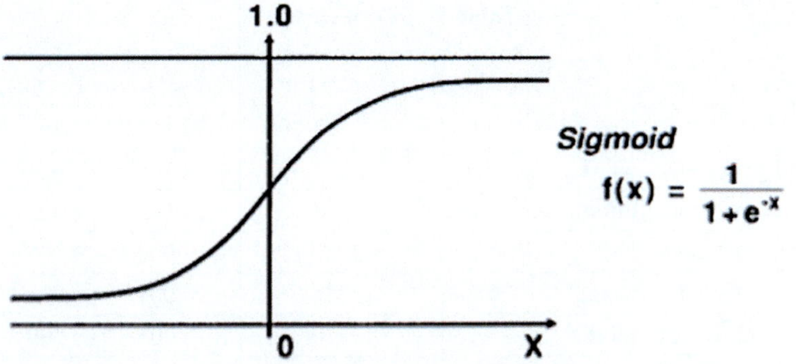

Fig. 5. Sigmoid function

e. *Training the data.* Logistic regression is a supervised learning classification algorithm that we have chosen for our categorical variables. The purpose of training the dataset is to develop a generalized learning model for dealing with future datasets of similar characteristics. To train the dataset we had to partition the dataset into 70% training set and 30% Test set. We then applied the generalized learning model (glm) of the Binomial family. Our target or dependent variable is the probability of being a 'cybercrime victim or not', which was modeled against other variables and their statistical significance determined.

4 Results and Discussion

The result of the training after 19 iterations showed that the following variables were not statistically significant: age range, type of device used, type of network used, and type of application used for the transaction. Only the mode of cybercrime target/attack that was significant.

4.1 Results

Table 1 shows that out of the five modes of financial cybercrime attacks identified by respondents, ATM and Phone modes were statistically significant, having p-values less than 0.001. The p-values with asterisks show that they are statistically significant. Figure 5 captures the visualization of the results of the model. We have an Akaike Information Criterion (AIC) value of 51.992 after 19 iterations when convergence occurred. The AIC is just the residual deviance adjusted for the number of parameters in the model. It can be used to compare one model with another. AIC having values less than 100 shows that the model performed well.

Table 1. Model results

Coefficients	Estimate	Std. error	Z-value	P-value
Intercept	−4.663	1.005	−4.642	3.46e−06 ***
Fin_cybercrime_mode: **ATM**	6.966	1.452	4.796	1.62e−06 ***
Fin_cybercrime_mode: **Phone**	6.609	1.179	5.605	2.08e−08 ***
Fin_cybercrime_mode: **SMS**	25.230	3780.128	0.007	0.995
Fin_cybercrime_mode: **Email**	25.230	10236.634	0.002	0.998
Fin_cybercrime_mode: **Bank**	25.230	17730.370	0.001	0.999

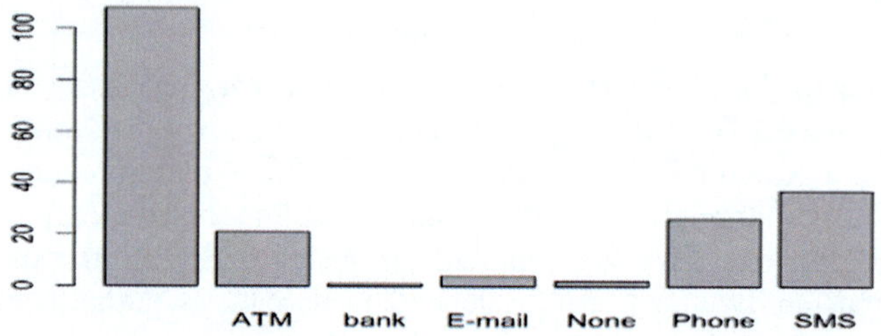

Fig. 6. Plot showing the distribution and significance of the modes of the cybercrime

From Table 1 the following machine learning logistic model in Eq. (5) is derived from the linear model of Eq. (4):

Linear model:

$$Y = -4.66 + 6.97 * ATM\_mode + 6.61 * Phone\_mode \ldots . \quad (4)$$

Logistic model:

$$\text{Logit(prob)} = 1/(1 + \exp(-(-4.66 + 6.97 * \text{ATM\_mode} + 6.61 * \text{Phone\_mode})) \ldots \quad (5)$$

If we assume that both ATM and Phone modes of attack do not occur, then the log (odds) of an attack happening is −4.66. However, if we are predicting the probability of a cybercrime attack occurring through ATM, we get Logit(prob) = −4.66 + 6.97*ATM. The second term indicates the increase in the log (odds) that ATM has for cybercrime to occur. In other words, the second term indicates the log (odds ratio) that ATM cybercrime will occur over the odds of it occurring without any mode of attack. When we add the Phone mode of attack, we provide an additional increase of 6.61 of the log (odds) of cybercrime attacks occurring. Odds are defined as the ratio of the probability of success and the probability of failure. A logit is defined as the log base e (log) of the odds, given as logit(p) = log(odds) = log(p/q), where p is the probability of an event occurring, and q is the opposite. The value of the logit can range from negative infinity to positive infinity.

4.2 Testing the Model

To evaluate the performance of the model, we make use of a confusion matrix to see if the classification was properly done. We shall also calculate the accuracy of the model.

Confusion Matrix (Error Matrix). A confusion matrix is a tabular description of the performance of a supervised machine learning model on the testing set, where true values are known. The output of a confusion matrix is a matrix. In the matrix, each column represents instances in an actual class (actual outcome) while each row shows instances in a predicted class (predicted outcome). Confusion matrix can be used for a classification problem where the classes of the output are two or more types. When there are 2 outcomes (classes), there will be 2 rows and 2 columns in the confusion matrix (see Table 2), 3 outcomes will be 3 rows and 3 columns, 20 outcomes will be 20 rows and 20 columns, etc.

Table 2. The confusion matrix for a two-class (binary) classification problem

		Predicted class	
		Class 1	Class 0
Actual class	Class 1	• True Positive (TP)	• False Negative (FN)
	Class 0	• False Positive (FP)	• True Negative (TN)

Important terms to note

a. *True Positives (TP).* Cases when YES was predicted and YES was also the actual output. When the classes in a confusion matrix exceed two, say three, four, and above, the diagonal outputs in a confusion matrix is seen as the true positive. They tell how many times the algorithm classified the outcome correctly.
b. *True Negatives (TN).* Cases when NO was predicted and NO was also the actual output. The total number of a true negative for a certain class will be the sum of all columns and rows excluding that present class's column and row.
c. *False Positives (FP).* Cases when YES was predicted and No was the actual output. The total number of false positive for a certain class is the sum of values in the corresponding column excluding the TP.
d. *False Negatives (FN).* Cases when NO was predicted and YES was the actual output. The total number of false negative for a certain class is the sum of values in the corresponding row excluding the TP.

The confusion matrix stands as the basis for other metrics here. To evaluate the quality of the classifications, four standard measures may be used, namely accuracy, precision, recall, and F1 score. Table 3 shows the confusion matrix of the model. From the table, TP = 81, while FP = 3, and TN and FN are 53 and 1, respectively. This result shows that there were minimal misclassifications of 3 and 1, respectively.

Table 3. Confusion matrix

Predicted	No	Yes
0	**81**	1
1	3	**53**

Testing Accuracy. Accuracy can be calculated from the confusion matrix as the sum of correct classification (TP) divided by the total number of classifications.

$$Accuracy = \frac{No\ of\ correct\ predictions}{total\ no\ of\ predictions\ made} \tag{6}$$

The calculated accuracy of the model from Eq. (6) is **81.38%**, which is good enough and promising for the generalization of the model, though could be improved on to above 90% if more dataset is collected.

Probability Accuracy. The values in Table 4 show the head section of the predicted results. The data in positions 5 and 7 are missing because they were assigned to the Test dataset before the training of the model.

Table 4. Sample of predicted results.

1	2	3	4	6	8
0.009345794	0.999999999	0.909090909	0.999999999	0.999999999	0.009345794

Applying the value in data point 6 (i.e. 0.99), for example into Eq. (4) and the result of Eq. (4) into Eq. (5), we can determine the probability accuracy of cybercrime occurring through ATM and Phone calls, respectively by making one zero while calculating for the other, in turn.

$$\mathrm{Y} = -4.66 + 6.97 * \mathrm{ATM\_mode} + 6.61 * \mathrm{Phone\_mode} \tag{4}$$

$$\mathrm{Logit(prob)} = 1/(1 + \exp\,(-(-4.66 + 6.97 * \mathrm{ATM\_mode} + 6.61 * \mathrm{Phone\_mode})) \tag{5}$$

Probability accuracy for ATM cybercrime is **90.37%,** while that of Phone calls is **87.76%**, respectively.

4.3 Discussion

Figure 7 shows that out of the 18% of respondents who accepted being victims of cybercrime, 61.5% were through SMS, 30.8% through phone calls, 15.4% were through ATM, and phone calls, respectively.

Comparing the result of the logistic model with the survey responses, Fig. 7 shows a very close correlation with our analysis above in Fig. 6. Even though SMS was favoured mostly as the most common method of attack followed by Phone, then ATM and Email, respectively in that order, the logistic regression model seems to show otherwise, placing ATM and Phone ahead of SMS as the most common mode of cybercrime attack as it observed from the dataset. The machine learning algorithm did not think that SMS was statistically significant enough to contribute to financial cybercrime as against what the respondents said.

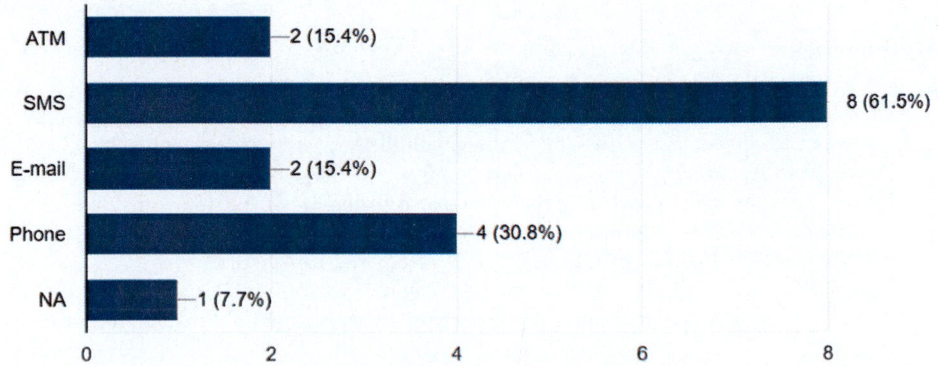

Fig. 7. Means of cybercrime

Means of Cybercrime Attacks. Ibikunle and Eweniyi [20] and Maitanmi et al. [21] both agree that one of the most prevalent modes of cybercrime in Nigeria is through identity theft, mostly perpetrated through debit/credit cards, which the respondents in our survey referred to as ATM cards. They observed that most banks do not report these incidences for fear of losing their customers. It was reported in 2015 that with stolen information from 3,000 accounts at a South African bank, criminals taking advantage of the 24 hours' service of the ATM, made 14,000 withdrawals from 1,700 ATMs in Japan amounting to about $17 Million in cash transactions within a space of 3 h on a Sunday morning [2].

Again, what the respondents referred to as Phone-based cybercrime includes generally, fraudulent text messages, phone calls from individuals who seem to have every information about the target victim, proposing a non-existent job, contracts, or feign to be someone known to the victim who is in critical need of paying a hospital bill, in a police case, etc., with the intention to extract money from the unfortunate victim driven either by the lust for money or empathy.

5 Conclusion

In this paper, it has been shown through a survey that many people have either been targets or victims of financial cybercrime, sometimes, they are oblivious of it. Cybercrime is present with us on a daily basis. This work has contributed in no small way in showing that economic losses due to cybercrime is huge, and maybe more according to projections by experts if cybercriminals and their activities are not check-mated. The paper has also contributed by showing that it is possible to model and predict the possibility of cybercrime occurrence and by what means it may occur. This model achieved an accuracy of more than 80%, which shows the possibility of generalization. We are optimistic that given more data, the model would achieve a higher percentage of accuracy, and be able to correctly and accurately predict cybercrimes. Our future work will be geared towards gathering industry-based cybercrime dataset with additional parameters for training using deep neural networks.

References

1. Clement, J.: Global financial cybercrime losses 2017, by victim country (2019). https://www.statista.com/statistics/799875/countries-with-the-largest-losses-through-cybercrime/. Accessed 15 Apr 2019
2. Lewis, J.: Economic Impact of Cybercrime-No Slowing Down (2018). https://www.mcafee.com/enterprise/en-us/forms/gated-form.html?docID=5fee1c652573999d75e4388122bf72f5&tag=ec&eid=18TL_ECGLQ1_CT_WW. Accessed 19 Apr 2019
3. Morgan, S.: A 2017 report from Cybersecurity Ventures sponsored by Herjavec Group (2017). https://cybersecurityventures.com/hackerpocalypse-cybercrime-report-2016/. Accessed 19 Apr 2019
4. Broadhurst, R., Chang, L.Y.C.: Cybercrime in Asia: trends and challenges. In: Liu, J., Hebenton, B., Jou, S. (eds.) Handbook of Asian Criminology, pp. 49–63. Springer, New York (2013). https://doi.org/10.1007/978-1-4614-5218-8_4
5. Kshetri, N.: Cybercrime and cybersecurity in Africa. J. Glob. Inf. Technol. Manag. **22**(2), 7781 (2019). https://doi.org/10.1080/1097198X.2019.1603527
6. Gercke, M.: Understanding cybercrime: phenomena, challenges and legal response, pp. 11–40. International Telecommunication Union, ITU (2012)
7. Shoba, P.V.: An analysis of cyber crime and internet security. World J. Pharm. Pharm. Sci. **6**(5), 304–308 (2017)
8. Smyth, S.M.: Measuring the extent of cyber - fraud: a discussion paper on potential methods and data sources. Public Safety Canada, no. 8, pp. 1–64 (2011)
9. Levi, M., Doig, A., Gundur, R., Wall, D., Williams, M.: Cyberfraud and the implications for effective risk-based responses: themes from UK research. Crime Law Soc. Change **67**(1), 77–96 (2016). https://doi.org/10.1007/s10611-016-9648-0
10. Omodunbi, B.A., Odiase, P.O., Olaniyan, O.M., Esan, A.O.: Cybercrimes in Nigeria: analysis, detection and prevention. J. Eng. Tech. **1**(1), 37–42 (2016)
11. Taiwo, O.C., Igbekoyi, E.K.: Prevalence and consequence of cybercrime perpetuated by students in public tertiary institutions in Ekiti State. Int. J. Arts Lang. Bus. Stud. **2**(1), 211–224 (2019)
12. Okeshola, F.B., Adeta, A.K.: The nature, causes and consequences of cyber crime in tertiary institutions in Zaria-Kaduna state, Nigeria. Am. Int. J. Contemp. Res. **3**(9), 98–114 (2013)
13. Adesina, O.S.: Cybercrime and poverty in Nigeria. Can. Soc. Sci. **13**(4), 19–29 (2017)
14. Udelue, M.C., Mathia, D.B.: Prevalence of cybercrimes among youths in Onitsha south local government area of Anambra state, Nigeria. Int. J. Health Soc. Inq. **5**(1), 82–106 (2019)
15. Igba, I.D., Igba, E.C., Nwambam, A.S., Nnamani, S.C., Egbe, E.U., Ogodo, J.V.: Cybercrime among university undergraduates: implications on their academic achievement. Int. J. Appl. Eng. Res. **13**(2), 1144–1154 (2018)

16. Tade, O., Aliyu, I.: Social organization of internet fraud among university undergraduates in Nigeria. Int. J. Cyber Criminol. **5**(2), 860–875 (2011). http://www.cybercrimejournal.com/tadealiyui2011julyijcc.pdf
17. Banerjee, A., Barman, D., Faloutsos, M., Bhuyan, L.N.: Cyber-fraud is one typo away. In: IEEE INFOCOM, pp. 66–70 (2008)
18. Oogarah-Hanuman, V., Ching, Y.L.L., Wan, A.Y.P., Chetty, D.: Assessing the prevalence of cybercrime in mauritius. Econ. Rev. **47**(4), 149–507 (2018)
19. Oluwadare, C.T, Oluwasanmi, L.A., Igbekoyi, K.E.: Prevalence and forms of cybercrime perpetrated by students in public tertiary institutions in Ekiti State. Int. J. Econ. Bus. Manag. Res. **2**(4) (2018). http://ijebmr.com/uploads2018/IJEBMR_02_263.pdf
20. Sulaiman, L.A., Ishowo, L., Muhammed, A.Y., Raji, A.: Cybercrime and Nigeria's external image: a critical assessment. Africology J. Pan Afr. Stud. **9**(6) (2016). https://pdfs.semanticscholar.org/6029/f903dd98cf7d96dc0ac234d2f6a2a742d31b.pdf
21. Abdulkabir Olaiya Suleiman, M.A.: The impulsive upsurge of yahoo-yahoo in the 21st century in Nigeria: islamic perspective. Afr. J. Criminol. Justice Stud. (AJCJS) **12**(1) (2019). https://www.umes.edu/uploadedFiles/_WEBSITES/AJCJS/Content/VOL12.1.%20SULEIMAN%20PDF.pdf
22. Fadare, O.A.: Impact of ICT tools for combating cyber crime in Nigeria online banking: a conceptual review. Int. J. Trade Econ. Finance **6**(5) (2015). http://www.ijtef.org/vol6/481-E304.pdf
23. Chauhan, N.S.: Real world implementation of logistic regression. Towards data science (2019). https://towardsdatascience.com/real-world-implementation-of-logistic-regression-5136cefb8125. Accessed 16 June 2020
24. Ibikunle, F., Eweniyi, O.: Approach to cyber security issues in Nigeria: challenges and solution. Int. J. Cogn. Res. Sci. Eng. Educ. (IJCRSEE) **1**(1) (2013). www.ijcrsee.com
25. Maitanmi, O., Ogunlere, S., Ayinde, S., Adekunle, Y.: Cyber crimes and cyber laws in Nigeria. Int. J. Eng. Sci. **2**(2), 19–25 (2013)

PEDAM: Priority Execution Based Approach for Detecting Android Malware

Olorunjube James Falana[1](✉), Adesina Simon Sodiya[1], Saidat Adebukola Onashoga[1], and Anas Teju Oyewole[2]

[1] Federal University of Agriculture Abeokuta, Abeokuta, Ogun State, Nigeria
{falanaoj,Sodiyaas,Onashogasa}@funaab.edu.ng

[2] Indigo Books and Music, 620 King Street W., Toronto M5V 1M6, Canada

Abstract. With the openness and growing popularity of Android Operating system all over the world, it has become a target of attack for Malware authors who are determined to take advantage of over 2.5 billion monthly active users of Android devices. Despite Google's various protection measures, android malware continues to grow in complexity and scope. In recent time, many research efforts have focused on detecting malware on the Android operating system using both static and dynamic approaches. Most of the existing techniques are still not perfect because of the problems of false positive, false negative and high detection time. In this work, a Priority Execution-based Approach for Detecting Android Malware (PEDAM) is proposed to solve some of these problems. In PEDAM, a two-phase dynamic analysis scheme is used for malware analysis. The first phase involves the use of a time-based filter for prioritizing the android application that will execute based on permissions and intents. Any suspected samples not captured in the first phase are further analysed in the second phase, which does behavioural analysis using Support Vector Machine classifier to analyse permissions, intent filters and Activity features set for effective detection. The evaluation of the proposed model on different Android malware families' shows that PEDAM outperformed another android-based malware detection system known as Iterative Classifier Fusion System (ICFS) with improved accuracy of 1.04%. These results indicated that the approach could be deployed for detection of android malware.

Keywords: Malware detection · Android · Dynamic analysis · Permission · Intent filter

1 Introduction

According to the mobile threat report released by McAfee in first quarter 2020, over 37 million android malware were detected in the fourth quarter of 2019 out of which nine hundred thousand were new mobile malware [28]. Mobile users have suffered from various degrees of security breaches as well as financial losses occasioned by malicious apps [24].

Android operating system (OS) has become the most popular smartphone platform with 72.26% of the market share in 2020, while iOS Apple accounts for 27.03% [11].

J. H. Abawajy et al. (Eds.): EATI 2020, LNNS 254, pp. 152–165, 2021.
https://doi.org/10.1007/978-3-030-80216-5_12

The popularity of Android OS has made it a targeted attack for Malware authors that want to take advantage of over 2.5 billion monthly active users of Android devices. Due to the Android open source development environment, developers can deploy their apps using android data centre or other data centres. This openness of the Android development environment has made it possible for malware authors to develop apps that abuse the platform features. Cybercriminals have continued to abuse Android's accessibility and permission to perpetrate attack [9, 18]. Android's accessibility features were proposed to assist individuals with a disability or other challenges in utilizing their gadgets. For instance, it is convenient to utilize voice commands rather than the touch screen. To prevent the abuse of accessibility features, Google has limited the permissions on the accessibility features by moving some of the functions to a new application programming interface. However, cybercriminal continues to take advantage of this feature to perpetrate an attack.

Existing android malware detection strategies are grouped into Static-based, Dynamic-based, signature-based and Hybrid-based detection. The signature-based uses a signature of known malicious sample to decide whether a program is malicious and can be built directly on any of the malware detection techniques [17, 19, 23]. Static investigation of a program can be performed without really executing the program. Some data that can be gotten from the static examination incorporates opcode sequence, control flow graph, API calls etc. Dynamic investigation necessitates that we execute the program, most often the programs are executed in a virtual environment [4] while Hybrid techniques combine both the features of static and dynamic analysis. The approach helps to detect malicious application early before it is executed and made available to the public.

Considering the highlighted problems, in this research, we develop a priority-based execution technique that focuses on extracting permissions, intent and activity to analyse the behaviour of Android malware using Support Vector Machine.

The rest of the paper is organized as follows: Related work is presented in Sect. 2. Proposed method and framework are described in Sect. 3. The system implementation, results and are contained in Sect. 4. The conclusion is presented in Sect. 5.

2 Related Work

Despite the recent security improvements in Android OS, its underlying open development environments remain vulnerable to novel exploits. Faruki et al. [10] proposed signature-based solutions to mitigate malicious apps threats in Android. The authors utilized statistical features to detect malicious applications. The result of the experiment shows that the approach can detect unknown, unseen suspicious apps from play store with an accuracy of 72.27%.

As of late, various researches have been made to address the security and privacy dangers occasioned by malware in the Android operating system, among which are static examination [19, 21]. Kate and Dhavale [19] proposed a two-stage static investigation utilizing bloom filters. The principal stage includes two distinctive bloom filters used to classify selected samples into malevolent and benign. Any dodged malicious samples from the first phase are further broken down by the second phase, which comprises of Naïve Bayes Classifier utilizing authorization and code-based blended mixed features set. The general accuracy of detection achieved is 87%.

Dai et al. [7] utilized the SVM-based malware recognition scheme for Android platform and utilized both dangerous API calls and dangerous permissions blends as features in the SVM characterization. Dimjašević et al. [8] proposed a dynamic analysis using an instinctive system call tracking. Malicious apps were executed in a virtual environment. However, the results of the experiment were not consistent.

In another related work by Meng et al. [22], the authors proposed a semantic approach in light of Deterministic Symbolic Automaton (DSA), for malware perception, detection and grouping. DSA captures the malevolent behaviours of malware families by extracting semantic features learned to classify according to attack patterns. The outcome shows that the method was able to distinguish 4583 new malware. Meng et al. [22] proposed a backtracking method for malware detection based on the fusion of a convolutional neural network using APK extracted features. The method used relevant feature sets that can contribute positively to the classification decision. The result of the experiment showed that the approach outperforms other existing approaches in time cost.

Li et al. [20] considered the relationship between permissions and API calls to adequately recognize malware in Android platform. Features are ordered depending on the level of segregation and the continuous pattern. The Naive Bayes classifier method is further used to enhance the weighted properties of successive patterns. Finally, the improved Naive Bayes algorithm and the features selection method are consolidated to actualize Android malicious application detection method. The experimental results showed that the component screening method and the progressive component configuration mining method could improve the detection performance of the algorithm in a specific way. Simultaneously, for the Naive Bayes algorithm, the weighted property can be improved to accomplish better outcomes. However, since the method depends on a static based method, it acquires the limitation of static detection procedures, which implies that it cannot detect obfuscated or encrypted malware.

Niu et al. [25] proposed a unique permission management database which records the state of consents for each app. Just the authorization, which is allowed condition in the database that can be used by the app. In the entire procedure, the method initially characterises the app using AI, at that point retrieves the comparing consent information from databases, and issues the risky authorization cautioning to clients. Finally, the method invigorates the dynamic administration database as demonstrated by the clients' decisions.

Due to the limitation of other detection methods, there is a pressing requirement for new detection methods that can deal with the massive number of polymorphic malware and the new malware innovations. In this manner, AI and data mining algorithms have been brought into the malicious applications detection space and these algorithms have exhibited its effectiveness [6].

In Xiao et al. [32], a dynamic analysis framework was created to detect the applications' malicious conduct. The technique depends on a deep learning design with Stacked AutoEncoders (SAEs) to characterise applications as malevolent or benevolent [13]. In Yuan et al. [33], a Deep Belief Networks (DBN) based deep learning model had been utilized to portray Android applications. In Huang [14], the classes.dex document, which contains a key part of the execution logic of Android application, has been changed over.

Huda et al. [16] presented a hybrid system to recognize run-time action for malware by utilizing more than one channel. The proposed system misused the highlights of inside channels and between covering channels by hybridizing discriminant, least repetitive, and greatest important channels with the covering way to deal with the information from the run-time conduct of malware. The design of the proposed discovery engine was made out of run-time social log collection, critical run-time highlight selector and recognition engine. A sum of 1500 benign executable and 2000 malware tests was utilized for the analyses. The proposed hybrid approaches were tried utilizing 10-fold cross-validation for 10-trials and the run-time information. The results indicated that the hybrid methodologies altogether improve the detection accuracies. However, the computational unpredictability of the proposed strategy will be extremely high.

Abawajy and Kelarev [3] built up an Iterative Classifier Fusion System (ICFS) that combined both static and dynamic highlights of android malware. The methodology consolidated three classifiers iteratively by exploiting the prescient abilities of the three classifiers. The authors performed trials by exploring the request for the three classifiers chose in the set J48, LibSVM, MP, NBTree, NNge, by employing feature selection methods Information Gain Feature Evaluator, Classifier Feature Evaluator and the Wrapper Subset Evaluator. However, some of the concerns are the size of the memory and the time-consuming learning phase will make the approach difficult for a mobile operating system.

In another related work by Abawajy et al. [1], an ensemble pruning method was proposed to address the problem of memory size and processing requirements associated with an ensemble classifiers. A Hybrid Consensus Pruning was employed to combine several classifiers into single. The experimental results showed that the approach performed better than K-means pruning scheme. Abawajy et al. [2] also provided a study that used a permission-based approach to determine the behaviour of application logic yet to execute in mobile IoT application. The classification was done based on 44 permission deleted through information gain features selection method. The study showed that J48 classifier performed better with a TPR of 95.2%

3 Methodology

The general architecture of PEDAM is shown in Fig. 1. The idea behind this proposed system is to analyse the permission of android application structure to predict the state of the file. The approach involves extracting the permission, intent filters and activity for each android application package. Each android application has androidmanifest.xml, classes.dex, resources and assets. The androidmanifest.XML contains a manifest file, all app components, permission etc. Before the installation or running of any application by an android framework, the manifest file is read for authorization required by the application, for example, read access to contacts, camera etc.

3.1 Phase 1

This phase consists of permissions and intent extraction stage to determine the priority level of an application. This involves analysing app manifest file to extract the permissions and intent attributes. One way of achieving this is to explicitly define the resources

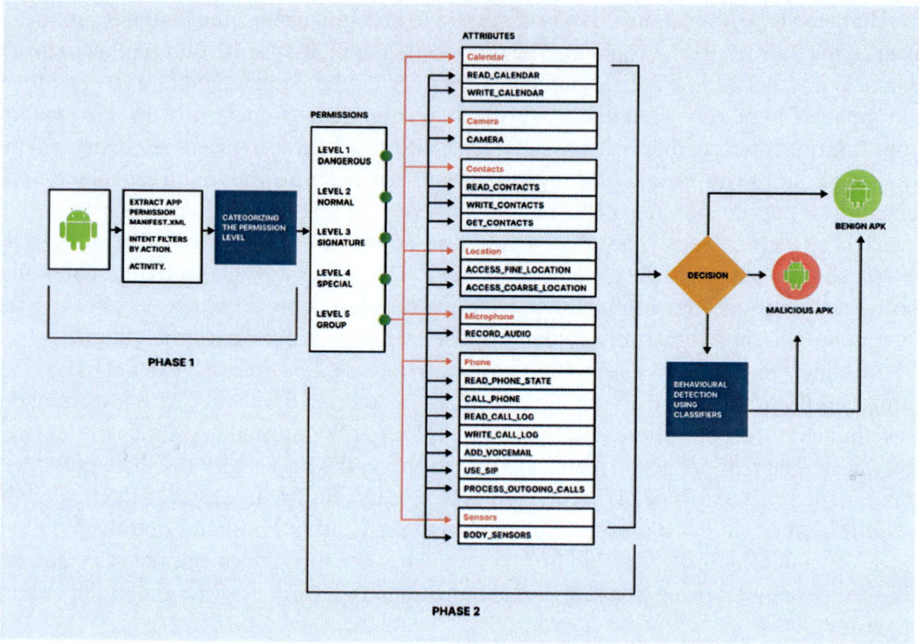

Fig. 1. Architecture android malware detection system

that can invoke the permission and map the resources to the permission level. Android Operating system requires an app to list all the resources they are going to use in the manifest.xml file [12, 29].

Given a list of risky permission R consisting of dangerous permission D and group permission P.

R can be defined as.

$$R = \{D \cup G\} \tag{1}$$

Let L be the list of all possible intent associate to a given permission.

Suppose an application A consisting of P permissions and I intents. Let P_i be the i-th permission sample and T_kP_i is the percentage of k-th permission in S.

In addition, let I_i be the i-th intent in a given intent sample and T_kI_i is the percentage of k-th intents in an application. The priority scheduling is computed as follows

$$T_kP_i = \frac{\sum_{i=1}^{m} p_i}{\sum_{i=1}^{n} S_i} \tag{2}$$

$$T_kI_i = \frac{\sum_{i=1}^{m} I_i}{\sum_{i=1}^{n} L_i} \tag{3}$$

Let $F = \min(T_kP_l, T_kI_l)$. If F is greater than the threshold defined by the user then the application is placed in lower priority otherwise, it is placed in High priority. Any application that is placed in lower priority will be subjected to behavioural detection.

```
Algorithm 1: Permission and Intent Filter Extraction
Input:  Risky_permission,whitelist_permission,method_call,
        parameter
Output: start_Activity
list_API ← requestPermission(Activity, permission, re_
  questCode)
list_activity ← Activity
RC <- requestCode
  if RC in Risky_Permission:
    Percentage_Permission <- Compute(Sum_permision/To_
    tal_per)
    Percentage_Intent <-Compute(Sum_Intentn/Total_Intent)
    T = Min(Percentage_Permission, Percentage_Intent)
    If T > threshold
      Goto Phase 2
    elif:
      list_permission ← permission
      list_intent ← intent
      start_Activity ← getCurrentMethod(list_permission,
      list_intent)
  return start_Activity
```

3.2 Phase 2: Behavioural Detection

The second phase consists of behavioural checks. Once an app is assigned to a low priority, a further check is carried out to ascertain the true state of the App. In essence, each analysis extracts permission, intent and activity from the given app. An activity is an important component of the android application that allows users to interact with the phone.

Let the device state be represented as δ in sliding window W.

δ = (P, I, A),

Where:
$P = \{P_1, P_2, P_3, \ldots, P_n\}$ is a set of requested permission by an application in W
$I = \{I_1, I_2, I_3, \ldots, I_n\}$ is a set of intents in W
$A = \{A_1, A_2, A_3, \ldots, A_n\}$ set of actions performed in W.

This can further be expressed as:

$$P = \bigcup_{t=1}^{t=n} \{set\ of\ permission\ request\ in\ \mathrm{W}\} \tag{4}$$

$$I = \bigcup_{t=start}^{t=stop} \{set\ of\ intents\ in\ \mathrm{W}\} \tag{5}$$

$$A = \bigcup_{t=1}^{t=n} \{set\ of\ actions\ performed\ in\ \mathrm{W}\} \tag{6}$$

The extracted features for all the samples is given

$$F = \{P_1, P_2, P_3, \ldots P_n; I_1, I_2, I_3, \ldots I_n; A_1, A_2, A_3, \ldots A_n\} \equiv f \in P \bigcup I \bigcup A \quad (7)$$

```
Algorithm 2: Permission and Intent Filter Extraction

Input: protection_Dangerous, whitelist_permission,
method_call, parameter
Output: S_array: a set of 3-tuple(P, I, A)

foreach RC in protection_Dangerous:
  list_activity ← t.remove(< name= '".ListAllActivity" >)
  list_permission ← t.remove(<uses-permission>)
  list_intent ← t.remove(android:name="android.intent.ac
    tion.MAIN)
   #Encoding the extracted features.
    features_list = {list_activity, list_intent,
    list_permission}
    S_array ← [ ]
    current_file ← [ ]
    status ← None
    if status== 'malware':
      labels = np.ones([compressed_data.shape[0],])
    elif status == 'benign':
      labels = np.zeros([compressed_data.shape[0],])
#Compress the rows of the one-hot-encoded columns by add-
ing them together for each malware/benign file.
    for i in range(len(features_list)):
      if i not in labels:
        if current_file == []:
          current_file = features_list[i]
        else:
        current_file += features_list[i]
        #Row contains attributes of new malware file.
       if (current_file != []) and (i != 0):
          S_array.append(current_file)
          current_file = features_list[i]
          i += 1
      continue
    S_array.append(current_file)
    return S_array
```

As stated in Eq. 4, n permissions were extracted from the samples out of which the ten frequently requested permissions from both malicious application and benign application were shown in Table 1.

Table 1. Ten frequently requested permissions

S/N	Malware PERMISSION	Benign PERMISSION
1	android.permission.INTERNET	android.permission.INTERNET
2	android.permission.READ_PHONE_STATE	android.permission.WRITE_EXTERNAL_STORAGE
3	android.permission.WAKE_LOCK	android.permission.ACCESS_COARSE_LOCATION
4	android.permission.SEND_SMS	android.permission.ACCESS_WIFI_STATE
5	com.google.android.c2dm.permission.RECEIVE	android.permission.ACCESS_NETWORK_STATE
6	android.permission.READ_SMS	android.permission.FOREGROUND_SERVICE
7	android.permission.RECEIVE_SMS	android.permission.READ_PHONE_STATE
8	android.permission.WRITE_EXTERNAL_STORAGE	android.permission.WAKE_LOCK
9	com.software.application.permission.C2D_MESSAGE	com.google.android.c2dm.permission.RECEIVE
10	android.permission.ACCESS_NETWORK_STATE	android.permission.BLUETOOTH

3.3 Classification Model

The feature sets generated from phase II, which consist of 3-tuple (Permission, Intent, Activity) together with labels (malware and benign) were used to build the support vector machine classifier.

Support Vector Machines (SVM)
Support vector machine is one of the most efficient supervised machine-learning models that can be used for data classification. The basic principle behind the working of Support vector machines is creating a hyperplane that separates the dataset into classes.

1. Training of Data

This is an initial set of data used to help the SVM learns and make a prediction. The steps in training our data are as follows:
Given training data (X_i, y_i).
Where:

$i = 1, 2, 3, \ldots,$ m
$y_{i \in}\{-1, +1\}$, which is the class label representing Malicious or Benign
$X_i \in R^b$, which represents the data value of the feature sets.

2. Kernel Mapping

Given any two vectors X_i and X_j as inputs into a kernel function, it produces an output defined by the inner of their images $\theta(X_i)$ and $\theta(X_j)$

$$K(X_i, X_2 = \theta(X_i)^T \theta(X_2) \tag{8}$$

To generate a learning Algorithm that functions in the kernel space. The values of all inner products are substituted into the newly formed kernel space

$$f(x) = \theta(X)^T w + b = \sum_{j=1}^{m} \alpha_j y_j K(X, X_j) + b \quad (9)$$

The parameter b can be derived from any support vector Xi.

$$b = yi - \theta XTw = yi - = 1m\alpha jyjK\theta XiT\theta Xj = yi - j = 1m\alpha jyjKXi, Xj \quad (10)$$

3. Hyperplane

Hyperplane helps to separate multidimensional data linearly by lines into the positive and negative hyperplane.
Given a straight line function.

$$Y = ax + b \quad (11)$$

Let x be x_1 and y be x_2
Then we have:

$$ax_1 - x_2 + b = 0 \quad (12)$$

if $X = (x_1, x_2)$ *and* $w = (a, -1)$
Then we have:

$$w.x + b = 0 \quad (13)$$

This is the equation of the hyperplane. Where w is normal to hyperplane.
The perpendicular distance to the hyperplane is given as $|b|/|w|$
Classifier
Once the hyperplane is derived, the hyperplane is then used to make predictions. as defined in equation

$$h(x_i) = \begin{cases} +1 \; if \; w.x + b \geq 0 \\ -1 \; if \; w.x + b < 0 \end{cases} \quad (14)$$

4 Dataset and Implementation

To show the performance of the work, the dataset collected is tested on Jupiter notebook running python 3.7.1. The dataset used for this experiment consists of 600 malicious Android packages in MD5 hash form and 550 benign Android applications (.apk). The malicious files came from Virushare [30], while the clean android programs were downloaded from Apkpure [5] and Google play store [26]. After the collection, both clean files and malicious files were scanned on Virustotal [31] containing 61 detection engines to ascertain the true state of the files.

4.1 Evaluation Measurements

The following metrics were used to assess the performance of PEDAM:

- Accuracy: This refers to the ratio of the total applications that are correctly classified whether as clean or malicious.

$$Accuracy = \frac{TP + TN}{TP + TN + FP + FN} \tag{15}$$

- Precision: This is the ratio of correctly predicted malicious application to the total applications that are classified as malicious.

$$Precision = \frac{TP}{TP + FP} \tag{16}$$

- Recall: This refers to the ratio of correctly classified malicious applications to the total number of correctly or incorrectly classified malicious and benign applications.

$$Recall = \frac{TP}{TP + FN} \tag{17}$$

- F-Measure: This model tells us how discriminative the model is.

$$F_{Measure} = 2 * \frac{Precision * Recall}{Precision + Recall} \tag{18}$$

Where:

True Positive (TP): It represents the malicious applications correctly classified.
True Negative (TN): It represents the benign applications correctly classified.
False Positive (FP): It represents the benign applications correctly classified.
False Negative (FN): It represents the number of malicious applications that are misclassified.

4.2 Results

The experiments were performed on different classifiers in a Jupter Notebook environment. The result shows that PEDAM had the best accuracy of 95.24% when compared with Naïve Baye, KNN, and Random forest with an accuracy of 82.3%, 71.4% and 76.5% respectively as shown in Table 2 and Fig. 2.

Table 2. Experimental result of PEDAM

Classifier		Precision	Recall	F1-score	Support
PEDAM (SVM)	0.00	1.00	0.89	0.94	9
	1.00	0.92	1.00	0.96	12
	Micro avg	0.95	0.95	0.95	21
	Macro avg	0.96	0.94	0.95	21
	Weighted avg	0.96	0.95	0.95	21
	Accuracy of SVM: **0.9524**				
Naïve Bayes					
	0.00	0.56	0.71	0.63	7
	1.00	0.83	0.71	0.77	14
	Micro avg	0.71	0.71	0.71	21
	Macro avg	0.69	0.71	0.70	21
	Weighted avg	0.74	0.71	0.72	21
	Accuracy of Naïve Bayes: **0.7143**				
KNN					
	0.00	1.00	0.57	0.73	7
	1.00	0.77	1.00	0.87	10
	Micro avg	0.82	0.82	0.82	17
	Macro avg	0.88	0.79	0.80	17
	Weighted avg	0.86	0.82	0.81	17
	Accuracy of Naïve Bayes: **0.8235**				
Random Forest					
	0.00	1.00	0.43	0.60	7
	1.00	0.71	1.00	0.83	10
	Micro avg	0.76	0.76	0.76	17
	Macro avg	0.86	0.71	0.72	17
	Weighted avg	0.83	0.76	0.74	17
	Accuracy of Random Forest: **0.7647**				

The result of the approach was also compared with other existing methods on malware detection as presented in Table 3. Rana et al. [27] achieved an accuracy of 94.33% with Random Forest. Abawajy and Kelarev (2017) used ICFS and had an accuracy of 94.2%. Abawajy et al. (2018) used a permission-based approach with J48 classifier and achieved an accuracy of 95.2%.

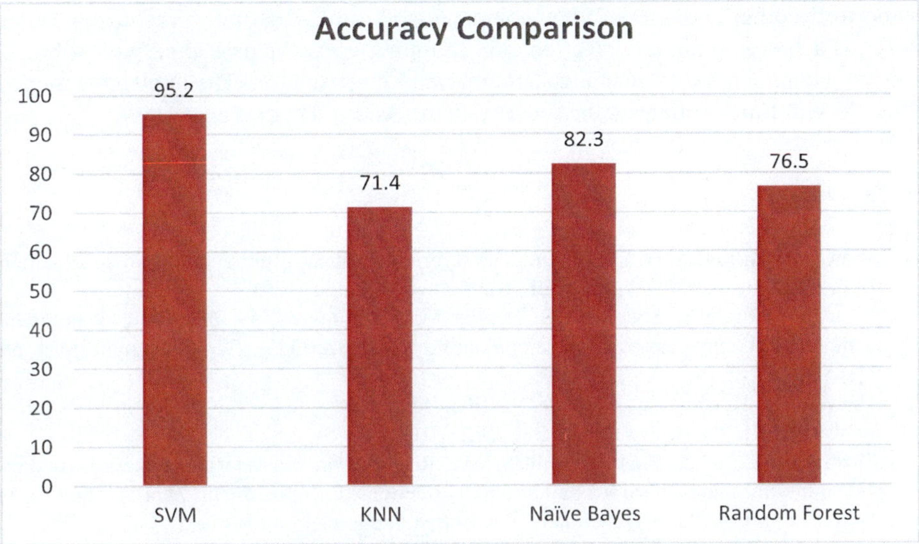

Fig. 2. An accuracy comparison for four different classifiers

Table 3. Result of the comparison with other existing research work

Ref	Features	Accuracy %	Detection method
Rana et al. [27]	API Calls, App Components, Filtered Intents, API Calls, Requested Permissions, Network Address, Restricted, Used Permissions	94.33	Random Forest
Abawajy and Kelarev [3]	Control Flow Graph, Call Graph, Permissions,	94.20	LibSVM with polynomial kernel, Multilayer Perceptron, NBtree, and iterative feature selection
Huda et al. [15]	Permission	95.20	J48
PEDAM	Permission, Intent, Activity	95.24	SVM

5 Conclusion and Future Work

This work presented a malware detection system on an Android Application using a two-phase approach. It was based on extracting permissions, intent filters and Activity from each android applications. These parameters were used to train the machine-learning algorithm. As shown from the experimental results, the detection accuracy of PEDAM outperforms other classifiers. When compared with other existing works, the proposed PEDAM achieved a higher detection rate. In future work, the procedure will be broadened to include both dynamic and static malware investigation. The implementation of PEDAM will further enhance the security of the Android operating system.

References

1. Abawajy, J., Chowdhury, M., Kelarev, A.: Hybrid consensus pruning of ensemble classifiers for big data malware detection. IEEE Trans. Cloud Comput. (2015)
2. Abawajy, J., Huda, S., Sharmeen, S., Hassan, M.M., Almogren, A.: Identifying cyber threats to mobile-IoT applications in edge computing paradigm. Futur. Gener. Comput. Syst. **89**, 525–538 (2018)
3. Abawajy, J., Kelarev, A.: Iterative classifier fusion system for the detection of Android malware. IEEE Trans. Big Data **5**, 282–292 (2017)
4. Al-rimy, B.A.S., Maarof, M.A., Shaid, S.Z.M.: Ransomware threat success factors, taxonomy, and countermeasures: a survey and research directions. Comput. Secur. **74**, 144–166 (2018)
5. Apkpure. Android Apps Download (2020). https://apkpure.com/
6. Bakour, K., Ünver, H.M., Ghanem, R.: The Android malware detection systems between hope and reality. SN Appl. Sci. **1**(9), 1–42 (2019). https://doi.org/10.1007/s42452-019-1124-x
7. Dai, G., Ge, J., Cai, M., Xu, D., Li, W.: SVM-based malware detection for android applications. In: Proceedings of the 8th ACM Conference on Security & Privacy in Wireless and Mobile Networks, pp. 1–2 (2015)
8. Dimjašević, M., Atzeni, S., Ugrina, I., Rakamaric, Z.: Evaluation of android malware detection based on system calls. In: Proceedings of the 2016 ACM on International Workshop on Security And Privacy Analytics, pp. 1–8 (2016)
9. Falana, O.J., Ebo, I.O., Tinubu, C.O., Adejimi, O.A., Ntuk, A.: Detection of cross-site scripting attacks using dynamic analysis and fuzzy inference system. In: 2020 International Conference in Mathematics, Computer Engineering and Computer Science (ICMCECS), pp. 1–6. IEEE (2020)
10. Faruki, P., Ganmoor, V., Laxmi, V., Gaur, M.S. Bharmal, A.: AndroSimilar: robust statistical feature signature for Android malware detection. In: Proceedings of the 6th International Conference on Security of Information and Networks, pp. 152–159 (2013)
11. Globalstat: Mobile Operating System Market Share Worldwide (2020). https://gs.statcounter.com/os-market-share/mobile/worldwide. Accessed June 2020
12. Google: Developers Documentation (2020). www.developers.android.com
13. Hou, S., Saas, A., Chen, L., Ye, Y.: Deep4maldroid: a deep learning framework for android malware detection based on linux kernel system call graphs. In: 2016 IEEE/WIC/ACM International Conference on Web Intelligence Workshops (WIW), pp. 104–111. IEEE (2016)
14. Huang, T.H.-D.: Hunting the ethereum smart contract: color-inspired inspection of potential attacks. arXiv preprint arXiv:1807.01868 (2018)
15. Huda, S., Islam, R., Abawajy, J., Yearwood, J., Hassan, M.M., Fortino, G.: A hybrid-multi filter-wrapper framework to identify run-time behaviour for fast malware detection. Futur. Gener. Comput. Syst. **83**, 193–207 (2018)
16. Huda, S., Islam, R., Abawajy, J., Yearwood, J., Hassan, M.M., Fortino, G.: A hybrid-multi filter-wrapper framework to identify run-time behaviour for fast malware detection. Future Gener. Comput. Syst. **83**, 193–207 (2018)
17. Idika, N., Mathur, A.P.: A survey of malware detection techniques. Purdue University **48**, 2007–2002 (2007)

18. Kang, B., Kang, B., Kim, J. Im, E.G.: Android malware classification method: Dalvik bytecode frequency analysis. In: Proceedings of the 2013 Research in Adaptive and Convergent Systems, pp. 349–350 (2013)
19. Kate, P.M., Dhavale, S.V.: Two phase static analysis technique for Android malware detection. In: Proceedings of the Third International Symposium on Women in Computing and Informatics, pp. 650–655 (2015)
20. Li, J., Wu, B., Wen, W.: Android malware detection method based on frequent pattern and weighted Naive Bayes. In: Yun, X., et al. (eds.) CNCERT 2018. CCIS, vol. 970, pp. 36–51. Springer, Singapore (2019). https://doi.org/10.1007/978-981-13-6621-5_4
21. Mahindru, A., Singh, P.: Dynamic permissions based android malware detection using machine learning techniques. In: Proceedings of the 10th Innovations in Software Engineering Conference, pp. 202–210 (2017)
22. Meng, G., Xue, Y., Xu, Z., Liu, Y., Zhang, J., Narayanan, A.: Semantic modelling of android malware for effective malware comprehension, detection, and classification. In: Proceedings of the 25th International Symposium on Software Testing and Analysis, pp. 306–317 (2016)
23. Ngamwitroj, S., Limthanmaphon, B.: Adaptive Android malware signature detection. In: Proceedings of the 2018 International Conference on Communication Engineering and Technology, pp. 22–25 (2018)
24. Nguyen, G., Nguyen, B.M., Tran, D., Hluchy, L.: A heuristics approach to mine behavioural data logs in mobile malware detection system. Data Knowl. Eng. **115**, 129–151 (2018)
25. Niu, S., Huang, R., Chen, W., Xue, Y.: An improved permission management scheme of android application based on machine learning. Secur. Commun. Netw. **2018** (2018)
26. Google PlayStore: Apps (2020). https://www.play.google.com/store/apps/category/GAME?hl=en
27. Rana, M.S., Gudla, C., Sung, A.H.: Evaluating machine learning models for android malware detection: a comparison study. In: Proceedings of the 2018 VII International Conference on Network, Communication and Computing, pp. 17–21 (2018)
28. Samani, R.: Mobile malware is playing hide and steal. McAfee Mobile Threat Report Q1 (2020)
29. Sheen, S., Anitha, R., Natarajan, V.: Android based malware detection using a multifeature collaborative decision fusion approach. Neurocomputing **151**, 905–912 (2015)
30. Virusshare: Malware Collections (2020). https://www.virusshare.com/
31. Virustotal: Free Online Virus, Malware and URL Scanner (2020). https://www.virustotal.com/
32. Xiao, X., Zhang, S., Mercaldo, F., Hu, G., Sangaiah, A.K.: Android malware detection based on system call sequences and LSTM. Multim. Tools Appl. **78**(4), 3979–3999 (2017). https://doi.org/10.1007/s11042-017-5104-0
33. Yuan, Z., Lu, Y., Xue, Y.: Droiddetector: android malware characterization and detection using deep learning. Tsinghua Sci. Technol. **21**, 114–123 (2016)

Mathematical Verification of Hybrid Model for Prime Decision-Making in Driving

Rabi Mustapha[1(✉)], Muhammad Aminu Ahmad[1], Mohammed Auwal Ahmed[2], and Muktar Hussaini[3]

[1] Department of Computer Science, Kaduna State University, Kaduna, Nigeria
{rabichubu,muhdaminu}@kasu.edu.ng
[2] Department of Computer Science, Kaduna Polytechnic, Kaduna, Nigeria
[3] Department of Computer Science, Hussaini Adamu Federal Polytechnic Kazaure, Kazaure, Nigeria

Abstract. This paper explains the verification of a hybrid model for prime decision-making in driving using mathematical analysis technique which defines the study problem. The hybrid model is the integration of Automaticity Recognition Prime Decision (RPD) training model and Situation Awareness (SA) model. Verification is a mechanism used to certify that the hybrid model simulation results, assumptions and specifications are correct and to also confirms that it is rightly built to achieve the design and implementation objectives. The verification of the model can be done in two phases thus; by automated and mathematical analysis. The first phase has been achieved. The mathematical analysis is describing the relevant system mathematically in the form of a partial differential equation (PDE) using the first order principle technique. It verifies the theoretical and structural correctness of the hybrid model. It also, determines the stability points that explains the presence or possibility of getting the equilibrium state of the system. This equilibrium occurs when for all states no changes occur. The presence of reasonable equilibrium is a sign of accuracy of the hybrid model. The results acquired evaluate the hybrid model validity and it also depicts the behavior of the driver on the road. Given that evaluation of the hybrid model for prime decision-making by verification using mathematical analysis is yet be achieved, this paper bridges the gap by providing the verification of the model by using the mathematical analysis to ensure that the hybrid model is rightly built, which is the novelty of this paper.

Keywords: Hybrid model · Prime decision-making · Mathematical verification · Driving

1 Introduction

Modelling is a concept used in understanding and predicting real phenomena or existing systems. However, driving a car involves a continuous process of perception, understanding, act choice, and act execution [1, 2]. Error in situation recognition might occur while driving a car, and the error can occasionally result in an 'erroneous' behaviour of

J. H. Abawajy et al. (Eds.): EATI 2020, LNNS 254, pp. 166–180, 2021.
https://doi.org/10.1007/978-3-030-80216-5_13

the" driver [2]. Poor decision-making skills of a driver that is human errors have been observed to be the major contributing factors of road crashes [3, 4].

The hybrid model is the integration of Automaticity Recognition Prime Decision (RPD) training model and Situation Awareness (SA) model [5–9]. Hence, a hybrid model for Prime Decision Making in driving improves on the knowledge and the skills required to drive safely and efficiently. The knowledge and skills that the driver needs to have must be known, for the training to be appropriate [10]. The training for critical decision-making is important for providing the learners "with experiences and instruction on cues, patterns, mental models, and actions that efficiently establish a collection of well-learned concepts that enable the driver to perform mainly at the skill-based level of processing, while providing adequate knowledge-based foundation to perform well in new" situations [11]. Having analysed this ability, it will provide a good perception towards safety driving. Training drivers is also essential "in recognizing situations, in communicating situation assessment in the driving environment, and in acquiring the experience to conduct mental simulation of options through the act of human cognitive unconscious decision-making, or automaticity" [12]. The training of driver is modelled to predict the behaviour of the driver in prime decision-making situations such as sudden swerve and panic stop to alternative direction [13]. To ensure that the developed model is built correctly both in terms of the design and implementation the mathematical analysis method known as logical proof procedures can be used and hence bridge the gap of this study.

Mathematical verification used the differential equation to explain the changes that occur in the hybrid model. This can be done in two phases and phase one which the automated analysis using temporal trace language (TTL) has already been achieved and the second phase can be achieved by mathematical analysis. This can be done to show when stability points or equilibria take place. It can be stated that when a certain state is increasing or decreasing the state is not instability point or equilibrium. Also, the equilibrium point shows the correctness of the model under consideration. Since "differential equation defines the dynamic of the system, then by equating all derivatives to zero, an estimation of the equilibrium" would be obtained [14–16]. Moreover, an equilibrium condition of a system can be considered stable only if the system upholds to its level of stability after being acted upon by a little disturbance. This paper presented a mathematical analysis of hybrid model for prime decision-making in driving. This integrates related dynamic factors based on the cognitive theory of Situation Awareness and naturalistic decision-making theory to describe basic training required by a driver in the driving domain.

Moreover, the remaining part of this paper is organized as follows. The discussion of the hybrid model, then the verification of the model using mathematical analysis and finally the conclusion.

2 Hybrid Model for Prime Decision-Making

The hybrid model is divided into two thus; the situation awareness model and the automaticity RPD training model. There are several of interactions that occur between the interrelated factors of the hybrid model as shown in Fig. 1. From this figure, it can be understood that the hybrid model consists of numerous interconnected nodes that shows some external, instantaneous and temporal factors that are interrelated with each other. Factors in this study are analogous to variables in social sciences research. The external factors are the independent variables (IV's) that determines the outcome of the whole processes in the model. They are independent factors that contribute to other factors, while instantaneous factors are also known as the mediators, they are dependent factors that are of time-bounded with fast response time with no delay contrary, to the temporal factors (also known as the dependent variables (DV's) are time-bounded with much delay. That is in temporal factors changes occur over accumulated period. The hybrid model has been formalized subsequent to when its structural relationships have been determined. The model formalization has been achieved in this studies [5, 7] based on [14–18]. However, the formalization of the automaticity recognition prime decision training model is reviewed in this study for easy understanding of the model concept. In the formalization process, all the nodes are connected using flow arrows to show the relationship between the factors involved. These relationships have been simulated using Mat lab software in our previous studies stated.

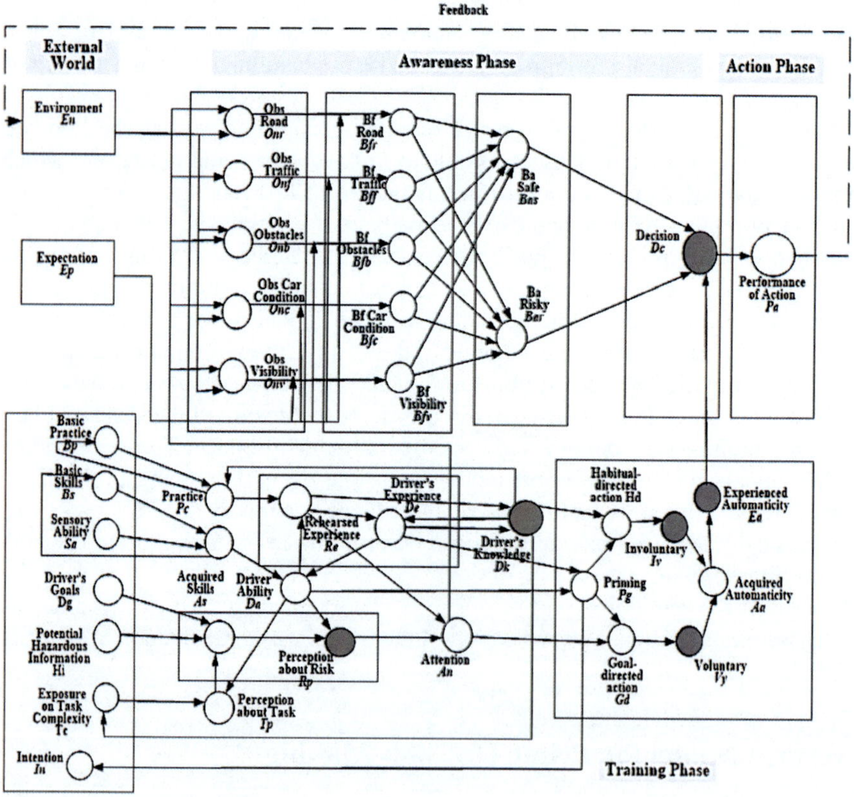

Fig. 1. Hybrid model for prime decision-making in driving.

In Fig. 1, there are various concepts (known as factors) used in the design of the model based on the review of related literatures. The concepts, their notations and their brief explanations will be given in a tabular form for easy understanding of the design procedures of the hybrid model and its mathematical formulations.

Table 1. Summary of external factors of the hybrid model

Concepts	Representation	Explanation
Environment	En	The surrounding in which car and driver operate
Expectations	Ep	Knowledge of possible consequences or expectancies of the future
Basic practice	Bp	The capacity to operate and control the vehicle
Basic skills	Bs	The operational competence of driver
Sensory ability	Sa	The ability of driver to have cognitive, physical and visual functions to manipulate the vehicle
Driving goal	Dg	Multiple driving aims that driver wants to achieve during the driving task
Potential hazardous information	Hi	Information acquired regarding potential threads that might need urgent respond in the traffic environment during driving
Exposure on task complexity	Tc	The complexity driver exposes to at the course of the interaction with the vehicle and environment
Intention	In	Driver's mental state that translates his goals into reality

The Table 1, summarizes the concepts, symbolic representations and explanations of the external factors of the hybrid model.

Table 2. Summary of instantaneous factors of the hybrid model

Concepts	Representation	Explanation
Observation	On	Ability to perceive elements in a driving environment
Belief formation	Bf	Ability to form certainty of the observation made
Belief activation	Ba	Ability to translate the certainty of the observations into activation values of beliefs, which can be safe or risky
Performance of action	Pa	Implementation of the decision taken by the driver
Practice	Pc	Method of developing the drivers' skills and knowledge that relates to specific useful competencies of the driving task
Acquired skills	As	Accumulated exposure of the basic skills

(continued)

Table 2. (*continued*)

Concepts	Representation	Explanation
Rehearsed experience	Re	Experiences acquired due to continuous driving routine that might decay overtime
Driver ability	Da	Capability driver possesses to manipulate /operate car
Driver's experience	De	Driver's accumulation of the reoccurrence of knowledge or skill acquired that result from direct participation in the driving activity
Hazard Perception	Hp	Driver's ability to anticipate potentially dangerous situations on the road ahead
Task perception	Tp	The way driver sees or experiences task in the potential traffic environment
Attention	An	The ability of driver to perceive multiple items in parallel accurately
Priming	Pg	The stimulus that makes driver initiates response sequence in driving
Habitual-directed action	Hd	Action initiated by driver as an act of unconsciousness while driving
Goal-directed action	Gd	Action initiated by driver as an act of conscious willing while driving
Acquired automaticity	Aa	Short-term automaticity

Table 2, give a summary of the concepts, symbolic representations and explanations of the instantaneous factors of the model.

Table 3. Summary of temporal factors of the hybrid model

Concepts	Representation	Explanation
Perception of risk	Rp	Subjective experience of risk in potential traffic hazards
Driving knowledge	Dk	Knowledge of traffic rules and regulations of the road
Involuntary automaticity	Iv	Unconscious and automatic behaviors experienced by driver
Voluntary automaticity	Vy	Conscious and non-automatic behaviors experienced by driver
Experienced automaticity	Ea	Long-term automaticity that denotes accumulated exposure of the acquired automaticity of driver
Decision	Dc	The internal processes by which the driver selects a course of action or inaction from a set of alternatives

Table 3, gives the summary of the temporal factors of the model.

2.1 Equations of the Hybrid Model

Instantaneous Relationships of the Model Factors:
Basic practice (*Bp*) of the driver increased with practice (*Pc*). Practice is influenced by basic practice (*Bp*) and driving knowledge (*Dk*). Therefore,

$$Bp(\mathrm{t}) = \beta_{bp}.Bp_{basic}(\mathrm{t}) + \left(1 - \beta_{bp}\right).Pc(\mathrm{t}) \tag{1}$$

$$Pc(\mathrm{t}) = \omega_{pc}.Bp(\mathrm{t}) + \left(1 - \omega_{pc}\right).Dk(\mathrm{t}) \tag{2}$$

Rehearsed Experience (*Re*) of the driver is affected by driver's practice (*Pc*) and ability (*Da*) by saying "with continuous practice any knowledge or skill is retained in short-term memory and later transfer to long-term memory otherwise it will decay". Subsequently, the driver's experience (*De*) is influenced by rehearsed experience (*Re*) and driving knowledge *(Dk)* of the driver. The concept of acquired skills (*As*) is computed by combining driver's basic skills, and the driver's sensory ability *(Sa)*. Therefore,

$$\mathrm{Re(t)} = \gamma_{re}.Pc(\mathrm{t}) + (1 - \gamma_{re}).Da(\mathrm{t}) \tag{3}$$

$$\mathrm{De(t)} = \lambda_{de}.\mathrm{Re(t)} + (1 - \lambda_{de}).Dk(t) \tag{4}$$

$$\mathrm{As(t)} = \beta_{as}.[\mathrm{w}_{as1}.Bs(\mathrm{t}) + \mathrm{w}_{as2}.Sa(\mathrm{t})] + (1 - \beta_{as}) \tag{5}$$

In the case of driver's ability (*Da*), this concept is affected by the skills acquired and experiences of the driver (*De*) during the training session. The combination of driver's experience, ability and intention (*In*) generates priming levels. Hazard Perception (*Hp*) is determined by combining concepts in driving ability (*Da*), perception about task (*Tp*) and potential hazardous information (*Hi*). Therefore,

$$Da(\mathrm{t}) = \mathrm{w}_{da1}.De(\mathrm{t}) + \mathrm{w}_{da2}.As(\mathrm{t}) \tag{6}$$

$$Pg(\mathrm{t}) = \left[\xi_{\mathrm{pg}}.Da(\mathrm{t}) + (1 - \xi\mathrm{pg}\ ).De(\mathrm{t})\right].In(\mathrm{t}) \tag{7}$$

$$Hp(\mathrm{t}) = \left[\mathrm{w}_{hp1}.Dg(\mathrm{t}) + \mathrm{w}_{hp2}.Tp(\mathrm{t})\right].Hi(\mathrm{t}) \tag{8}$$

Attention (An) is generated combining a proportional ratio of rehearsed experience (Re) and Risk perception (Rp). Next, the proportional contribution between exposure on task complexity (Tc) and driving ability (Da) provides a computational concept of perception about task. Therefore,

$$An(\mathrm{t}) = \left[\xi_{an}.Rp(\mathrm{t}) + (1 - \xi_{an}).Re(\mathrm{t})\right] \tag{9}$$

$$Tp(t) = \left[\eta_{tp}.Da(t) + \left(1 - \eta_{tp}\right).Tc(t)\right] \quad (10)$$

Another important concept is the exposure on task complexity (Tc). This concept is positively correlated with the knowledge of the driver. Habitual-directed action (Hd) is affected by driving knowledge (Dk) of the driver and priming (Pg). Using the same computational concept as in habitual directed action, priming (Pg) and attention (An) generates goal-directed action (Gd). Similarly, it also the case of acquired automaticity (Aa) as it is influenced by weight contribution of involuntary (Iv) and voluntary (Vy).

$$Tc(t) = \beta_{tc}.Tc_{basic}(t) + (1 - \beta_{tc}).Dk(t) \quad (11)$$

$$Hd(t) = w_{hd1}.Pg(t) + w_{hd2}.Dk(t) \quad (12)$$

$$Gd(t) = w_{gd1}.An(t) + w_{gd2}.Pg(t) \quad (13)$$

$$Aa(t) = w_{aa1}.Iv(t) + w_{aa2}.Vy(t) \quad (14)$$

Equations (1) to (14) are as a result of the relationship that shows the interrelated connectivity of the nodes in the hybrid model in Fig. 1. βbp, βas, ωpc, γre, ξpg, ξan, ηtp, βtc are known as proportional parameters while λde is a decay function. Moreover, was1, was2, wda1, wda2, whp1, whp2, whd1, whd2, wgd1, wgd2, waa1 and waa2 are weight parameters with $\sum w = 1$. While (1 – βbp), (1 – βas), (1 – ωPc), (1 – γre), (1 – λde), (1 – ξpg), (1 – ξan), (1 – ηtp) and (1 – βtc) are regulating function in the equation that regulate the equations.

Temporal Relationship of the Model Factors:

Driving knowledge (Dk) of the driver "primarily contributed to the accumulation of rehearsed experience (Re) and driver's experience (De). Risk Perception of the driver (Rp) is affected by hazard perception of the driver (Hp) and driver's ability to handle vehicle (Da)," while the involuntary decision (Iv) is contributed through habitual-directed action (Hd). The positive change in goal-directed action (Gd) improves voluntary (Vy) level. Experienced automaticity is influenced by acquired the automaticity of the driver.

$$Dk(t + \Delta t) = Dk(t) + \gamma_{dk}.[(Pos((\omega_{dk1}.Re(t) + \omega_{dk2}.De(t)) - Dk(t)).(1 - Dk(t))) - Pos(-(\omega_{dk1}.Re(t) + \omega_{dk2}.De(t)) - \lambda_{dk}).Dk(t)].\Delta t \quad (15)$$

$$Rp(t + \Delta t) = Rp(t) + \gamma_{rp}.[\left(Pos\left((\omega_{rp1}.Hp(t) + \omega_{rp2}.Da(t)\right) - Rp(t)\right).(1 - Rp(t))\right) - Pos(-\left(\omega_{rp1}.Hp(t) + \omega_{rp2}.Da(t)\right) - Rp(t)) - \lambda_{rp}).Rp(t)].\Delta t \quad (16)$$

$$Iv(t + \Delta t) = Iv(t) + \beta_{iv}.\left[(Pos((Hp(t) - Iv(t))).(1 - Iv(t))) - Pos(-(Hd(t) - Iv(t))).Iv(t)\right].\Delta t \quad (17)$$

$$Vy(t + \Delta t) = Vy(t) + \beta_{vy}.\left[(Pos((Gd(t) - Vy(t))).(1 - Vy(t))) - Pos(-(Gd(t) - Vy(t))).Vy(t)\right].\Delta t \quad (18)$$

$$Ea(t+\Delta t) = Ea(t) + \beta_{ea}.[(Pos((Aa(t) - Ea(t))).(1 - Ea(t))) - Pos(-(Aa(t) - Ea(t))).Ea(t)].\Delta t \quad (19)$$

$$Dc(t+\Delta t) = Dc(t) + \gamma_{dc}.((Bas(t) - Bar(t)) - Dc(t)).(1 - Dc(t)).\Delta t \quad (20)$$

Equations (15)–(20) were obtained based on the concepts of the differential equation (DE). The change process in these equations is "measured in a time interval between t and $t + \Delta t$. Moreover, the rate of change for all temporal specifications is determined by flexibility rates" of γ_{dk}, γ_{rp}, β_{iv}, β_{vy} and β_{ea} which are change rate parameters. The derivation of all the equations in this paper follows the same concepts used in our papers [5, 9] and in this study [20].

2.2 Verification of the Hybrid Model

Verification is a mechanism used to certify that the hybrid model simulation results, assumptions and specifications are correct and to also ensure that it is rightly built to achieve the design and implementation objectives. The verification of the model can be done in two phases thus; by automated and mathematical analysis. Therefore, to build the model right, there are processes involved as identified in Fig. 2. In this Figure, the evaluation processes of the hybrid model are illustrated to ensure that the model is the actual representative of the phenomenon under investigation. The automated analysis (using Temporal Trace Language (TTL)) of the hybrid model has been achieved in these studies [5, 6, 9]. Hence, the verification using the mathematical analysis can be achieved using ordinary differential equations [18–20].

The evaluation of the hybrid Model for Prime Decision-Making from Fig. 2 can be achieved in stages starting from the development of the model based on theory driven. In this case the model was developed from the cognitive theory of Situation Awareness [22] and naturalistic decision-making theory [12, 23] based on literature. Followed by model simulation, the model is simulated in a simulation environment using MATLAB [15]; Then, the model evalauation that aims to ensure that the computational model is the actual representative of the phenomenon under investigation. It is divided into two sub-stages, thus; first verification and second, validation [24]. The model verification is also subdivided into automated analysis and mathematical analysis. The automated analysis of the hybrid model is performed by using Temporal Trace Language (TTL), used in the analysis of properties of simulated traces of specified properties and states against generated traces. The matthematical analysis can be achieved using ordinary differential equations [23, 24] to ensure that the model is built right to achieve its designed and implementation objectives. Both automated and mathematical analyses are used to verify the hybrid model. The second form of the hybrid model evaluation is the model validation and it is to be achieved emperically using human experiment and it is the future work of this study.

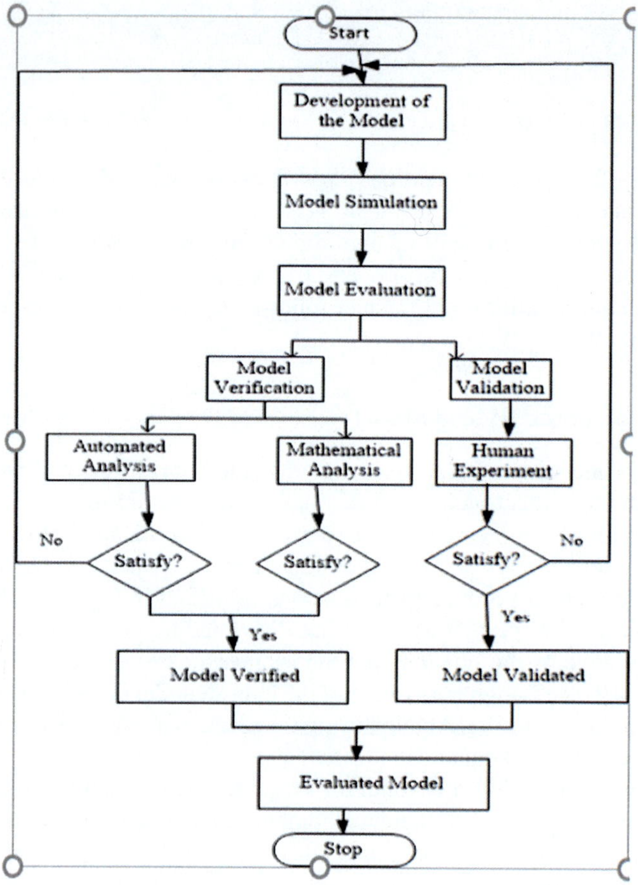

Fig. 2. Evaluation processes of the hybrid model for prime decision-making.

2.3 Mathematical Analysis of the Hybrid Model

The mathematical analysis is conducted to verify the structural and theoretical correctness of the model. This study performs equilibrium analysis, which describes conditions in which a stable state has been reached. That is, if the dynamics of a system are described by a differential equation, then equilibrium levels can be estimated by setting a derivative (or all derivatives) to zero. It is important to note that an equilibrium condition is considered stable if the system always returns to it" state after little disturbances. Mathematical analysis can be used to analyze the dynamic properties of dynamic models [27] like the hybrid model for prime decision-making in driving in order to understand the structural and accuracy of the model through theoretical concept. Examples of such properties are as follows:

i. Values of the variables for which no change occur (stationary points or equilibria);
ii. "Certain variables in the model converge to some limit value (equilibria)";

iii. "Certain variables will show monotonically increasing or decreasing values over time (monotonicity)";
iv. "Situations occur in which no convergence occur, but in the end a particular sequence of values is repeated all the time (limit cycle)."

These equilibrium conditions indicate the accuracy of the hybrid model, that is revolved on the model concept. Therefore, the existence of reasonable equilibria is a sign of the accuracy of the model [20]. The concept of mathematical analysis used in this paper is derived from the concepts of the differential equation based on [18–20]. To obtain possible equilibrium values for the temporal factors, the study first describes the temporal equations presented in Eqs. 15 to 20 in differential equations, 21 to 26. These six differential Eqs. 21 to 26 provide the differential values for Driver's Knowledge (*Dk*), Risk Perception (*Rp*), Involuntary Automaticity (*Iv*), Voluntary Automaticity (*Vy*), Experience Automaticity (*Ea*) and Decision (*Dc*), respectively.

$$\mathrm{dDk/dt} = \gamma_{\mathrm{dk}}.((\omega_{\mathrm{dk1}}.\mathrm{Re} + \omega_{\mathrm{dk2}}.\mathrm{De}) - \mathrm{Dk}).\ \mathrm{Dk}.(1 - \mathrm{Dk}) \tag{21}$$

$$\mathrm{dRp/dt} = \gamma_{\mathrm{rp}}.\big((\omega_{\mathrm{rp1}}.\mathrm{Hp} + \omega_{\mathrm{rp2}}.\mathrm{Da}\big) - \mathrm{Rp}\big).\mathrm{Rp}.(1 - \mathrm{Rp}) \tag{22}$$

$$\mathrm{dIv/dt} = \beta_{\mathrm{iv}}.(\mathrm{Hd(t)} - \mathrm{Iv}).\mathrm{Iv}.(1 - \mathrm{Iv}) \tag{23}$$

$$\mathrm{dVy/dt} = \beta_{\mathrm{vy}}.(\mathrm{Gd} - \mathrm{Vy}).\mathrm{Vy}.(1 - \mathrm{Vy}) \tag{24}$$

$$\mathrm{dEa/dt} = \beta_{\mathrm{ea}}.(\mathrm{Aa} - \mathrm{Ea}).\mathrm{Ea}.(1 - \mathrm{Ea}) \tag{25}$$

$$\mathrm{dDc/dt} = \gamma_{\mathrm{dc}}.((\mathrm{Bas} - \mathrm{Bar}) - \mathrm{Dc}).\mathrm{Dc}.(1 - \mathrm{Dc}) \tag{26}$$

The symbols $\gamma_{dk}, \gamma_{rp}, \beta_{iv}, \beta_{vy}, \beta_{ea}, \gamma_{dc}$ are proportional parameters (speed factors) while $\omega_{dk1}, \omega_{dk2}, \omega_{rp1}, \omega_{rp2}$ are weight parameters, and both parameters are nonzero. From Eq. 21 to 26, for any possible simulation results, the following cases can be distinguished:

$$\begin{aligned}
(\mathrm{Re} + \mathrm{De}) - \mathrm{Dk}).(\mathrm{Dk} - 1).\ \mathrm{Dk} &= 0\\
(\mathrm{Hp} + \mathrm{Da}) - \mathrm{Rp}).(\mathrm{Rp} - 1).\ \mathrm{Rp} &= 0\\
(\mathrm{Hd} - \mathrm{Iv}).(1 - \mathrm{Iv}).\ \mathrm{Iv} &= 0)\\
(\mathrm{Gd} - \mathrm{Vy}).(1 - \mathrm{Vy}).\mathrm{Vy} &= 0\\
(\mathrm{Aa} - \mathrm{Ea})(1 - \mathrm{Ea}).\ \mathrm{Ea} &= 0\\
(\mathrm{Bas} - \mathrm{Bar}) - \mathrm{Dc}).(1 - \mathrm{Dc}).\ \mathrm{Dc} &= 0
\end{aligned}$$

These cases can further be distinguished into the following:

$$(\mathrm{Re} + \mathrm{De} = \mathrm{Dk}) \vee (\mathrm{Dk} = 1) \vee (\mathrm{Dk} = 0)$$
$$(\mathrm{Hp} + \mathrm{Da} = \mathrm{Rp}) \vee (\mathrm{Rp} = 1) \vee (\mathrm{Rp} = 0)$$
$$(\mathrm{Hd} = \mathrm{Iv}) \vee (\mathrm{Iv} = 1) \vee (\mathrm{Iv} = 0)$$
$$(\mathrm{Gd} = \mathrm{Vy}) \vee (\mathrm{Vy} = 1) \vee (\mathrm{Vy} = 0)$$
$$(\mathrm{Aa} = \mathrm{Ea}) \vee (\mathrm{Ea} = 1) \vee (\mathrm{Ea} = 0)$$
$$(\mathrm{Bas} - \mathrm{Bar} = \mathrm{Dc}) \vee (\mathrm{Dc} = 1) \vee (\mathrm{Dc} = 0)$$

Thus, it can be concluded that these are the points where the equilibrium can occur. That is, when Re + De = Dk, or k = 1, or Dk = 0. "By combining these three conditions, it can be re-written as a set of relationship in a" form $(A \wedge D) \vee (A \wedge G)$:

$$((Re + De = Dk) \vee (Dk = 1) \vee (Dk = 0)) \wedge$$
$$((Hp + Da = Rp) \vee (Rp = 1) \vee (Rp = 0)) \wedge$$
$$((Hd = Iv) \vee (Iv = 1) \vee (Iv = 0)) \wedge$$
$$((Gd = Vy) \vee (Vy = 1) \vee (Vy = 0)) \wedge$$
$$((Aa = Ea) \vee (Ea = 1) \vee (Ea = 0)) \wedge$$
$$((Bas - Bar = Dc) \vee (Dc = 1) \vee (Dc = 0))$$

These expressions can be elaborated using distributive law as:

$$(\mathrm{A} \wedge \mathrm{D}) \vee (\mathrm{A} \wedge \mathrm{G}) \vee (\mathrm{A} \wedge \mathrm{J}) \wedge (\mathrm{A} \wedge \mathrm{M}), \ldots, \vee(\mathrm{C} \wedge \mathrm{R}).$$
$$(\mathrm{Re} + \mathrm{De} = \mathrm{Dk} \wedge \mathrm{Re} + \mathrm{De} = \mathrm{Dk} \wedge \mathrm{Hd} = \mathrm{Iv} \wedge \mathrm{Gd} = \mathrm{Vy} \wedge \mathrm{Aa} = \mathrm{Ea} \wedge \mathrm{Bas} - \mathrm{Bar} = \mathrm{Dc}) \vee (\mathrm{Dk} = 1 \wedge \mathrm{Rp} = 1 \wedge \mathrm{Iv} = 1 \wedge \mathrm{Vy} = 1 \wedge \mathrm{Ea} = 1 \wedge \mathrm{Dc} = 1) \vee \ldots \vee$$
$$(\mathrm{Dk} = 0 \wedge \mathrm{Rp} = 0 \wedge \mathrm{Iv} = 0 \wedge \mathrm{Vy} = 0 \wedge \mathrm{Ea} = 0 \wedge \mathrm{Dc} = 0)$$

The temporal factors are used to determine the possible combinations from the concepts of DE, $F(x) = x^n$ where x is the possible number of equilibrium points that can occur in a particular temporal factor (e.g., (Hp + Da = Rp, Rp = 1, Rp = 0) and n is the number temporal factors in the hybrid model. This is resulting into 3^6 (729) possible equilibria points.

However, due to the large size of possible combinations, it makes it problematic to provide a complete classification of equilibria and typical cases were analyzed further as follows:

Case 1: (Hp + Da = Rp)

Hp = Rp − Da or Da = Rp − Hp

First considering the factors that receive Rp as an input from the causal relationship. For example, An receives Rp as an input. Therefore, substituting (Hp + Da = Rp) in Eq. (27),

$$An = \xi_{an} * Rp + (1 - \xi_{an}) * Re \tag{27}$$

The result is as follows:

$$= \text{Hp} + \text{Da} + (1-1) * \text{Re}, \text{ making assumption } \xi_{\text{an}} = 1$$
$$\text{An} = \text{Hp} + \text{Da}$$

Changing the values of $\xi_{\text{an}} = 0.5$
Then,

$$\begin{aligned} \text{An} &= 0.5 * \text{Rp} + (1-0.5) * \text{Re}, \\ &= 0.5 * \text{Rp} + 0.5 * \text{Re}, \\ &= 0.5\,(\text{Rp} + \text{Re}), \\ &= 0.5\,(\text{Hp} + \text{Da} + \text{Re}), \end{aligned}$$

And when $\xi_{\text{an}} = 0$, An = Re.

Case one demonstrates that risk perception is affected by the combination of drivers' hazard perception on the road and driver's ability. From the mathematical analysis, when the case is tested with the parameter Xi of attention $\xi_{\text{an}} = 1$, (i.e., high) the result indicated that driver's level of attention is affected by the driver's hazard perception combined with the driver's ability. This implies that the level of attention of the driver depended on the driver's task perception level and driver's ability and vice versa. When the case is tested with $\xi_{\text{an}} = 0.5$ parameter, (i.e., considered moderate), the result shows that doubling the attention level of the driver on the road depended on the driver's hazard perception on the road combined with the driver's ability and rehearsed experience. Similarly, when tested with parameter $\xi_{an} = 0$, (i.e., considered to be low) the result indicated that driver's level of attention on the road depended on the driver's rehearsed experience.

Case 2: (Rp = 0)

Case 2 can be analyzed by substituting (Rp = 0) in Eq. (28),

$$\text{An} = \xi_{\text{an}} * \text{Rp} + (1 - \xi_{\text{an}}) * \text{Re} \tag{28}$$

This gives the following results:

$$\begin{aligned} &= 1 * 0 + (1-1) * \text{Re}, \text{ assuming } \xi_{\text{an}} = 1 \\ \text{An} &= 0 \\ \text{An} &= \xi_{\text{an}} * \text{Rp} + (1 - \xi_{\text{an}}) * \text{Re}, \\ &= 0.5 * 0 + (1 - 0.5) * \text{Re}, \text{ assuming } \xi_{\text{an}} = 0.5 \\ &= 0.5\text{Re} \\ \text{An} &= \xi_{\text{an}} * \text{Rp} + (1 - \xi_{\text{an}}) * \text{Re}, \\ &= 0 * 0 + (1 - 0) * \text{Re}, \text{ assuming } \xi_{\text{an}} = 0 \\ \text{An} &= \text{Re}. \end{aligned}$$

Case two shows that the perception of risk of the driver is low. From the mathematical analysis, when the case is tested with parameter Xi of attention $\xi_{\text{an}} = 1$ (i.e., high) the result suggested that the attention of the driver also became low and when tested with ξ_{an}

= 0.5 (i.e., moderate) the result indicated that doubled the driver's level of attention on the road depends on the driver's rehearsed experience. Similarly, when tested with ξ_{an} = 0 (i.e., low) the result showed that driver's level of attention on the road is determined by driver's rehearsed experience.

From the analysis of cases one and two, when tested with different parameter values of (0, 0.5 and 1), the results obtained suggested that risk perception of driver on the road is influenced by the increase in his level of attention. More so, when the risk perception of the driver is low, it indicates that his attention level is also low and vice versa. These findings are in line with these studies [26, 27]. It implies that the equilibria state determined from the temporal equations also portrays the behavior of driver on the road.

3 Conclusion

This paper presented a verification of the hybrid model for prime decision-making in driving domain using mathematical analysis in order to ensure that the hybrid model is correctly built. The mathematical analysis of the model has been achieved in this study. To prove the existence of equilibria states, it is essentially important to define the convergence and stability states of the hybrid model. The mathematical analysis also proves that the hybrid model syntax and semantics presentations is mathematically compatible. Its model properties are also compatible with its specifications as well as its theoretical formulations. These implies that the equilibria states generated from the temporal equations also depicts the behavior of the driver on the road. Hence, the future work of this paper is to evaluate "the applicability of the model in real life domain empirically, using human experiment that is, the model validation.

References

1. Inagaki, T., Itoh, M.: Human's overtrust in and overreliance on advanced driver assistance systems: a theoretical framework. Int. J. Veh. Technol. **2013**, 8 (2013)
2. Inagakin, T.: To what extent may assistance systems correct and prevent 'erroneous' behaviour of the driver? In: Cacciabue, P., Hjälmdahl, M., Luedtke, A., Riccioli, C. (eds.) Human Modelling in Assisted Transportation, pp. 33–42. Springer, Milano (2011). https://doi.org/10.1007/978-88-470-1821-1_5
3. Mashadi, B., Majidi, M.: Two-phase optimal path planning of autonomous ground vehicles using pseudo-spectral method. Proc. Inst. Mech. Eng. Part K J. Multi-body Dyn. **228**, 426–437 (2014)
4. Stanton, N.A., Salmon, P.M., Walker, G.H., Salas, E., Hancock, P.A.: State-of-science: situation awareness in individuals, teams and systems. Ergonomics **60**, 449–466 (2017)
5. Mustapha, R., Yusof, Y., Aziz, A.A.: A hybrid model for prime decision making in driving. J. Telecommun. Electron. Comput. Eng. **9**, 95–99 (2017)
6. Mustapha, R., Yusof, Y., Aziz, A.A.: Computational model of situation awareness for decision making in driving. Adv. Sci. Lett. **24**, 1244–1248 (2018)
7. Mustapha, R., Yusof, Y., Aziz, A.A.: Computational-Rabi's driver training model for prime decision-making in driving. J. Theor. Appl. Inf. Technol. **97**, 3540–3563 (2019)
8. Mustapha, R., Yusof, Y., Aziz, A.A.: A computational agent model of automaticity for driver's training. Int. Res. Innov. Summit. 1–7 (2017)

9. Mustapha, R., Aziz, A.A., Yusof, Y.: Computational model of situation awareness for action performed in driving. Malaysian J. Hum. Factors Ergon. **1**, 1–8 (2016)
10. Vegvesen, S.: Driver training in Norway: foundations for the revisions of the regulations and curricula. NPRA Directorate of Public Roads, Norway (2014)
11. Greitzer, F.L., Podmore, R., Robinson, M., Ey, P.: Naturalistic Decision Making for power system operators. Int. J. Hum. Comput. Interact. **26**, 278–291 (2010)
12. Klein, G.: Naturalistic decision making. Hum. Factors Ergon. Soc. **50**, 456–460 (2008)
13. Leland, F.: Critical Decision Making under pressure. Homel. Secur. Rev. **3**, 43–74 (2009)
14. Treur, J.: Relating emerging network behaviour to network structure. In: Aiello, L.M., Cherifi, C., Cherifi, H., Lambiotte, R., Lió, P., Rocha, L.M. (eds.) COMPLEX NETWORKS 2018. SCI, vol. 812, pp. 619–634. Springer, Cham (2019). https://doi.org/10.1007/978-3-030-05411-3_50
15. Mohammadi Ziabari, S.S., Treur, J.: A modeling environment for dynamic and adaptive network models implemented in Matlab. In: Yang, X.-S., Sherratt, Sn., Dey, N., Joshi, A. (eds.) Fourth International Congress on Information and Communication Technology. AISC, vol. 1041, pp. 91–111. Springer, Singapore (2020). https://doi.org/10.1007/978-981-15-0637-6_8
16. Sahand Mohammadi Ziabari, S., Treur, J.: An adaptive cognitive temporal-causal network model of a mindfulness therapy based on humor. In: Davis, F.D., Riedl, R., vom Brocke, J., Léger, P.-M., Randolph, A., Fischer, T. (eds.) Information Systems and Neuroscience. LNISO, vol. 32, pp. 189–201. Springer, Cham (2020). https://doi.org/10.1007/978-3-030-28144-1_21
17. Aziz, A., Klein, M., Treur, J.: An agent model of temporal dynamics in relapse and recurrence in depression. In: Chien, B.-C., Hong, T.-P., Chen, S.-M., Ali, M. (eds.) IEA/AIE 2009. LNCS (LNAI), vol. 5579, pp. 36–45. Springer, Heidelberg (2009). https://doi.org/10.1007/978-3-642-02568-6_4
18. Treur, J.: Dynamic modeling based on a temporal-causal network modeling approach. Biol. Inspired Cogn. Archit. **16**, 131–168 (2016)
19. Treur, J.: Network-oriented modeling and its conceptual foundations. In: Treur, J. (ed.) Network Oriented Modeling. UCS, pp. 3–33. Springer, Cham (2016). https://doi.org/10.1007/978-3-319-45213-5_1
20. Treur, J.: Verification of temporal-causal network models by mathematical analysis. Vietnam J. Comput. Sci. **3**(4), 207–221 (2016). https://doi.org/10.1007/s40595-016-0067-z
21. Aziz, A.A., Ahmad, F., Yusof, N., Kabir, A.F., Azmi, M.Y.S.: Formal analysis of temporal dynamics properties in anxiety states and traits. J. Telecommun. Electron Comput. Eng. **8**, 133–138 (2016)
22. Endsley, M.R.: Designing for Situation Awareness: An Approach to User-Centeres Design. CRC Press and Taylor & Francis Group, New York (2016)
23. Klein, G.: A naturalistic decision making perspective on studying intuitive decision making. J. Appl. Res. Mem. Cogn. **4**, 164–168 (2015)
24. Mustapha, R., Ahmad, M.A., Daniel, A., Ahmed, M.A., Hussaini, M.: Validating measures of driver behavior's training factors for prime decision-making. Sci. World J. **15**, 102–112 (2020)
25. Treur, J.: Mathematical details of specific difference and differential equations and mathematical analysis of emerging network behaviour. In: Treur, J. (ed.) Network-Oriented Modeling for Adaptive Networks: Designing Higher-Order Adaptive Biological, Mental and Social Network Models. SSDC, vol. 251, pp. 375–403. Springer, Cham (2020). https://doi.org/10.1007/978-3-030-31445-3_15
26. Ziabari, S.S.M., Treur, J.: Decision making under acute stress modeled by an adaptive temporal-causal network model. Vietnam J. Comput. Sci. 123–142 (2020)

27. Sijm, M., Exel, C., Treur, J.: Using a temporal-causal network model for computational analysis of the effect of social media influencers on the worldwide interest in veganism. In: Yang, X.-S., Sherratt, S., Dey, N., Joshi, A. (eds.) Fourth International Congress on Information and Communication Technology. AISC, vol. 1027, pp. 129–140. Springer, Singapore (2020). https://doi.org/10.1007/978-981-32-9343-4_12
28. Deery, H.A.: Hazard and risk perception among young novice drivers. J. Safety Res. **30**, 225–236 (1999)
29. Rosenbloom, T., Shahar, A., Elharar, A., Danino, O.: Risk perception of driving as a function of advanced training aimed at recognizing and handling risks in demanding driving situations. Accid. Anal. Prev. **40**, 697–703 (2008)

A Review on Unmanned Aerial Vehicle Energy Sources and Management

Ibrahim Abdullahi Shehu[1](✉), Musa Mohammed[1], Sulaiman Haruna Sulaiman[1], Abubakar Abdulkarim[1], and Ahmad Bala Alhassan[2]

[1] Department of Electrical Engineering, Ahmadu Bello University, Zaria, Nigeria
iashehu@abu.edu.ng

[2] School of Mechanical Engineering, Xi'an Jiaotong University, Shaanxi 710049, People's Republic of China

Abstract. Unmanned Aerial vehicle (UAV) systems have an insufficient amount of onboard energy which is being shared for mobility, transmission, data processing, control and payload related applications. Different energy sources have been investigated and applied to solve unmanned aerial vehicle energy limitations. These energy sources were either used as single sources or hybrid for the reason of UAV weight budget or miniaturization. To achieve the aforementioned designed criteria, researchers around the globe are continuously investigating different energy sources and hybridization techniques while others focus on the Energy Management System (EMS) such to achieve optimal energy sources utilization. Such energy sources and supplying techniques are battery swapping, wireless Power Transfer (WPT), tethering methods, Fuel Cells (FC), and SuperCapacitors (SC). While, the hybridization technique is based on the combination of single energy source with either renewable energy sources, Internal Combustion Engine (ICE) to save fuel or WPT. Example of such hybridization techniques are solar and battery storage, Fuel Cell (FC) and Battery Storage (BS) etc. This article presents an in-depth review of UAV energy sources, hybridization techniques and EMS. Their challenges and methods adopted by researchers to mitigate such UAV energy challenges.

Keywords: Aeronautics · Energy · EMS · Hybrid · UAV

1 Introduction

Due to the rapid advancement of technology and manufacturing industries, Unmanned Aerial Vehicle (UAVs) originally used for military applications, recently demonstrated the potentials for many new applications and investment opportunities, such as advanced cargo distribution [1], surveillance and aerial photography [2], disaster management [3], search and rescue operations, data gathering and agricultural-related applications [4] etc. This is due to their flexibility, level of autonomy, and broad domains of applications. Based on the kind of applications, technology or mission, UAV system can be categorized

J. H. Abawajy et al. (Eds.): EATI 2020, LNNS 254, pp. 181–204, 2021.
https://doi.org/10.1007/978-3-030-80216-5_14

according to weight, mission, characteristics or configuration. Some of the features that differentiate UAV types are the range coverage, altitude, take-up, and endurance.

Moreover, to handle cooperative teamwork, multi-agent systems are needed. This implies that multi-agent systems have greater advantages compared to a single UAV agent. Multi-agents systems can localize themselves more efficiently, have faster search and exploration task and carry out missions within short periods as compared to a single UAV system. In this context, communication in multi-agents system plays an important role. It enables the agents to overcome many real-world issues and applications. In almost all real multi-agents UAV applications, such as geological exploration, search-and-rescue operation, connected multi agents have to cooperate to complete assign tasks effectively while maintaining communication links among themselves. Recent researches have demonstrated that, UAVs systems if carefully deployed and operated can be used to enhanced or provide efficient and wireless services for a wide range of applications scenarios [5]. In similar way, the potentials of deploying UAV as Flying Base Stations (FBSs) for wireless applications to provide temporary connectivity after a disaster or during temporary event has been reported in [6].Thus, to efficiently achieve all the above mentioned applications and potentials, the system must be supported with efficient and long endurance energy sources. The limitations of this energy sources remains a challenge stagnating UAV technology and deployment in various real applications scenarios.

This energy limitation and payload trade-up make the flight duration of UAV becomes limited and insufficient for achieving long term related application mission. The performance of onboard IoT devices is also affected by the performance of the UAV power source devices used. Furthermore, the flight duration as a function of its energy sources lifespan, depends on the weight of the system, speed, and flight trajectory of the UAV. Therefore, power allocation to energy-consuming devices is a trade-off between energy minimization and achieving added services other than the target mission of the drone. Base on the above-mentioned challenges, it is very important and necessary while designing UAV systems for any kind of mission to put into consideration the energy and flight time constraint. In addition, the above-mentioned UAV challenges attracted the attention of many researchers to investigate different energy sources, efficient energy utilization as well as hybridizing the energy sources for UAV successful deployment. The paper is organized as follows: UAV technology, Flight profile, Energy source and consumption is presented in Sect. 2. Section 3 discusses UAV hybrid techniques. In Sect. 4. UAV EMS are discussed. Section 5 contain challenges and future research while Sect. 6 gives concluding remarks.

1.1 Motivation

Although, a lot of publications have been made on the review of different aspects of UAV technology such as (1) UAV applications (2) path planning and energy efficiency (3) UAV technology (4) Modelling and control etc. only a few articles critically reviewed and addressed energy sources and optimal energy management in the UAVs. Table 1 presents a summary of recent review articles related to UAV sources of energy and energy management.

Table 1. UAV published review article related to energy, energy management and power supply techniques

Ref/Year	Focus	Contribution
[7] **2014**	Power supply techniques	Discuses and review the different power supply architecture for UAV
[8] **2017**	Renewable energy for aerial and non-aerial robotic system	Addresses the application of RES in robotic system's uninterrupted power supply and compare with the traditional techniques
[9] **2018**	Storage technology	Presents a critical review on available energy storage technology and their real life application
[10] **2018**	More electric aircraft system (MOEAS) onboard micro grid review	Presents and discussed the MOEAS onboard micro grid technology, challenges and feature research direction
[11] **2019**	Energy management	(1. Review of the topics on energy optimal management strategies of UAV (2. Hybrid UAV system power architecture (3. Discuses and summarizes the new energy sources for UAV
[12] **2019**	Power supply and energy management	Critical review on UAV power supply configurations and energy management
This article	Energy sources and management	(a. Review of the available energy source and their performance evaluation (b. Energy sources hybridization, interfacing techniques and EMS

2 UAV Technology, Flight Profile, Energy Sources and Consumption

UAV System basic building blocks as illustrated in Fig. 1, consist of propulsion system, onboard flight control system, energy management block, actuators and ground control system.

The flight control system is tasked with aerial data collection and analysis, guidance, control and navigation, mission planning as well as communicating to the ground station. Other UAV component are the sensors for autonomous flight, cameras and radar system.

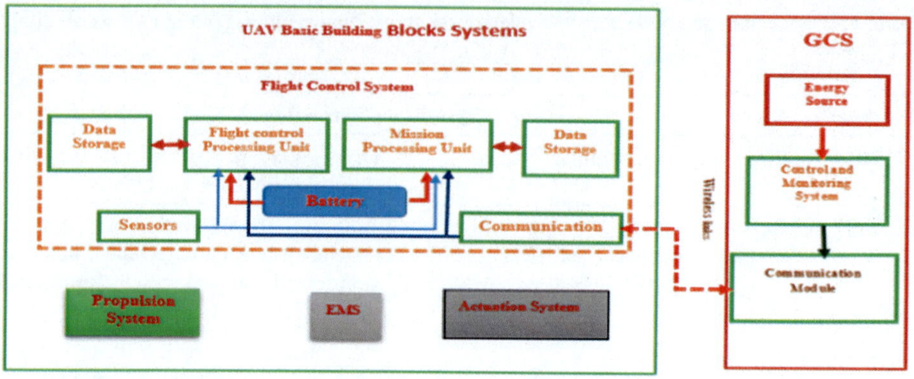

Fig. 1. UAV building blocks

2.1 Propulsion System Technology

The UAV propulsion system converts electrical energy into equivalent mechanical power generated by the motor-propeller system thereby allowing the motion of UAV system. Certainly, the propulsion system is the main power consuming part. The propulsion system component as shown in Fig. 2 includes the onboard Electronic Speed Controller (ESC), power sources, Electronics Power Converters (EPC), EMS, motors and propellers units. Furthermore, this constitutes the heaviest units in the UAV systems. The communication techniques between the propulsion system components is to allowed the onboard energy sources supply power to the bus system through either unidirectional or bidirectional electronics power converters as the case may be. These electronics power converters are employed to enable proper and efficient power flow control, storage systems interfacing, stepping down voltages to feed low voltage avionics, power conversion and power quality supply [4]. The EPC usually receives control signals from EMS unit responsible for power splitting and sharing between the energy sources [13]. Basically, small scale UAV systems uses Brushless Direct Current (BLDC) motors due to high speed and efficiency, high power density, good torque characteristics, control flexibility, long lifespan, and reliability [14–16]. While long endurance, large weight aircraft systems burnt fuel through chemical energy conversion in the aircraft engine to generate the required thrust, usually in the range of tens of megawatt. This constitutes the largest percentage of the output power of the engine. Furthermore, the aircraft engine is usually coupled to various kind of gearboxes, which then drive the electrical generators, the hydraulic pump for the actuators and the fuel pump for the engine [10, 16].

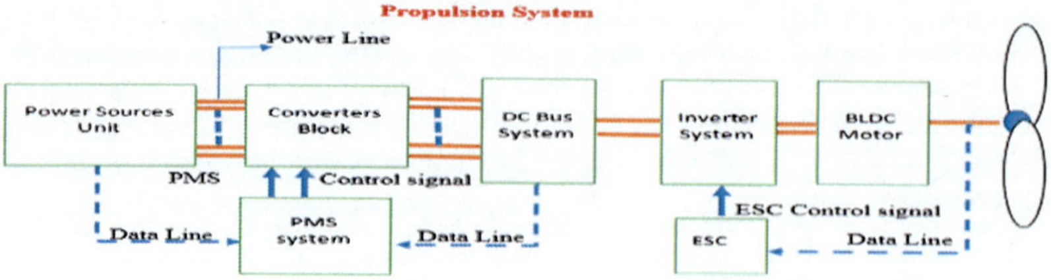

Fig. 2. UAV propulsion system technology

One of the basic and important building block of long endurance, large weight aircraft propulsion system is the Internal Combustion Engine (ICE) [17]. As compared to the Electric Propulsion System (EPS) Table 2, ICE allows long missions, large payload to be achieved as two important challenges of deploying medium to micro-scale UAV systems [16]. Basically, ICE systems involve many steps of producing energy which reduces its efficiency [17, 18]. Due to the above mentioned limitations, electric propulsion system stands the best choice due to the advantages of its minute thermal and acoustic signatures, automation and control flexibility, self-starting characteristics, cost effectiveness and a very high reliability as well as low chances of crash due to motor or power failure [17, 19, 20].

Table 2. ICE and EPS comparison analysis [10, 12, 14, 17, 21]

Technology	Advantage	Disadvantage
ICE	a. Large power to weight ratio b. Allow long mission and large payload c. Highly efficient	a. Not adequate for medium to micro scale UAV system b. Reduced efficiency due to many steps of converting energy c. High thermal and acoustic signature d. Control and automation complexity
EPS	a. Low thermal and acoustic signature b. Control and automation flexibility c. Cost effective and reliability d. Allow component redundancy	a. Short mission and low payload b. Inadequate for high power range

2.2 Flight Profile and Energy Consumption

Flight and energy consumption profiles are the important aspects that need to be given careful consideration when designing UAV system for a particular mission. For fixed wing UAV system, the flight profile involves five stages each with peculiar mode of energy consumption. These flight profile includes take off, climbing, cruising, descending and landing as illustrated in Fig. 3. The energy demand during each profile stages need to be stable, sequential and uninterrupted for the reason of safety [22]. Moreover, it can be clearly observed that, the load power demand varies at each stage of the flight profile mission. Typical power profile of UAV shows that, the load power demand during takeoff and climbing is the largest, hence power need to be shared between the sources as it can exceeds the primary source power level. The next high power consuming stage is the descending and landing flight profile with significant power fluctuation.

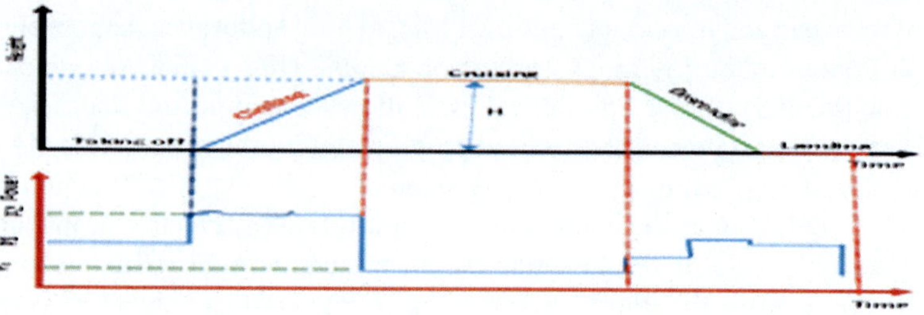

Fig. 3. UAV flight and energy consumption profile

2.3 Sources of Energy

Now, the EPS alternatives must be supported with a power source which becomes one of the UAV challenge [23]. This is due to fact that, flight duration of the system is a function of the energy source. Generally, UAV systems utilizes either liquid fuels as source of power reported by [24], BS system [25, 26], renewable sources of energy, FC or SC [8, 27, 28] to power all electrical and other avionics components. Recently, the FC, solar cells, BS, and SC have significantly increased their practical applications in UAV systems technology [24, 28, 29]. Several FC-powered UAV systems, SC, Solar system and BS powered UAV have been experimentally flight-tested. Based on the available published article and experimentally flight tested result, UAV energy sources can be summarized and presented as in Table 3. Progressively, Wireless Power Transfer (WPT) has been actively investigated in the cause of addressing UAV energy challenge. Authors in [6] proposed WPT technique for UAV Mobile Edge Computing (MEC). The proposed technique involves energizing two mobile devices wirelessly. Moreover, reference [30] investigation was on UAV enable WPT and optimization.

Table 3. UAV sources of energy research progress

Ref/Year	Energy source	Contribution
[31] **2011**	Battery Storage (BS)	Evaluate the range and endurance performance of BS powered UAV
[2] **2013**	Power line Tethering (PLT)	Experimental design of PLT UAV
[32] **2016**, [33] **2015**	Photovoltaic (PV)	Aerodynamic design of solar HALE UAV system Experimental design of solar powered UAV
[34] **2018**	Fuel Cells (FC)	Experimental design, cost analysis and environmental impact of FC-powered octocopter for mobile crane surveillance application
[35] **2018**	Super Capacitor (SC)	Comparative study on SC, FC and BS using hardware in the loop (HIL)
[30] **2018**	Wireless Power Transfer (WPT)	Energy optimization and trajectory design of UAV WPT

Battery Storage (BS)

Battery Storage (BS) system is the most frequent used source of energy for small to micro scale UAV systems [36]. This is due to the enhanced flexibility and simplicity of the propulsion systems. Furthermore, its usage can satisfy users in terms of flight duration, cost, and flexibility [12]. However, due to the weight budget and cost tradeoff, the BS systems limit the UAV to a very short endurance compared to the time required for its application [37]. There exist a lot of BS technologies available such as Ni-Cd, Ni-Mn and Li-S etc. [26, 38, 39]. Performance evaluation of the available BS technology proves less advantageous to LiPo battery technology in terms of energy density, power and weight budget [31]. This is the reason why LiPo battery system is mostly preferred for small to micro scale UAVs as it gives the highest flight mission of about 1 h 30 min and relatively saves weight and indeed high energy [40, 41]. Furthermore, [42] investigates and evaluate the available batteries technology state of charge for a particular mission while [31] carried his research on parameters that significantly affect the performances of a small UAV powered by battery and come up with a mathematical model to predict the range and endurance of the UAV taking into account the battery discharge stages.

Though, there exist a significant research improvement in batteries technology, yet the specific energy of current batteries technology still constraint the UAV endurance and range mission. The technology is not even adequate for high power range and long mission UAV system. As such, they will not reach many UAV real world applications. Additionally, the stability and safety levels are highly affected as a result of energy density improvement [12]. Now, to support UAV long flying missions, hovering and extreme weather conditions, the systems need to have a very dense and stable sustainable source of energy. Thus, different solutions were developed in the literature to deal with this battery energy limitations. Examples of such solution are WPT or laser beam aerial recharging, battery swapping, path planning and energy optimization. It can be clearly observe that, aerial drone technology is driving fast the advancement of batteries technologies that are chief, light and safer with long lives cycle, high-temperature tolerance, and higher energy density.

Battery Swapping

To extend the mission of UAVs and batteries long range limitation, researchers demonstrated the technique of *swapping* to recharge UAV's drained batteries. The swapping technique is either semi-autonomous assisted by an operator or full autonomous [43]. Two techniques of battery swapping have been reported in the literature, hot-swapping and cold-swapping [43, 44]. Hot swapping involves swapping UAV drained batteries with fully charged batteries while keeping the UAV powered on. As shown in Fig. 4 the hot swapping technique has the advantage of allowing UAV agents to activate and rejoin its hotspot again. Thus, by deploying multiple number of UAV agents, uninterrupted service can be achieve through coordination and hotspot handshake services between multiple agents. While cold swapping techniques involves UAV agents identifying Ground Charging Station (GCS) and connecting to the charging point. However, cold swapping technique is constraint with the long charging waiting time compared to operational time frame of the UAV agents. Hence, this can results to agents queuing for large number of UAV agents.

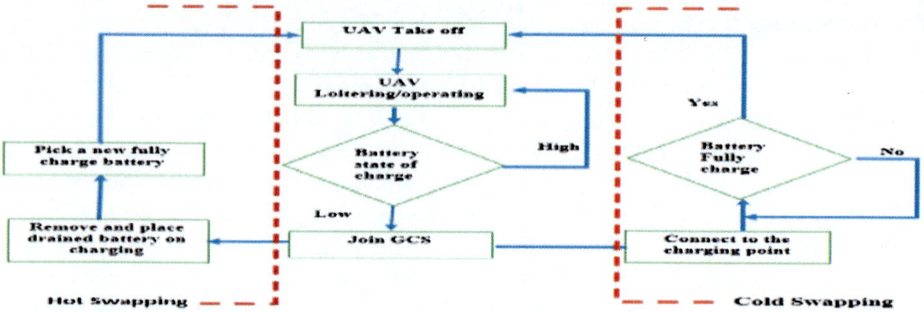

Fig. 4. UAV cold and hot swapping algorithm [12]

Basically, the technique of battery swapping must comprises three distinct components; (a) Charging station (b) Swarm agents for continuous operation (c) Management and control systems for swarm coordination. To maintained extended services and continuous operation of the UAV agents, GCSs must be situated in a strategic location along the route connecting the agent path. This GCSs can be carefully positioned on roofing, telecom towers, street light towers, poles, or designed standalone mast [45]. This GCSs consists of a platform for landing combined with other related electronic components. At the GCSs the battery can be charged through inductive coupling or contact enabling path. Furthermore, the GCSs sources of power can be either grid connected, solar systems or battery banks to have a full autonomous remote station.

Research works that have investigated the technique of battery swapping technology to achieve long-endurance missions are the example of [46]. The author developed four quadrotor aerial system models. State of charge estimation were adopted to continuously operate the system and achieve long mission. The idea was to operate one quad-copter at a time up to the threshold level of the battery before relinquishing its services to a fully charged quad-copter while allowing the drained quad-copter to join the GCSs to put the battery on charging and get new fully charged one before next stage of operation. This idea allows the system to be continuously recycled until a mission is achieved. But one of the limitations of this technique is that the time frame of operation of the systems depends solely on the number of batteries to be swapping with, charging and discharging time of the batteries. Therefore, this system does not depend on the number of UAV agents deployed for a particular mission.

An economic evaluation of recharging and refilling was proposed by [47]. The authors present analytical critics on the cost, coverage and system complexity. The conclusion drawn by the author was to adopt refilling stations for short endurance missions and battery swapping for long-endurance missions. Other research works on battery swapping in the literature are the work of [48, 49] where both proposed a GCS and a model to limit the charging time of the battery. For long mission enable UAV systems, authors in [48, 50] proposed and present a hot-swapping techniques based on online energy management, state of charge, battery health monitoring and precise landing techniques. The swapping technique is fully autonomous by the use of servo lift system to swap the batteries. Furthermore, the authors in [51] proposed an active infrared imaging to design GCS. The design mechanism allows the agents to be operated 24 h daily and incorporated a camera and infrared filter for the accurate and precise landing of the UAVs. An efficient hot-swapping technique was presented in [46]. Dynamic battery swapping stations (DBS) for delivery services drone to improve agents recycling was proposed by [52, 53]. The UAV agents were designed in such a way it moves to DBS stations according to a predefined schedule. The proposed technique allows the UAV to join a chosen station for changing storage or product delivery and loading [52, 53]. However, critical evaluation of battery swapping technology results to the following drawbacks. (a) Increased system complexity and cost (b) Increased number of batteries and swapping agents (c) Only applicable to specific mission and applications (d) Charging and discharging time of the batteries (e) Risk of flight and landing for charging and changing batteries (f) GCS feasibility.

Wireless Power Transfer (WPT)
Battery swapping technology was adopted by researchers as an alternative to enhanced UAV long mission and overcome energy limitations. However, the technology still faced with entirely different limitations. Such battery swapping drawbacks attracted researchers to investigate the viability of WPT. This technique of WPT was adopted by researchers as a choice to prolong the operational time of UAV systems or continuously operated to avoid the risk of flight and landing for charging and battery swapping [30]. In this context, a laser generator is used to transmit a beam of light from the Ground Wireless Power Source (GWPS) to the UAV system. Attached to the onboard UAV electronics components is an optical receiver for receiving and converting the received light to electric energy for charging or powering the UAVs electrical and electronics components Fig. 5. Using state of charge estimation (SCE), the UAV is autonomously directed to join an aerial power link for charging, hence safety of landing and takeoff is improved. Top of long story buildings, telecommunication mast, power line mast or dedicated mast, are used to stations the laser beam transmitters to avoid obstruction. Employing this mechanism, a continuous link of radiation will be establish between the transmitter and embedded receiver onboard UAV. The authors in [51] investigates the feasibility of the laser beam WPT technique by showcasing the first-ever 12 h quad-copter flight time. Also, an experimental justification of the investigation both mechanical and control aspect design were provided. Furthermore, the research work considered restricted UAV mission for this particular technique. Other contributing researches on UAV assisted wireless power transfer are the works of [54, 55] on magnetic resonant power transfer. However, WPT technique provides a solution to only some UAV energy constraint but another obstacle arises as result. Such obstacles are the UAV restricted area of operation to maintained wireless power link, reduced altitude and must significant is one to one transmitter and receiver WPT. A typical UAV WPT experimental demonstration is presented in Fig. 5. Therefore, WPT technology limits the number of agents to be deployed for a particular mission and hence increased operational cost [56].

Fig. 5. UAV Wireless Charging System with directional LED reported by [51, 55]

Photovoltaic (PV) Cell

Photovoltaic Cells (PV) as an energy generation system is receiving tremendous attention among researchers for UAV related application missions. The mechanisms used by many researchers and industries is to installed PV modules on UAV wings or surface in such a way that the modules will be supplying both UAV and battery bank during optimum solar light and have the battery bank as switching alternative in the night or when sunlight is not available [57]. In such an arrangement, the UAV can fly for a very long period or even stay indefinitely to the user requirement.

A lot of research works have been published by researchers around the globe on the analysis and design of solar-powered UAV systems. A review and feasibility study on solar-powered UAV is presented in [8, 33] and the history of solar-powered UAV is discussed in [32, 58]. Research trend and active participation in solar-powered UAV technology have indicated an extensive research value for technology development and commercial applications. The authors in [32, 58] developed a prototype model of solar-powered UAV. The basic idea was to optimize the airframe efficiency and improve fuel-saving apart from reducing the system weight. The investigations successfully proved the sufficiency of the solar energy to even carry additional load and presented about 60% of fuel saving apart from prolonging endurance. Progressively, the authors presented an experimental case study EMS using the developed solar-powered UAV with battery banks. The experimental model consists of Maximum Power Point Tracking Controller (MPPTC) to maximize module power under variation of temperature and solar radiation. The main drawback encountered with solar-powered UAV technology is the wing size for the maximization of the received solar radiation. Also, MPPTC is necessary which consists of a DC-DC converter, controllers, voltage sensor and current sensor hence increased system complexity. Unpredictable nature of solar radiation and solar irradiances fluctuations is a great challenge to solar-powered UAV system. Finally, the techniques is completely inadequate for small to micro scale UAV system.

Power Line Tethered (PLT) UAV

Alternatively, some researchers investigate the viability of tethering UAVs to power lines and continuously supplied the UAVs with the required energy. The technique is expected to be safe as well as assisting the UAV with continuous power and efficient data communication to Ground Station (GS). The tethering technology employed fiber cable as alternatives to copper conductor being used in power lines. The technology allows maximum energy in the range of kilowatt to be transferred optically using fiber optic conductor. The main advantages of this technology are reduced UAV payload, light power line systems and elimination of electrical signature. Authors in [2, 59] developed and proposed tethered UAVs one for data collection and surveillance applications while the latter for maritime pollution monitoring and heavy offshore contamination. However, critical examination of the PLT technique shows that, users can be constrained with limited area of application, low altitude operation and even restricted from moving far away from the GS. Though, some researchers have used Ground Vehicle GV as carrier for the prime power source in order to have a wide coverage but still complexity and financial budget constraint remain. In addition, the authors in [60] Proposed and demonstrated experimentally a long endurance mission tethered UAV for nuclear power plant surveillance. In that study, the UAV is expected to be operated indefinitely as long as the tethered cable supplied the required energy. Furthermore [61] proposed control of tout cable tethered UAV with restricted domain of applications and [62] present an experimentally evaluated tethered UAV for construction site application as shown in Fig. 6.

Fig. 6. A case study of tethered UAV for construction site [62]

Fuel Cell (FC)

UAV energy efficient utilization is very important for normal operation of the systems often required to achieve system operational goals of range, endurance, and other target requirements. Due to the limitations of the space available and the mass budget on the UAV, it is often a delicate balance between onboard energy sources (e.g. batteries) and the achievement of operational goals [37]. To enhance UAV endurance and achieved long mission researchers push the boundary of battery technology to higher efficient sources such as FC. Comparing the performance of BS technology and FC technology, it was found that a typical UAV LiPo batteries can have energy up to about 250 Wh/kg while compressed FC systems can reach up to 1000 Wh/kg about 75% increase in power [35]. Unlike batteries recharging and swapping technology, refueling in the case of FC is instant and efficient [63]. A lot of technologies have been used in the FC industries. Mostly based on chemicals, catalyst, electrolytes criteria, or temperature characterization. Sharaf *et al.* in [64] Present a comparative studies and general performance evaluation of the most frequent use of FC in UAV technology such as Proton Exchange Membrane Fuel Cell (PEMFC), Methanol Fuel Cell (MFC), Solid Oxide Fuel Cell (SOFC). It can be concluded based on most research works that PEMFC is the most frequent use of FC for UAV applications [64]. Some of the leading research works on UAV powered by FC are Homet UAV from Vironment in 2003, Georgia Tech University demonstrated an experimental UAV model powered by a 500 W FC, while the Office of Naval Research (ONR) break a world record with UAV powered by FC having the highest endurance of about 48 h [65]. Moreover, Korean aerospace research institute in 2017 demonstrated UAV powered by FC with hovering, monitoring, vertical takeoff and landing capability [66]. Other researches employing FC as a source of energy for UAV are the case of SKYCORP in 2018 developed e-Drone Zero, BATCAM and MetaVista in 2019 as reported by [12]. FC technology though has numerous advantages such as reduced carbon dioxide emission or total elimination, reduced power transmission losses but the FC technology is constraint with sluggish dynamic response that leads to voltage fluctuations and uncertainty when a sudden change in power is encountered [67]. Furthermore, FC cannot absorb regenerative and excess energy and hence have difficulty in cold start, large storage system for UAV long endurance mission, the safety of storing some FC under very high pressure and low temperature such as hydrogen FC. Finally, a details comparison of FC storage mechanism was presented in [64].

Super Capacitor (SC)

Batteries powered UAV are constrained with limited lifespan, sluggish response and high energy density. While FC powered UAV is constraint with FC sluggish dynamic response that lead to voltage instability and uncertainty when encountered with sudden change in power. Therefore, SC technology could stand the best choice for UAV related application. This is due to the fact that, SC is characterized with a higher energy, fast response, low energy density, wide operating temperature range, charging tolerance, cheap installation and maintenance cost, higher lifespan and reduced voltage instability [9]. The SC technology is expected to offer added advantages in terms of supplying architecture, rapid response of the EMS and reinforced power density. Based on the above mentioned benefits of SC storage system researchers have extensively investigate on integrating an SC storage systems as an alternative source of energy for UAV application. Some leading researches related to SC technology for UAV application are the work of [35, 68] where the authors presented an evaluation and critical analysis on hybrid UAV propulsion system consisting of battery bank, SC and FC. The authors tested and investigated the significance of SC compared to FC and BS system. The obtained result indicated a very good performance of SC on the system dynamic response and load stability. The works of [15, 16] developed a hybrid UAV test model. The result also shows SC significant contribution in supplying peak power, absorbing power instability during dynamic flight with instantaneous changes in power load. A summary and comparison analysis on the reviewed available article on UAV energy sources is presented in Table 4.

Table 4. Performance evaluation of UAV energy sources

Sources	Advantage	Disadvantage
ICE	Higher Power to Weight ratio than that of other types, Better speed (propulsion energy density) values, long endurance	High fuel consumption, not practically adequate for small scale UAVs, high noise level and environmental pollution
BS	Enhanced Simplicity and Flexibility, Enhanced Safety and Cost Effectiveness	Limited Endurance and Range, Reduced Autonomy, Limited lifespan and Increased Weight Budget
SC	Fast response, Low energy density, wide operating temperature range, charging tolerance, cheap installation and maintenance cost, and higher lifespan	Increased cost and complexity, EMS becomes integral part of UAVs system and voltage regulation is necessary

(continued)

Table 4. (*continued*)

Sources	Advantage	Disadvantage
FC	Instant refueling, higher energy compared to other sources, reduced or zero exhaust emission, negligible power transmission losses, zero or very low noise	Sluggish dynamic response, large storage system hence increased weight, difficulty in cold start, safety of storing some fuel cell under very high pressure and low temperature, Heavy and complex power supply system
PV	Clean source of energy, reduced system weight cheaper operational cost	Very large wings, MPPT controller is necessary, initial installation cost, not applicable to small scale rotary UAV due to space budget

3 UAV Hybrid Techniques

Due to the requirement of efficient energy sources for UAVs real world applications scenarios such as endurance, long mission target and mass budget, there is often a need to trade up between the available energy sources and achieving applications target [37]. Based on the limitations and constraints attached to all the available energy sources, it could be generally concluded that none of the available energy source can satisfy the performance of UAV energy requirement [10, 23]. To overcome such constraint, hybridization alternative that combined energy storage systems such as batteries and SC, batteries and PV or FC in combination with either one of the alternative source which has higher power density, is considered in most literature for improve dynamics, response and stability of power systems [35, 69, 70]. Hybrid systems techniques involve the use of two or more energy sources to create a very efficient source of power for unmanned aerial vehicle applications. Among the most important criterions considered in energy sources hybridization are low power consumption, system higher efficiency, robustness, DC bus power quality and increased lifespan of the energy sources [9, 10, 38]. Additionally, sources hybridization techniques can involves combining either one of the energy sources such as renewable energy sources or internal combustion engine (ICE) with auxiliary short-term boost capacity. Such auxiliary sources of energy can be any of batteries, FC or SC. The auxiliary source of energy is aim to enhance the UAV to achieve its aerial goals, which usually requires a significant amount of energy for small segment of the flight profiles and instant response to energy fluctuations [11, 28, 68]. A typical hybrid systems, in which a FC is combined with auxiliary short-term boost capacity, such as that provided by BS or SC, may significantly enhanced the UAV to achieve its missions, which usually requires a large power for a short fraction of the flight and a fast response to fluctuations in power demand [11, 28, 68]. On the hybrid system, Korean Aerospace Research Institute (KARI) in [66] demonstrated a case study on electric-powered UAV and developed hybrid EAV-2 with three distinct sources of power (solar cells, fuel cells, and battery pack). The power from the three sources can either be supplied to the propulsion system using an active or passive method.

3.1 Interfacing Techniques for Hybrid Energy Sources

The techniques adopted to connect the individual sources of energy is usually through either of the single unidirectional power converters and an auxiliary switches or bi-directional power converters based architecture as shown in Fig. 7 [71]. The converter unit act to control the output power from the auxiliary sources and maintained DC level voltage. The techniques that employ single power conveter and an auxilliary switch involves connecting the primary sources of energy through a single power converter while the auxiliary energy sources through an auxiliary switches. The auxilaiary energy sources can be any of FC, SC, BS or combination of the three [35]. The main function of the auxiliary switch is to control individual energy sources by either switching ON or OFF the sources [12, 15, 72]. While the HP system that employ multi converters as illustrated in Fig. 7b involves replacing the auxilliary switch with EPC. Hence, each individual source of the auxilliary unit and the primary unit is connected to the DC bus through EPC. However, this techiniques results in an increase system payload, size and complexity in design but has the advantage of improve fuel economy [73], charging the auxilliary power sources and improve safety.

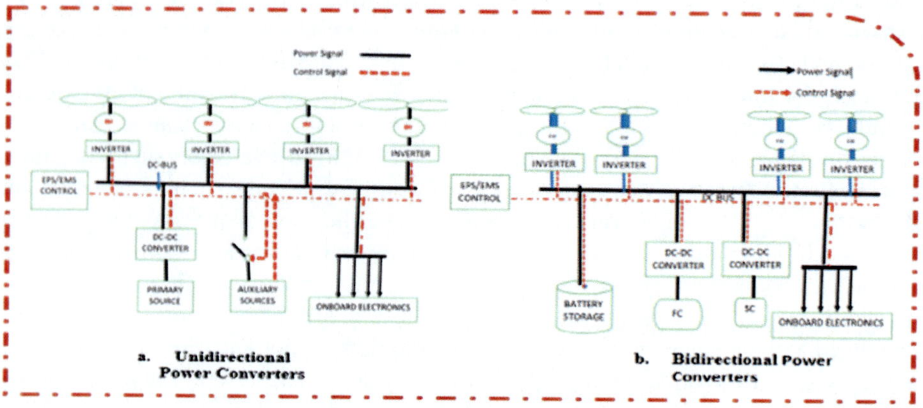

Fig. 7. Hybrid energy sources interfacing techniques

4 UAV Energy Management System (UAV-EMS)

Research works have proved the feasibility of power sources hybridization choice to power UAV propulsion systems and other avionics components [8, 74]. However, the design should not be just swapping the original energy sources with the hybrid system but a conceptual design approach needs to be strictly ensured. This hybrid system has the advantage of combining different energy sources performances and balancing their drawbacks. Thus, the energy from individual sources combined to form a hybrid system has to be optimally shared to attain high performance, efficient energy usage and long lifespan of the energy sources [75]. Hence, a well-designed EMS installed to optimally share the power from the individual sources is a good solution to UAV hybrid propulsion systems. This lead to passive and active EMS for hybrid propulsion system.

4.1 Active and Passive Energy Management System

Active Energy Management System (AEMS) involves the real-time sharing of power between the available sources through EPC by the management system block. While Passive Energy Management System (PEMS) which is the most frequently used techniques for UAVs applications involves connecting the available energy sources directly to the UAV propulsion systems without any EMS controller as used in the case of AEMS [27, 76]. PEMS when compared to AEMS, has the advantages of reduced complexity, weight, and power losses associated with the power conversion unit [33]. Conversely, AEMS has the advantage of full control of the sources through power converters which leads to the high efficiency of the energy sources and increases safety [69]. Reference [12] summarizes and provides details comparison of AEMS and PEMS. Therefore, hybrid energy sources can either be actively or passively shared to power UAV propulsion system and related avionics components based on the flight profile and the energy requirement of each stage [11, 23]. Thus, in conventional passive control techniques, output power of the individual sources relies on their own power characteristics, which can lead to significant amount of energy loss and low security of the system. These conventional passive control methods drawbacks can be curtail through an active EMS techniques. While designing AEMS for hybrid UAVs, minimized energy consumption, Improve system efficiency, lifespan of the power sources, robustness of the hybrid system and DC power quality are the most important points to be considered [70].

4.2 UAV-EMS Research Advances

A lot of research works have been carried out by researchers on EMS using either PV/BS system [58], FC/BS [69], or combination of either of the available energy sources, PV, FC, SC and BS [10, 35, 77]. While for large weight UAV utilizing ICE as a primary source of power and PV, BS or SC as secondary sources have been reported in [15, 78]. The authors in [79] developed and proposed a hybrid PV/BS EMS. The test model consists of an MPPT, EPC system and AEMS. Lee *et al.* in [68] extend the model storage capability to PV, FC and BS. Authors in [73, 80] investigates the feasibility of AEMS in a hybrid FC/BS with electronic DC-DC converters. The authors successfully proved AEMS as excellent alternatives to manage FC energy under maximum power and efficiency strategy. Other research works on the hybrid energy system are the work of [81] on modelling and control of PV, FC and BS hybrid system while [11, 66] reviewed the available EMS techniques and comparative analysis between AEMS and PEMS. The simulation results indicated that, PEMS could not maintain the battery minimum State of Charge (SC) and hence affect the lifespan of the battery while AEMS recorded 4.7% increase in energy management however a lot of flight-related conditions were neglected in the test. Authors in [30, 31] presented a critical review studies on the most recent researches on the power split system techniques and EMS for UAV applications. To generalize the EMS techniques, four classes of AEMS are the most frequently discussed from the vast amount of literatures namely intelligent method, rule-based method, optimization methods and control based method. The subsequent subsections briefly reviewed those management techniques.

Rule Based EMS: Rule based EMS technique is based on either machine tool control or power following techniques. The method is applied to EMS due to its simplicity and reliability. The strategy involves a set of rules predefined over the control variable (CV). The defined rules are deterministic and easy to realize with less computational stress [22]. Authors in [66] proposed an AEMS to control a hybrid PV, FC and BS for UAVs applications. The study is based on the level of power required by the UAV flight profile and battery state of charge to assign contributing power to each energy source output. FC was chosen only to serve the UAV in a predefined power intervals. This is to enable keep the system working in its nominal conditions. Furthermore, rule base AEMS driven by conventional PI controllers and DC-DC converters was presented in [11] though the strategy threshold setting was based on heuristic and previous knowledge of the designer. The authors in [81] developed a model rule-based AEMS for PEMFC and battery system based on the battery state of charge and using a DC-DC buck-boost converter. Rule-based EMS though considered as very simple and reliable with very little computational stress but still faced with very poor dynamic adaptability and the design is based on the level of past experience and investigative knowledge of the designer [82, 83]. Furthermore, the designed involves the use of DC-DC converter which may fail or malfunction [19, 84].

Intelligent Based EMS: The intelligent based method is a technique developed to imitate human beings or animals' intelligent systems for the purpose of automation and control. The intelligent base method has played a vital role in hybrid EMS. One such intelligent based method with high robustness and fast response is the Fuzzy Logic Control Technique (FLC) which is based on membership function and set of rules [85]. FLC has the advantage of reduced modelling complexity and nonlinearity compared to other control methods, not necessarily dependent on historical data as in the case of rule base method. Thus, this FLC technique could serves as best choice for on-line control with less computational stress [80, 86]. Some of the leading research on AEMS based on FLC is the work of [87] which proposed an experimental test of FLC based AEMS for UAV. The result of the study indicated that FC consumption in FLC is less as compared to the RB technique. But the main drawback of the research was that the design of the FLC rule depends on the level of experience and knowledge of the designer or expert [85]. Progressively, other research works have combined the potential of advanced optimization techniques such as genetic algorithm [88], particle swarm optimization [22] or an evolutionary technique [89] with FLC to achieve better energy management for UAV. Recently, [29] proposed a state machine strategy for hybrid FC/BS UAV application. In that study, the author uses a control logic to partition the decision area into five states according to flight profile and power consumption demand by the system units. A neural network as an intelligent based energy management system has the advantage of self-learning, good optimization performance, compared to FLC as such, becomes a very good choice for an optimal solution [90]. Though, a neural network may be faced with a peculiar problem of paralysis caused by complex objective function and could led to the algorithm failure. Some representative research study on AEMS based on NN is the work of [74] where the authors proposed a NN backpropagation AEMS with four layers for hybrid FC and battery system.

Control and Optimization Based EMS: An optimization-based technique can be classified into causal and non-causal methods. Example of non-causal methods includes Dynamic Programming (DP) [88] Pontryagin's Minimum Principle (PMP) and Linear Programming (LP) all belong to optimal control techniques and were successfully applied in UAV EMS [91]. While the examples of causal are Model Predictive Control (MPC) and Equivalent Consumption Minimization Strategy (ECMS) [92, 93]. The causal and non-causal techniques were all applied to AEMS. MPC has been applied to solve AEMS in four operational steps as predictive, control, application and update steps [94]. The performance of the control scheme was based on the model quality, sampling technique and the length of the prediction horizon. Path planning and optimization is the techniques of finding an energy efficient path that covers all the point of a particular area [95]. Energy efficiency and optimization problem have been extensively investigated from different perspective by researchers worldwide. The goals of optimization in most literatures are (a) to cover a specific area effectively (b) to find an efficient and fastest route to cover a particular area [96] (c) to determine an energy efficient path [97] (d) collision avoidance path for team of multiple agents [95].

5 UAV Energy Challenges and Future Research Direction

5.1 Long Endurance, Space and Payload Budget

Long endurance, space and payload budget are the major UAV design criteria for any real world applications. However, all the available energy sources and EMS techniques are attached to either one of those constraint (Long endurance, space and weight budget).The future research will aim at investigating UAV ultralight, high efficiency energy sources and EPC. This is due to the fact that, the energy sources and EPC are contributing largely to UAV space scale and weight budget. Hence control of the energy storage system and EPC holds a great impact to the possibilities of weight reduction and range extensions.

5.2 Multi-agents UAV-EMS

Energy hybridization have demonstrated a very good promising result for powering UAV propulsion system and other avionics components. Certainly, most EMS design which form part of efficient and optimal energy sources hybridization are considered at single UAV level and hence the space scale is limited. Moreover, the advent of WPT, smart agents UAV and efficient communication links, multi-agents UAV system are expected to attain collaborative sensing and exploration, coordinated motion planning, and formation to improve the flight endurance and energy efficiency. There exist a need to investigate and develop multi agents UAV cooperative EMS. This is due to the fact that EMS considered at single UAV level could be entirely different from that for multi agents. The technique to achieve multi-agents UAV-EMS remain a research challenge. Thus, future research will aim at investigating more deeply on different multi-agents communication techniques and ways of agent to agent information sharing about agent internal state, energy level and consumption profile [98]. For example, the UAV agent could directly exchange information about their energy states. Moreover, the agents must associate a confidence level to the information they are communicating using efficient communication links. Advanced control techniques, artificial intelligent methods and machine learning techniques can be used to predict energy consumption and energy level among UAV agents, hence help immensely to design efficient and coordinated EMS.

5.3 Weather, Energy Forecast and Future Energy Consumption

Flight profiles and weather condition are part of the factors contributing largely to energy consumption and limitations for many real UAV applications. Therefore, it is expected for the future UAV-EMS to have high efficient and ultra-light EMS that can forecast weather and energy consumption at each flight profile stages and manage the energy consumption in advance. Prediction algorithm using advanced artificial intelligent system, machine learning can be incorporated to UAV-EMS for efficient energy sources utilization and optimization.

5.4 Energy Sources Fault Isolation and Voltage Stability Control

The goal of all UAV system design is the avoidance of critical situations that may likely compromise the safety of the system as well as structure and people on the ground. Since energy sources hybridization technique supported with efficient EMS have been accepted as the best energy supplying technique for UAV long range applications. Most EMS design techniques considered only optimal and efficient energy sharing while energy source fault isolation is neglected. Furthermore, EMS must ensure power quality supplied to the load and other critical avionics components. Very few research works have been carried out related to energy source fault isolation and power quality. Thus, there is need for deep research investigation on EPC, smart EPC and EMS with fault tolerant capability, high quality DC bus voltage supply with DC circuit breakers. The technique is to have EMS algorithm incorporate energy source fault detection and prediction capability so that it can take necessary action before occurrence of any dangerous situation [14].

5.5 Multi-objective Optimization Techniques

Optimization and control based EMS though regarded as a good solution to hybrid EMS in an aerial vehicle but still faced with some constraints, drawbacks and limitations. First and famous, critical examination of vast researches on UAV EMS control and optimization indicates concentration on single objective optimization such as energy consumption minimization, optimal path planning while other objectives are neglected which may affect constrained or unconsidered objectives functions. Control scheme such as MPC is faced with the vulnerability to uncertainties in the model of hybrid systems and the disturbances result in a poor EMS. Hence, there is need for future research to deeply investigate multi-objectives optimization techniques for UAV energy sources efficient utilization. Finally, summary of some leading research on hybrid UAV EMS performance evaluation and challenges/limitations of each techniques is presented in Table 5.

Table 5. Assessment of EMS performance and challenges

Sources	EMS	Contribution	Limitation/Challenge
FC/BS/PV [71]	Logic-gate switching	Integration of fuel cell and solar system to form UAV hybrid system and achieve a flight duration of 1.5 h	Strategy algorithm missed, UAVs vibration affect the solar system due to vibrational motion of the wing hence cannot supply constant power to meet the load variation. $NaBO2$ precipitation issue
FC/BS/PV [99]	Fuzzy state machine	Integration of fuzzy logic and state machine strategies through simulation	Lack of experimental validation, expert knowledge or designer experience is essential
FC/BS/PV [66]	Rule-based strategy	Experimental validation of the integrated model and achieved flight duration of 3.8 h	Night flight was not tested, designer experience is essential, wing vibration and space budget
FC/BS [82, 99]	State machine strategy	Experimental validation of the technique and flight duration of 15 min	Only ground test, limited flight time compared to complexity and increased cost
FC/BS [87]	Online fuzzy logic strategy	Experimental validation for long and short endurance mission of the hybrid system	No flight test, lack reliability assessment of the system, designer experience is essential, increased complexity and cost, limited endurance
ICE/BS [80]	Fuzzy-ECMS	Integrate fuzzy and ECMS which improved system safety and fuel consumption	The study lack experimental validation. Designer experience is essential and increased weight budget of the system
PV/BS/[100, 101]	Control based	Integration of PV/BS, provide sizing and conceptual design of PV-UAV system	The study lack experimental validation, the system could be affected by solar radiation fluctuation and PV space scale

6 Conclusion

A critical review of UAV energy sources and management is presented. Based on the UAV electrification trends, the expected feature of hybrid UAVs systems, and transmuting the energy grid by widespread adoption of RES will still require new innovative ideas for the hybrid EMS, energy storage systems and reliability oriented control system. As it can be clearly observed from the review, there exist quite a small number of researches on UAV-EMS with experimental flight validation. This could be due to some specific UAV challenges.

References

1. Cord, T.: Skytote advanced cargo delivery system. In: AIAA International Air and Space Symposium and Exposition: The Next 100 Years (2003)
2. Woodworth, A., Peverill, J.: Tethered aerial system for data gathering. Google Patents (2013)
3. Lin, Z.L., Liu, H.H.T., Wotton, M.: Kalman filter-based large-scale wildfire monitoring with a system of UAVs. IEEE Trans. Ind. Electr. **66**(1), 606–615 (2018)
4. Valavanis, K.P.V., George, J.: Handbook of Unmanned Aerial Vehicles, pp. 2639–2641. Springer, Dordrecht (2015)
5. Mozaffari, M., et al.: A tutorial on UAVs for wireless networks: applications, challenges, and open problems (2019)
6. Hu, X., Wong, K.-K., Yang, K.: Wireless powered cooperation-assisted mobile edge computing. IEEE Trans. Wireless Commun. **17**(4), 2375–2388 (2018)
7. Cwojdziński, L., Adamski, M.: Power units and power supply systems in UAV. Aviation **18**(1), 1–8 (2014)
8. Iqbal, J., Khan, Z.H.: The potential role of renewable energy sources in robot's power system: a case study of Pakistan. Renew. Sustain. Energy Rev. **75**, 106–122 (2017)
9. Aneke, M., W.M.: Energy storage technologies and real life applications–a state of the art review. Appl. Energy **179**, 350–77 (2018)
10. Buticchi, G., et al.: On-board microgrids for the more electric aircraft—technology review. IEEE Trans. Industr. Electron. **66**(7), 5588–5599 (2018)
11. Lei, T., et al.: The state of art on energy management strategy for hybrid-powered unmanned aerial vehicle. Chinese J. Aeronaut. (2019)
12. Nadir, B.M., Zhibin, Z., Benbouzid, M.: A critical review on unmanned aerial vehicles power supply and energy management: solutions, strategies, and prospects. Appl. Energy **255**, 113823 (2019)
13. Chung, P.-H., Ma, D.-M., Shiau, J.-K.: Design, manufacturing, and flight testing of an experimental flying wing UAV. Appl. Sci. **9**(15), 3043 (2019)
14. Freeman, P., Balas, G.J.: Actuation failure modes and effects analysis for a small UAV. In: 2014 American Control Conference. IEEE (2014)
15. Hung, J., Gonzalez, L.F.: On parallel hybrid-electric propulsion system for unmanned aerial vehicles. Prog. Aerosp. Sci. **51**, 1–17 (2012)
16. Jansen, R., et al.: Overview of NASA electrified aircraft propulsion (EAP) research for large subsonic transports. In: 53rd AIAA/SAE/ASEE Joint Propulsion Conference (2017)
17. El-Sayed, A.F.: Aircraft Propulsion and Gas Turbine Engines. CRC Press, Boca Raton (2017)
18. Fahlstrom, P., Gleason, T.: Introduction to UAV Systems. Wiley, Hoboken (2012)
19. King, D.W., Bertapelle, A., Moses, C.: UAV failure rate criteria for equivalent level of safety. In: International Helicopter Safety Symposium, Montreal (2005)

20. Petritoli, E., Leccese, F., Ciani, L.: Reliability assessment of UAV systems. In: 2017 IEEE International Workshop on Metrology for AeroSpace (MetroAeroSpace). IEEE (2017)
21. Bongermino, E., et al.: Hybrid aeronautical propulsion: control and energy management. IFAC-PapersOnLine **50**(2), 169–174 (2017)
22. Huang, Y., et al.: Energy-optimal path planning for solar-powered UAVs monitoring stationary target. In: Proceedings of the 4th ACM SIGSPATIAL International Workshop on Safety and Resilience. ACM (2018)
23. Chin, C.K.: Extending the endurance, missions and capabilities of most UAVs using advanced flexible/ridged solar cells and new high power density batteries technology. Naval Postgraduate School Monterey CA (2011)
24. Adam, P., Leachman, J.: Design of a reconfigurable liquid hydrogen fuel tank for use in the Genii unmanned aerial vehicle. In: AIP Conference Proceedings. AIP (2014)
25. Saha, B., et al.: Battery health management system for electric UAVs. In: 2011 Aerospace Conference. IEEE (2011)
26. Sastry, A.M., et al.: Energy storage system. Google Patents (2019)
27. Koster, J., et al.: Hybrid electric integrated optimized system (HELIOS)-design of a hybrid propulsion system for aircraft. In: 49th AIAA Aerospace Sciences Meeting including the New Horizons Forum and Aerospace Exposition (2011)
28. Zandi, M., Payman, A., Martin, J.P., Pierfederici, S., Davat, B.: Energy management of a fuel cell/supercapacitor/battery power sources for electric vehicular applications. IEEE Trans. Veh. Technol. **60**, 433–443 (2011)
29. Yang, Z., et al.: The testing platform of hybrid electric power system for a fuel cell unmanned aerial vehicle. In: IEEE International Conference on Electrical Systems for Aircraft, Railway, Ship Propulsion and Road Vehicles & International Transportation Electrification Conference (ESARS-ITEC), pp. 1–8. IEEE, Nottingham (2018)
30. Xu, J., Zeng, Y., Zhang, R.: UAV-enabled wireless power transfer: trajectory design and energy optimization. IEEE Trans. Wireless Commun. **17**(8), 5092–5106 (2018)
31. Traub, L.W.: Range and endurance estimates for battery-powered aircraft. J. Aircraft **48**(2), 703–707 (2011)
32. Hwang, S.-J., Kim, S.-G., Kim, C.-W., Lee, Y.-G.: Aerodynamic design of the solar-powered high altitude long endurance (HALE) unmanned aerial vehicle (UAV). Int. J. Aeronaut. Space Sci. **17**(1), 132–138 (2016)
33. Morton, S., D'Sa, R., Papanikolopoulos, N.: Solar powered UAV: design and experiments. In: IEEE/RSJ International Conference on Intelligent Robots and Systems (IROS). IEEE (2015)
34. Belmonte, N., et al.: Fuel cell powered octocopter for inspection of mobile cranes: design, cost analysis and environmental impacts. Appl. Energy **215**, 556–565 (2018)
35. Gong, A., et al.: Analysis of a fuel-cell/battery/supercapacitor hybrid propulsion system for a UAV using a hardware-in-the-loop flight simulator. In: 2018 AIAA/IEEE Electric Aircraft Technologies Symposium (EATS). IEEE (2018)
36. Gong, A., Verstraete, D.: Role of battery in a hybrid electrical fuel cell UAV propulsion system (2014)
37. Lubkowski, S., et al.: Trade-off analysis of regenerative power source for long duration loitering Airship. In: 2010 IEEE Systems and Information Engineering Design Symposium. IEEE (2010)
38. Hasvold, O., Johansen, K.H., Mollestad, O., Forseth, S., Størkersen, N.: The alkaline aluminium/hydrogen peroxide power source in the Hugin II unmanned underwater vehicle. J. Power Sources **80**(1–2), 254–260 (1999)
39. Thackeray, M.M., Wolverton, C., Isaacs, E.D.: Electrical energy storage for transportation—approaching the limits of, and going beyond, lithium-ion batteries. Energy Environ. Sci. **5**(7), 7854–7863 (2012)

40. Mansouri, S.S., et al.: Remaining useful battery life prediction for UAVs based on machine learning. IFAC-PapersOnLine **50**(1), 4727–4732 (2017)
41. Morbidi, F., Cano, R., Lara, D.: Minimum-energy path generation for a quadrotor UAV. In: 2016 IEEE International Conference on Robotics and Automation (ICRA). IEEE (2016)
42. Donateo, T., et al.: A new approach to calculating endurance in electric flight and comparing fuel cells and batteries. Appl. Energy **187**, 807–819 (2017)
43. Jacobsen, R., Ruhe, N., Dornback, N.: Autonomous UAV Battery Swapping (2018)
44. Wang, M.: Systems and methods for UAV battery exchange. Google Patents (2016)
45. Gentry, N.K., Hsieh, R., Nguyen, L.K.: Multi-use UAV docking station systems and methods. US Patent App. 15/159,859 (2018)
46. Michini, B., et al.: Automated battery swap and recharge to enable persistent UAV missions. In: Infotech@ Aerospace 2011, p. 1405 (2011)
47. Kemper, F.P., Suzuki, K.A., Morrison, J.R.: UAV consumable replenishment: design concepts for automated service stations. J. Intell. Rob. Syst. **61**(1–4), 369–397 (2011)
48. Swieringa, K.A., et al.: Autonomous battery swapping system for small-scale helicopters. In: 2010 IEEE International Conference on Robotics and Automation. IEEE (2010)
49. Ure, N.K., et al.: An automated battery management system to enable persistent missions with multiple aerial vehicles. IEEE/ASME Trans. Mechatron. **20**(1), 275–286 (2014)
50. Liu, Z.-N., et al.: QUADO: an autonomous recharge system for quadcopter. In: 2017 IEEE International Conference on Cybernetics and Intelligent Systems (CIS) and IEEE Conference on Robotics, Automation and Mechatronics (RAM). IEEE (2017)
51. Achtelik, M.C., et al.: Design of a flexible high performance quadcopter platform breaking the MAV endurance record with laser power beaming. In: 2011 IEEE/RSJ International Conference on Intelligent Robots and Systems. IEEE (2011)
52. Sun, B., Tan, X., Tsang, D.H.: Optimal charging operation of battery swapping stations with QoS guarantee. In: 2014 IEEE International Conference on Smart Grid Communications (SmartGridComm). IEEE (2014)
53. Tan, X., et al.: Optimal scheduling of battery charging station serving electric vehicles based on battery swapping. IEEE Trans. Smart Grid **10**(2), 1372–1384 (2017)
54. Junaid, A., et al.: Autonomous wireless self-charging for multi-rotor unmanned aerial vehicles. Energies **10**(6), 803 (2017)
55. Griffin, B.: Automated resonant wireless power transfer to remote sensors from an unmanned aerial vehicle (2012)
56. Kausar, A.Z., et al.: Energizing wireless sensor networks by energy harvesting systems: scopes, challenges and approaches. Renew. Sustain. Energy Rev. **38**, 973–989 (2014)
57. Vega-Garita, V., et al.: Integrating a photovoltaic storage system in one device: a critical review. Progr. Photovol. Res. Appl. **27**(4), 346–370 (2019)
58. Shiau, J.-K., et al.: Design of a solar power management system for an experimental UAV. IEEE Trans. Aerosp. Electron. Syst. **45**(4), 1350–60 (2009)
59. Muttin, F.: Umbilical deployment modeling for tethered UAV detecting oil pollution from ship. Appl. Ocean Res. **33**(4), 332–343 (2011)
60. Gu, B.W., et al.: Novel roaming and stationary tethered aerial robots for continuous mobile missions in nuclear power plants. Nuclear Eng. Technol. **48**(4), 982–996 (2016)
61. Nicotra, M.M., Naldi, R., Garone, E.: Taut cable control of a tethered UAV. IFAC Proc. Vol. **47**(3), 3190–3195 (2014)
62. Kiribayashi, S., Yakushigawa, K., Nagatani, K.: Design and development of tether-powered multirotor micro unmanned aerial vehicle system for remote-controlled construction machine. In: Hutter, M., Siegwart, R. (eds.) Field and Service Robotics. SPAR, vol. 5, pp. 637–648. Springer, Cham (2018). https://doi.org/10.1007/978-3-319-67361-5_41
63. Kim, T., Kwon, S.: Design and development of a fuel cell-powered small unmanned aircraft. Int. J. Hydrogen Energy **37**(1), 615–622 (2012)

64. Sharaf, O.Z., Orhan, M.F.: An overview of fuel cell technology: fundamentals and applications. Renew. Sustain. Energy Rev. **32**, 810–853 (2014)
65. Swider-Lyons, K., et al.: Hydrogen fule cell propulsion for long endurance small UVAs. In: AIAA Centennial of Naval Aviation Forum. 100 Years of Achievement and Progress (2011)
66. Lee, B., et al.: Active power management system for an unmanned aerial vehicle powered by solar cells, a fuel cell, and batteries. IEEE Trans. Aerosp. Electron. Syst. **50**(4), 3167–3177 (2014)
67. Williams, M.C., Suzuki, A., Miyamoto, A.: Assessment of the performance of fuel cells and batteries. ECS Trans. **51**(1), 183–192 (2013)
68. Lee, B., Park, P., Kim, C.: Power managements of a hybrid electric propulsion system powered by solar cells, fuel cells, and batteries for UAVs. In: Handbook of Unmanned Aerial Vehicles, pp. 495–524 (2015)
69. Gao, L., Jiang, Z., Dougal, R.A.: Evaluation of active hybrid fuel cell/battery power sources. IEEE Trans. Aerosp. Electron. Syst. **41**(1), 346–355 (2005)
70. González, E.L., et al.: Experimental evaluation of a passive fuel cell/battery hybrid power system for an unmanned ground vehicle. Int. J. Hydrogen Energy **44**(25), 12772–12782 (2019)
71. Gang, B.G., Kwon, S.: Design of an energy management technique for high endurance unmanned aerial vehicles powered by fuel and solar cell systems. Int. J. Hydrogen Energy **43**(20), 9787–9796 (2018)
72. Rapinett, A.: Zephyr: a high altitude long endurance unmanned air vehicle. Doctor, Department of Physics, University of Surrey (2009)
73. Donateo T., Spedicato, L.: Fuel economy of hybrid electric flight. Appl. Energy **206**, 723–38 (2019)
74. Khayyam, H., Bab-Hadiashar, A.: Adaptive intelligent energy management system of plug-in hybrid electric vehicle. Energy **69**(5), 319–35 (2014)
75. Sikeridis, D., et al.: Wireless powered Public Safety IoT: A UAV-assisted adaptive-learning approach towards energy efficiency. J. Netw. Comput. Appl. **123**, 69–79 (2018)
76. Larsson, V., Johannesson, L., Egardt, B.: Analytic solutions to the dynamic programming sub-problem in hybrid vehicle energy management. IEEE Trans. Veh. Technol. **64**(4), 1458–1467 (2014)
77. Chen, Q., Gao, L., Dougal, R.A., Quan, S.: Multiple model predictive control for a hybrid proton exchange membrane fuel cell system. J. Power Sources. **191(2)**, 473–82 (2009)
78. Glassock, R., et al.: Design, modelling and measurement of a hybrid powerplant for unmanned aerial systems. Austr. J. Mech. Eng. **6**(2), 69–78 (2008)
79. Peng, L., et al.: A novel tangent error maximum power point tracking algorithm for photovoltaic system under fast multi-changing solar irradiances. Appl. Energy **2018**(210), 303–316 (2018)
80. Xie, Y., Savvaris, A., Tsourdos, A.: Fuzzy logic based equivalent consumption optimization of a hybrid electric propulsion system for unmanned aerial vehicles. Aerosp. Sci. Technol. **85**, 13–23 (2019)
81. Karunarathne, L., Economou, J.T., Knowles, K.: Model based power and energy management system for PEM fuel cell/Li-Ion battery driven propulsion system (2010)
82. Zhang, X., Liu, L., Xu, G.: Energy management strategy of hybrid PEMFC-PV-battery propulsion system for low altitude UAVs. In: 52nd AIAA/SAE/ASEE Joint Propulsion Conference (2016)
83. Kuroki, Y., Young, G.S., Haupt, S.E.: UAV navigation by an expert system for contaminant mapping with a genetic algorithm. Expert Syst. Appl. **37**(6), 4687–4697 (2010)
84. Todd, R., Forsyth, A.: DC-bus power quality for UAV systems during generator fault conditions (2010)

85. Logic, F.: Foundations of fuzzy logic and semantic web languages (2014)
86. Smith, J.F., Nguyen, T.: Fuzzy logic based resource manager for a team of UAVs. In: NAFIPS 2006–2006 Annual Meeting of the North American Fuzzy Information Processing Society. IEEE (2006)
87. Zhang, X., et al.: Experimental investigation on the online fuzzy energy management of hybrid fuel cell/battery power system for UAVs. Int. J. Hydrogen Energy **43**(21), 10094–10103 (2018)
88. Wang, A., Yang, W.: Design of energy management strategy in hybrid electric vehicles by evolutionary fuzzy system part II: tuning fuzzy controller by genetic algorithms. In: 2006 6th World Congress on Intelligent Control and Automation. IEEE (2006)
89. Wang, A., Yang, W.: Design of energy management strategy in hybrid vehicles by evolutionary fuzzy system part I: fuzzy logic controller development. In: 2006 6th World Congress on Intelligent Control and Automation. IEEE (2006)
90. Lakhmi Jain, P.D., Anna Maria Fanelli, P.D.: Recent Advances in Artificial Neural Networks Design and Applications, vol. 315. CRC Press (2000). ed. P.D. L.C. Jain, M.E., B.E. (Hons), Fellow I.E. (Australia)
91. Xu, L., et al.: Multi-objective component sizing based on optimal energy management strategy of fuel cell electric vehicles. Appl. Energy **157**, 664–674 (2015)
92. Eren, U., et al.: Model predictive control in aerospace systems: current state and opportunities. J. Guid. Contr. Dyn. **40**(7), 1541–1566 (2017)
93. Geng, B., Mills, J.K., Sun, D.: Energy management control of microturbine-powered plug-in hybrid electric vehicles using the telemetry equivalent consumption minimization strategy. IEEE Trans. Veh. Technol. **60**, 4238–4248 (2011)
94. Han, J., Kum, D., Park, Y.: Synthesis of predictive equivalent consumption minimization strategy for hybrid electric vehicles based on closed-form solution of optimal equivalence factor. IEEE Trans. Veh. Technol. **66**, 5604–5616 (2017)
95. Choi, Y., Jimenez, H., Mavris, D.N.: Two-layer obstacle collision avoidance with machine learning for more energy-efficient unmanned aircraft trajectories. Robot. Auton. Syst. **98**, 158–173 (2017)
96. Bellingham, J., et al.: Multi-task allocation and path planning for cooperating UAVsl. In: Butenko, S., Murphey, R., Pardalos, P.M. (eds.) Cooperative Control: Models, Applications and Algorithms, vol. 1, pp. 23–41. Springer, Boston (2003). https://doi.org/10.1007/978-1-4757-3758-5_2
97. Amoiralis, E.I., Tsili, M.A., Spathopoulos, V., Hatziefremidis, A.: Energy efficiency optimization in UAVs: a review. Mater. Sci. Forum **792**, 281–286 (2014)
98. Leonard, J., Savvaris, A., Tsourdos, A.: Energy management in swarm of unmanned aerial vehicles. J. Intell. Rob. Syst. **74**(1–2), 233–250 (2013). https://doi.org/10.1007/s10846-013-9893-8
99. Zhang, X., Liu, L., Dai, Y.: Fuzzy state machine energy management strategy for hybrid electric UAVs with PV/fuel cell/battery power system. Int. J. Aerosp. Eng. **2018** (2018)
100. Panagiotou, P., Tsavlidis, I., Yakinthos, K.: Conceptual design of a hybrid solar MALE UAV. Aerosp. Sci. Technol. **53**, 207–219 (2016)
101. Hajianmaleki, M.: Conceptual design method for solar powered aircrafts. In: AIAA Aerospace Sciences Meeting including the New Horizons Forum and Aerospace Exposition (2011)

An Efficient Multi-sensor Positions Human Activity Recognition: Elderly Peoples in Rural Areas in Focus

Haruna Abdu[1,3](✉), Mohd Halim Mohd Noor[2], and Rosni Abdullah[2]

[1] National Advanced IPv6 Center of Excellence, Universiti Sains Malaysia, George Town, Malaysia
haruna.abdu@fulokoja.edu.ng, haruna.abdu@student.usm.my
[2] School of Computer Sciences, Universiti Sains Malaysia, George Town, Malaysia
{halimnoor,rosni}@usm.my
[3] Department of Computer Science, Federal University Lokoja, Lokoja, Kogi State, Nigeria

Abstract. The increase in the availability of wearable devices offers a prospective solution to the increasing demand for elderly human activity monitoring, in the essence of improving the independent living standard of the growing population of elderly humans. With all the availability of the wearable devices fully embedded with sensors that are being used for human health monitoring, a lot of techniques are been proposed and used in the process. However, most of the publicly available datasets in use today, are collected in a fully controlled or semi-natural settings. Also, elderly peoples from rural areas and transitional activities are mostly not considered, which will cause a lack of generalization of the dreamed HAR models. The purpose of this research is to collect a new dataset from elderly peoples in a rural area and find the best sensor position among ankle and waist by subjecting the newly collected datasets to different machine learning classifiers. Sliding window technique with 50% overlapping was used to segment the sensor data collected from the elderly subjects. Relevant features were extracted, and selected using the wrapper method. From the results obtained, it has shown that the sensor attached at the waist position yield a better result compared to the ankle position on the newly collected elderly data. KNN algorithm has the highest accuracy level in both cases compared to the remaining tested classifiers.

Keywords: Human-activity-recognition · Elderly · Waist · Sensor and position

1 Introduction

Nowadays we are seeing ever-longer life expectancy. Over a long time, fear and vulnerability within the development can have an extraordinary effect on day-to-day activities. In the process of supplying the elderly with a satisfactory quality of life indeed at a better age, the necessities forced by wellbeing care must be taken into consideration when designing and preparing the living and care offices and homes of the elderly. As reported by world population prospects 2019 [1], one in six people in the world will be more

J. H. Abawajy et al. (Eds.): EATI 2020, LNNS 254, pp. 205–219, 2021.
https://doi.org/10.1007/978-3-030-80216-5_15

than 65 years of age, up from one in eleven in 2019. As the population of the elderly is increasing, therefore, there is a need for assisted living technology, to assist the elderly in their independent living due to their physical impairment. Human activity recognition (HAR)HA is beneficial to different domains such as academia and health industries. It's also an area of great importance [2] for health and well-being applications that include Ambient Intelligent (AmI) spaces, Ambient Assisted Living (AAL) environments, and wearable healthcare systems. It can also be used in e-health applications [3] such as monitoring elderly humans who are vulnerable in different aspects as a result of their physical impairment and cognitive that is seriously needing assistance for their standard daily living [4]. With the help of HAR, the elderly activity level can be automatically monitored and generate [5] timely activity reports such as; daily, weekly, and/or monthly activity, which could also be automatically emailed to the user or saved for some other purposes like self-health monitoring. The generated reports would keep track of an overall activity level of the user, which could be used to measure if the elderly need to improve on exercise and determine the number of calories expended. The activity report generated could be utilized to support sound practices and may caution some users on how inactive they are. The activity data can be used to configure the mobile phone's actions automatically. For example, changing the call status from a normal call to voicemail or even reject the call.

Human Activity Recognition is a promising and interesting area of research that aims at identifying what activity a person is engaged at a time. HAR is a very challenging and important field of research that can support many novel ubiquitous applications. Nowadays the availability of sophisticated wearable devices such as smartphones and smartwatches is fully embedded with several powerful and useful sensors. Due to the small size of these mobile devices, their substantial computing power, their ability to send and receive data, and their nearly ubiquitous use in our society, these devices [6] open up exciting new areas for data mining research and data mining applications. These sensors include motion sensors, light and sound sensors have made it easy for researchers in collecting data for activity recognition. Accelerometer and gyroscope are the most common examples of motion sensors that are meant to provide meaningful information that will aid in detecting and recognizing the activity engaged by a user; such as walking, running, sitting, standing, etc. The light and sound sensor include ambient light, microphone, and proximity, these are responsible for detecting the light around the wearable device [7].

A lot of human activity recognition datasets are publicly available for researchers or interesting individuals. Most of these datasets are collected in a laboratory setting, not in a natural setting. In the laboratory setting the subject is asked to perform some physical activity under the supervision of the researcher, and the subject doesn't have any right to perform any activity without instruction given from the researcher, in short, data collection in laboratory settings is instruction dependent. Secondly, transitional activities are usually ignored by researchers due to its shorter period, and if not dealt with properly it will become noise to the classifiers and leads to low accuracy or poor classification. Thirdly, to the best of our knowledge no publicly available datasets for human activity recognition purely collected in a natural setting, from elderly peoples in a rural area. If such data (from elderly peoples in a rural area) is not collected and integrated into the big picture of the dreamed HAR, the models will not be generalized and will be biased at all times. Also, the sensors are mostly placed at a different part of the subject body without considering which position gives the best sensor data, for activity detection or recognition.

Triaxial accelerometer and gyroscope have been playing a very important role in elderly human activity recognition, and become a good choice for that purpose, due to its smaller size, less energy consumption, low cost, easy integration, etc. Besides, they have strength in providing the best relevant features to recognize human body movement and/or posture. Single or multiple sensors can be used for elderly human activity recognition depending on the researcher's choice and the activity kind. The sensor position contributes to the overall accuracy of the activity classification, especially low-level activity that is particularly difficult to recognize among elderly humans. In this manner, sensor positioning, even though depends on the sort of activity being watched, plays a vital role in assessing the pervasiveness and wearability of the device used [9].

An elderly healthy individual or a patient can participate in activity recognition to improve their standard of livings. However, the choice of a position on the elderly body (where to place the sensor) that will give or produce a better signal, that is expected to capture the whole-body movement or posture is a factor to be considered. Placing the sensor device on some part of the elderly body like ear [8] will not capture body acceleration, chest [9] will completely produce noisy signals on elderly peoples as they can't stand erect, that will make the devices to be swinging in an uncontrol direction, back [10] and lower-back [11, 12] will distort the device orientation and the produced signals as mostly the elderly body shakes. Also, carrying multiple sensors by elderly humans is disturbing and cumbersome, and will end up causing incompatibility to their living standards. Waist and ankle are the most common positions that are usually always in contact with some other materials such as belts and socks respectively, and also, it's easy to attach the device on those positions without disturbing the wearer as it will not be in contact with their body. Besides, the waist position can capture the major body motion but may impede daily living activities. While the ankle is not, it won't be able to capture the body motion.

Therefore, this research aims to collect a new dataset from elderly peoples in a rural area and to investigate and determine the best sensor position among ankle and waist by subjecting the newly collected datasets to different machine learning models. This research work is divided into subsections; Section II of this paper provides an overview of the HAR model architecture. Section III explains some recent HAR research literature, publicly available HAR datasets, their strength, and weaknesses. Section IV highlight the materials and methods used. Section V gives out the results and conclusion.

1.1 Overview of HAR

The sensing approaches for HAR can be classified into two; a sensor-based and a vision base [13–17]. Sensor-based human activity recognition depends on the received signals from the sensor worn by the target audience, triaxial accelerometer and gyroscope are majorly used in this category, these sensors generate data along three axes, *x*, *y*, and *z*-axis. While vision-based activity recognition involves using cameras to capture images and videos as input to the activity recognition models. However, many researchers [16, 18, 19] have made it clear that the sensor-based is more appropriate, due to its availability and easy affordability, and also the privacy of the participants is well respected, unlike in vision-based systems, where the participant is video recorded intentionally or unintentionally. Besides, the use of cameras in vision-based is affected by angle, location, obstruction and some other environmental factors, such as light, non-target audience, etc.

For ordinary people recognizing human activities appears to be simple without complications or confusion, but it involves a series of complicated tasks of sensing, learning, and inference. The process of wearable sensor-based activity recognition starts by collecting the raw sensor data from the wearable devices, that are embedded with a variety of sensors before the other data analytic phases. In the process of data collection, the sensors or wearable devices are placed in any part of the body chosen by the subject or the researcher, and the sensor will then start sending signals of the active or current activity engaged. The data is recorded with various signal patterns based on different activities. For example, the static activity signal is different from dynamic and transitional activity signals and vise vasa.

Typically, the process of creating a HAR system is divided into five (5) steps: 1). *Data Collection*: the data is collected from the sensors attached to the participant and the rate at which the data will be collected is specified at this stage. 2). *Data Pre-processing*: this is a very important step and will lead to terrible results if not handled properly. In this step, the whole data collected by the single or multiple sensors must be merged and labeled to correspond to an activity. Checking for missing values, resampling, noise reduction, balancing of the dataset are all to be done in this step. 3). *Signal Segmentation*: the sensor signals received are segmented into groups called windows, instead of treating the data by individual samples. 4). *Relevant Feature Extraction and Selection*: a lot of relevant features are extracted and selected from the windows of step three. 5). *Activity Classification*: this is the last step; the extracted features are then used to train the classifiers or algorithms to be able to classify which activity is by mapping the features and identify the corresponding activity class. Figure 1 shows a complete architecture of human activity monitoring from start to the end of the process. Among all the steps of activity signal segmentation recognition and relevant feature vector extraction are the most critical and important stages of all. However, the quality of the relevant feature vector extracted depends on the excellent signal segmentation which will lead to better performance and generalization of the model used or is to be used.

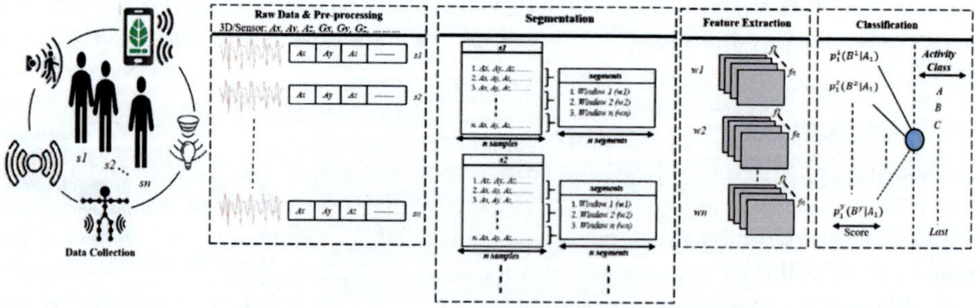

Fig. 1. Human activity recognition architecture

2 Related Works

Many kinds of research have been conducted in the field of human activity recognition in general, but very few researchers considered looking at recognizing physical activities using the data collected from elderly peoples. Meanwhile, there are many publicly available sensor-based human activities recognition datasets, that are been used to test most of the developed models for human activity recognition. Some of the most commonly used and publicly available datasets are summarized below, and some of their shortcomings are highlighted.

- *WISDM Dataset* [20] is a publicly available dataset collected in a laboratory setting from fifty-one (51) different subjects using two sensors, accelerometer, and gyroscope from smartwatch and smartphone. The subjects performed eighteen (18) different diverse daily living activities repeatedly (specifically, X3). The activities performed include basic activities (such as jogging, walking, climbing upstairs, etc.), eating activities (such as eating soup, eating pasta, eating chips, etc.), hand-base activities (such as folding cloth, brushing teeth, etc.). However, In the documentation of this dataset, no age range was specified, and no transitional activity was recorded.
- *UCI HAR and HAR Datasets* [21] In these datasets, the data was collected from thirty (30) subjects with ages ranging from 19 to 48 years. Only two different kinds of activities namely; static (sitting, standing and lying) and dynamic (walking, walking-upstairs, walking downstairs) was recorded, in total six different activities were performed by the subjects. Accelerometer and gyroscope sensors were used in the cause of the data collection. In this dataset, not even one, elderly person was selected/volunteered among the subjects and no transitional activity was recorded.
- *OPPORTUNITY Dataset*: Twelve (12) subjects were asked to perform up to seventeen (17) kitchen-related activities. The subjects have access to a fridge, dishwasher, etc. and can perform the activities repeatedly, and without restrictions. Also, multiple sensors were attached to both the subjects and the objects they interact with. However, no age range was specified in the dataset documentation, and no transitional activity was recorded.
- *PAMAP2 Dataset* this dataset contains eighteen (18) different physical activities involving static and some complex dynamic activities. The data were collected in a laboratory setting from nine (9) different subjects aged 27.22 ± 3.31 years, wearing three multiple inertial sensors and heart rate monitors. In this dataset, transitional activities are ignored, and the subject's age range couldn't seem to cause any challenge with the dataset.
- *USCHAD Dataset* in this dataset fourteen (14) subjects with age range 21–49 have participated in the process. The subjects perform some physical activities that include both static and dynamic activities. However, dynamic activities constitute most of the percentage of the dataset with zero transitional activity recorded.
- *mHEALTH Dataset:* In the process of making this dataset, ten (10) subjects were volunteered to perform several physical activities with sensors attached to a different part of their bodies, which include the right wrist, left ankle, and chest.

In summary, all the above mentioned most commonly used and publicly available datasets for sensor-based human activity recognition didn't put elderly peoples into consideration talk less of their geographic locations, that might probably affect there body posture as a result of their nutritional diets, life style and culture. Therefore, there is a need for a benchmark dataset collected from elderly peoples, especially from rural areas.

In previous years, some research focuses on investigating the best sensor positions. [9] tried to address two issues related to the sensor placement, that is, optimal accelerometer sensor best position with other activities, and feature relevance of the received signals, by using multiple sensors up to seven attached at different positions of the human body. In this regard, the number of sensors is cumbersome to carry by an elderly individual, the features generated are much and will make the system computationally expensive.

[22] in their research aimed at identifying the best sensor position amongst the wrist and ankle. A quite large number of subjects were used (up to thirty-three) for that purpose, and each subject was asked to perform twenty-six different daily activities with the sensors attached at four different parts of the body ankle, thigh, hip upper arm, and wrist. However, the data collection process is instruction dependent, the participants are instructed to perform the intended daily activities, not in a natural setting. Besides, the participants are adults not elderly and are wise enough to perform the activities as instructed. From the results obtained, it has shown that the sensor attached to the wrist position is the best.

In [23], research challenges concerning automated elderly activity monitoring using wearable sensors were perspicuously highlighted that opened doors for researchers in that aspect. Six wireless triaxial accelerometers were used by [24] to investigate the optimal sensor position, by fusion the sensor data collected from only eight (8) samples. The positions included in this research are; chest, thigh, foot, lower back, hip, and wrist. The activities included in this research are only static and dynamic activities, and the transitional activities were neglected. From the raw sense data, 11 features were extracted from each window and used. Among the six different algorithms (Decision tree (J48), Neural Network (NN), Naïve Bayes (NB) and Support Vector Machine (SVM)) used by the author, SVM stands out to be the best in yielding a better accuracy. Based on the result obtained by the authors from the few classifiers, the hip position is the best sensor position.

Transitional activities are usually abandoned due to their shorter duration [4], which in turn causes a lot of misclassification. [25, 26] in their research, they considered only static and dynamic activities and left aside transitional activities on the data collected internally and other external. In the internal aspect, the author uses the accelerometer, gyroscope, gravity with no specification of the positions where the sensors were attached, accelerometer, gravity, linear acceleration, magnetometer were utilized for the external data collection.

A systematic human activity recognition was proposed and implemented by [13] to recognizes basic activities and transitional activities using SBHAR public dataset. The dataset contains 30 samples collected from individuals between the ages range of 18 to 60 years all-inclusive. Different kinds of basic activities and falls are included that constitute a huge amount of data. Also, the author has proposed a good model for the classification of these basic and transitional activities. However, the dataset used includes peoples with a young age that are not having any difficulty in performing any basic or complex activity. The newly proposed model should be subjected to a purely elderly dataset and see its behavior/performance.

Physical human activity recognition research targeting elderly peoples was conducted by [25], the data was collected from aged human beings using a smartphone. This research is limited to recognizing only static and dynamic activities. The transitional activities are ignored. However, during the data collection process, the authors make use of three (3) different sensors to collect the data; the sensors include; Accelerometer, Gyroscope, and Magnetometer.

An activity monitoring and classification were proposed by [19, 27] for the detection of Activities of Daily Livings (ADLs) of an elderly person using wrist-worn sensors that are small, low-cost, non-intrusive and non-stigmatize. However, the dataset collected is limited to static and dynamic activity, in which no transitional activity was considered. In this research, multiple sensors were used in the data collection process, and only twelve (12) volunteers were considered between the age of 65 and 78 years.

3 Materials and Methods

In this section, we explain the methodology used in this research, right from the data collection up to the activity classification stage.

3.1 Dataset Collection

For this research, data were collected using inertial sensors (accelerometer and gyroscope) from fifteen (15) elderly peoples, and validated using the recorded videos during annotation. The inertial sensor is configured to generate samples at 50 Hz. The mean age of the subjects is 64.8 with a standard deviation of 8.79. Each subject wore two inertial sensors, one on the right waist and one on the right ankle. The subjects were asked to perform a set of activities (walking, standing, sitting, lying down and the transitions between two postures) in their preferred style and pace as shown in Table 1. No specific instructions were given about how to perform the activities. The activities were performed continuously for a single trial in the elderly house. Written informed consent was obtained before data collection in accordance with the approval by the human research ethics committee of Universiti Sains Malaysia (USM/JEPeM/18040205).

3.2 Pre-processing and Signal Segmentation

The sensor data collected from the elderly participants look noisy, so preprocessing is necessary. The data was pre-processed using different python programming language libraries to help in improving the classification accuracy, a median filter was used to remove the noise in the received sensor data. Missing values in the data were checked and then the data were normalized, to scale the raw sensor data to be between 0 and 1 to improve the classification model's accuracy.

Table 1. Activities involved

Activity name	Activity category		
	Static	*Dynamic*	*Transition*
• Walking		✓	
• Standing	✓		
• Stand-to-Sit			✓
• Sitting	✓		
• Sit-to-lie			✓
• Lying down (face-up)	✓		
• Lie-to-Sit			✓
• Sit-to-Stand			✓

Signal segmentation is one of the most important steps in human activity recognition. After the pre-processing stage, the represented signals are then segmented into different sizes of samples called windows. In this research, the window size used is one (1) second at a hundred (100) samples per window. The sliding window was used to segment the signals. To reduces the risk of missing the transitional activities due to their short time duration and correctly classify them, we use 50% windows overlap. In the process of segmenting the data, each window is ensured to contain a hundred samples. In the case of where the remaining samples of the current activity are not up to the windows' sample size, we complete it by fetching some part of the next activity signal data. That means the last window at the end of any activity might probably contain some activity data of the next activity next to it.

3.3 Feature Extraction and Selection

After the signals are segmented into a specific sliding window size; relevant features are extracted from each window that represents the number of samples. Some of the most common statistical time-domain features used include; average, standard deviation, minimum, maximum, skewness, the mean absolute deviation, etc. these features are extracted to form a representative feature vector [28]. Table 2 summarizes the features extracted and their formulas.

The relevant features vector was extracted from the three axes, *x, y, and z-axis*. For instance, the average feature we extracted three-dimensional accelerometer data; *x-mean, y-mean*, and *z-mean*, likewise for standard deviation*; x-std, y-std,* and *z-std,* and so on. Twenty-four (24) features were extracted and then relevant very important features were selected using wrapper-based feature selection methods.

Table 2. Summary of features extracted

S/No.	Time domain feature	Formula
1	Average	$\frac{1}{N}\sum_{i=1}^{N} x_i$
2	Standard deviation	$\sqrt{\frac{1}{N}\sum_{i=1}^{N} x_i - \mu}$
3	Median	$\left\{\frac{(N+1)}{2}\right\}$
4	Variance	$\sigma^2 = \left(\frac{1}{N}\sum_{i=1}^{N} x_i\right) - \mu^2$
5	Maximum	$x_i : x_i \geq x_j, i \neq i \forall i, j \in n$
6	Minimum	$x_i : x_i \leq x_j, i \neq i \forall i, j \in n$
7	Skewness	$\frac{[(N-1)(N-2)]}{N}\sum_{j=1}^{N}\frac{\left(x_j-\overline{\overline{x}}\right)^3}{\sigma_x^3}$
8	Mean absolute deviation	$\frac{\sum_{i=1}^{n}\lvert x_i-\mu\rvert}{n}$

3.4 Activity Classification

After the whole process of preprocessing, signal segmentation, feature extraction, and selection, the data was divided into 80% training and 20% testing. The data collected from the first twelve (12) subjects constitute 80% of the training set, and the remaining 20% was gotten from the remaining three subjects. Four different machine learning classification algorithms have been utilized for the activity classification that includes; K-Nearest Neighbor (KNN), Decision Tree, Logistic Regression, and Naïve Bayes Classifier.

4 Results and Discussions

The performance result is classified into two parts, ankle and waist. Four different machine learning classification algorithms were used in classifying the activities based on the extracted features from the sensor signals. Figure 2 shows the confusion matrix of the four machine learning classifiers used on the data collected from the sensor attached at the waist position. Looking at the best result obtained from the sensor attached to the waist position; KNN with an accuracy of 97% still some misclassification happened. 29% of the lie-to-sit data were misclassified to sit-to-lie and 23% of the stand-to-sit was misclassified to walking. Figure 2a is a confusion matrix of the activities involved and their classification results. In the decision tree classifier, sit-to-stand has the highest misclassification percentage of 43%, which was misclassified as stand-to-sit, then followed by sit-to-lie with 35% that was misclassified as the opposite of the activity i.e. lie-to-sit, and so on.

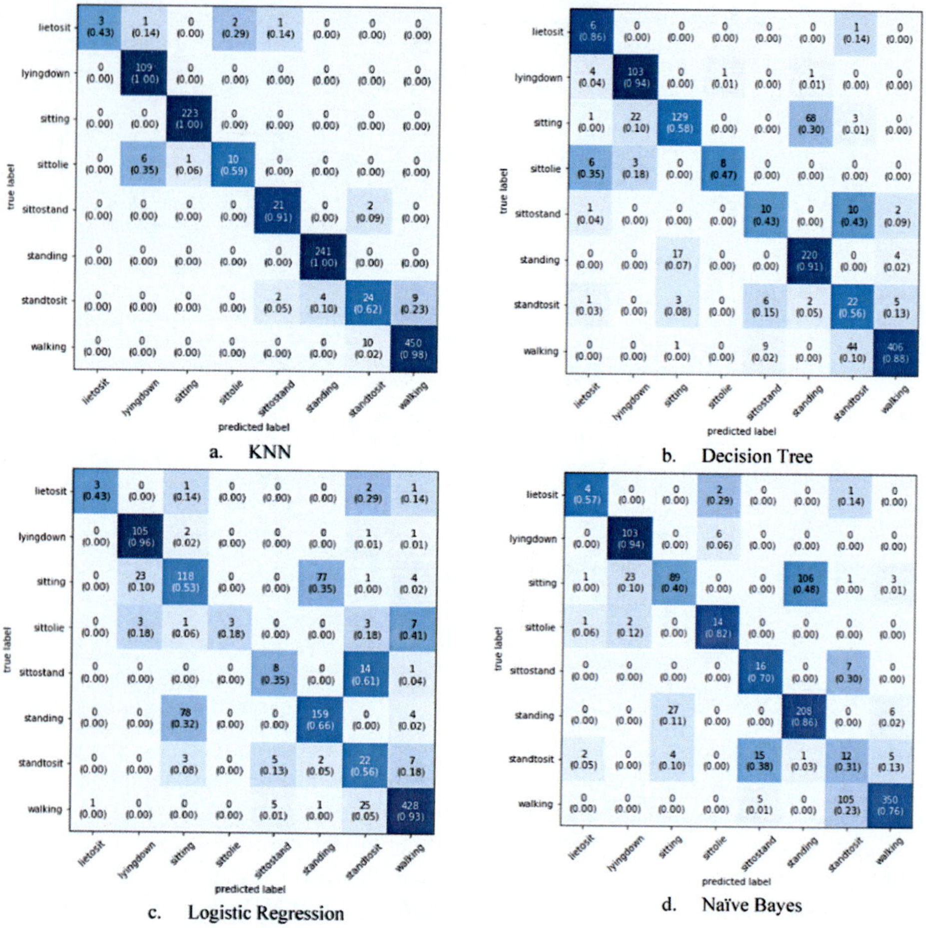

Fig. 2. Confusion Matrix of the four classifies used on sensor data attached at a waist position

The second part of the result is from the sensor data attached to the ankle position. Figure 3 shows the confusion matrix of the result obtained from the classifiers used on the data obtained from the sensor attached at the ankle position. Misclassifications also happened especially on the transitional activities after testing the data obtained from the sensor attached to the ankle position. In this case, some activities were almost 100% misclassified. All this misclassification probably happened because data from the ankle does not capture the major motion of the body. That is why HAR accuracy is significantly lower. E.g. for standing and sitting, the signal patterns might be the same. Similarly, for stand-to-sit and sit-to-stand, the pattern would be the same since the motion could not be captured by the sensor is attached to the ankle.

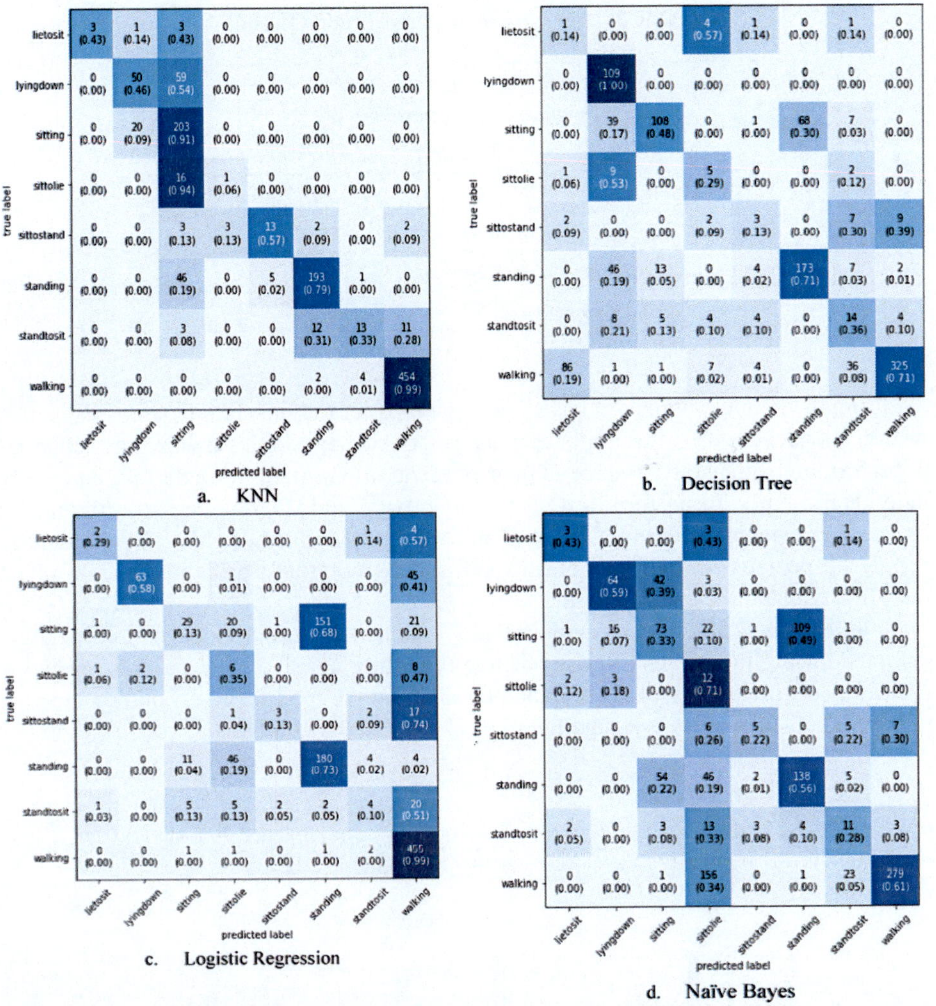

Fig. 3. Confusion Matrix of the four classifies used on sensor data attached at an ankle position

The performance result is classified into two parts, ankle and waist. Table 3 shows the machine learning algorithms/classifiers used and their classification accuracy when the sensor is placed in the two different parts of the participating agent's body.

Table 3. Summary of the Classification Results

Classifiers	Accuracy at different sensor positions (%)	
	Ankle	*Waist*
K-Nearest Neighbor (KNN)	**83**	**97.00**
Decision Tree	66	81
Logistic Regression	66	**75**
Naïve Bayes Classifier	52	71

From the results obtained it has shown that most of the misclassifications happened on transitional activities, this might be as a result of elderly peoples are having challenges in performing some activities due to their physical impairment. It might take an elderly individual a while to perform a transitional activity, and it's not easy to differentiate between the signal patterns of most of the activities performed. Most of the elderly peoples, especially in the rural areas cannot stand erect, that was caused as a result of their lifestyle and the environment.

Figures 4 and 5 summarizes the results obtained from the recall or true positive computed using the values of false positive (FP), true positive (TP) and false-negative (FN). Recall (true positive) is computed by dividing the true positive value by the sum of the true positive and false negative.

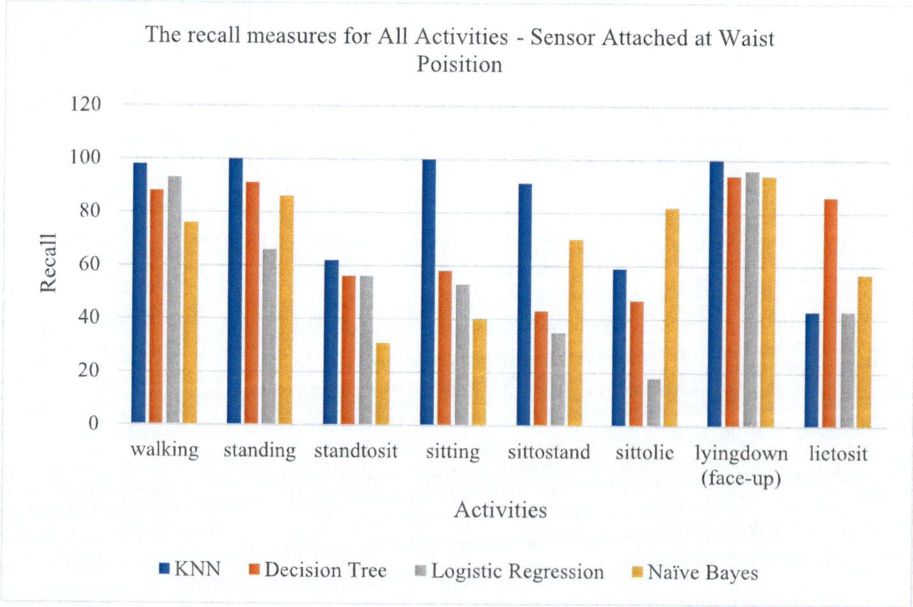

Fig. 4. Classification accuracy of all activities - sensor attached at waist position

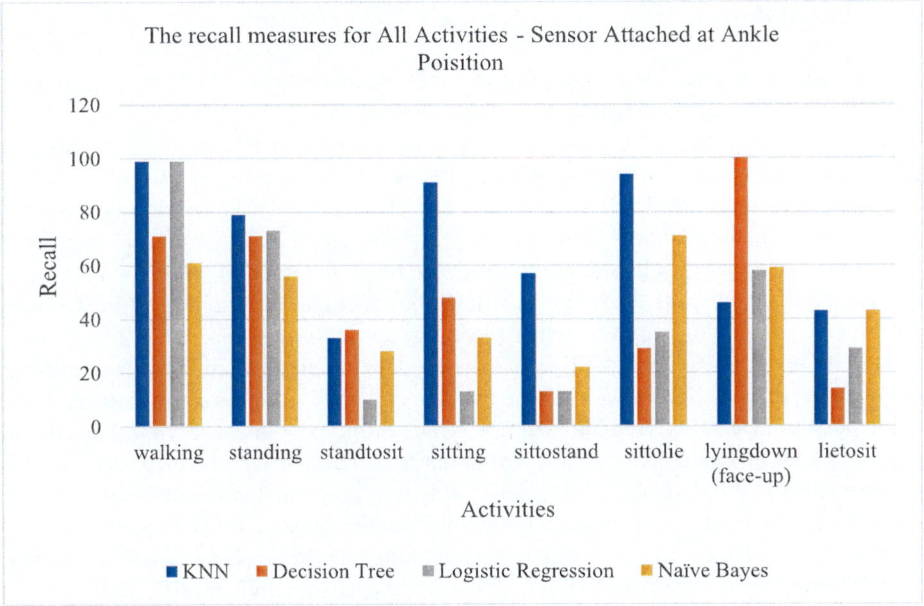

Fig. 5. Classification accuracy of all activities – sensor attached at ankle position

In summary, in our attempt to find the best sensor position among waist and ankle positions we collected our new datasets from elderly peoples in rural areas. In the data collected from fifteen elderly individuals; the data was first pre-processed, a median filter was used to remove the noise in the received sensor data. Missing values in the data were checked and then the data were normalized, to scale the raw sensor data to be between 0 and 1 to improve the classification model's accuracy. The pre-processed sensor data was segmented at 100 samples per second using a sliding window with 50% overlap. Relevant features were extracted and selected from each window. The selected features were used to train and test the four selected classifies in both cases (sensor attached at the waist and ankle positions). The results obtained from the sensor data attached at the waist position using the KNN classifier is 97% accurate, with some little misclassifications at when recognizing transitional activities due to the inability of the elderly peoples to perform some activities correctly. 83% accuracy was achieved using the same algorithm KNN at the sensor data attached at an ankle position, which is 14% less accurate than when attached at the waist position. Decision Tree algorithm was the second-best classifier in showing a promising result with 81% accuracy from the sensor attached at waist position and 15% less accurate when the sensor attached at an ankle position. The remaining two tested classifiers resulted in less than 80% accuracy in both cases. In all the cases, the tested classifiers show a higher level of accuracy from the sensor attached at waist position, than the data obtained from the sensor attached at an ankle position. Based on the findings sensor attached at the waist position produces better data for activity classification among elderly individuals in rural areas.

5 Conclusion

Human activity monitoring or recognition is dependent on the kind of sensor signal data collected and the relevant feature vector extracted from the correctly segmented data. Sensor signal data collected from the elderly peoples especially in the rural areas are quite challenging due to its noisy nature, that is caused as a result of the inability of the elderly peoples to perform some certain activities correctly. That inability makes the attached sensor automatically change the orientation, and make the signals look different and in the end cause misclassification. From the result obtained, it has shown that when the sensor is attached at the waist position, the machine learning algorithms show a remarkable result, compared to when placed at an ankle position. KNN has shown a promising result (97% accuracy), tested on the data collected from the sensor attached in a waist position and 83% accuracy on the data collected from the sensor attached in an ankle position. Decision Tree algorithm was the second-best in classifying the activities correctly with 81% and 66% accuracy from the sensor attached at the waist and an ankle positions, while the remaining classifies (Logistic regression and Naïve Bayes) have an accuracy less than 80% in both data obtained from the sensors attached at an ankle and waist positions. Based on our findings, we conclude that the sensor attached at the waist position provides better/correct signals for activity classification among elderly peoples in rural areas. That can be as a result of the ability of waist position to capture body acceleration better than the ankle position.

Acknowledgment. This work has been was supported in part by the Universiti Sains Malaysia under Short Term Grant 304/PKOMP/6315206.

References

1. UN DESA's Population Division, World Population Ageing 2019 (2019)
2. Woznowski, P.R., King, R., Harwin, W., Craddock, I.: A human activity Recognition framework for healthcare applications: Ontology, labelling strategies, and best practice. In: IoTBD 2016 – Proceedings of International Conference Internet Things Big Data, no. IoTBD, pp. 369–377 (2016)
3. Wang, Z., Meng, F., Yuan, G., Yan, Q., Xia, S.: An overview of human activity recognition based on smartphone. Sensor Rev. (2018)
4. Noor, M.H.M., Salcic, Z., Wang, K.I.K.: Adaptive sliding window segmentation for physical activity recognition using a single tri-axial accelerometer. Pervasive Mob. Comput. **38**, 41–59 (2017)
5. Wang, X., Kim, H.: Detecting user activities with the accelerometer on android smartphones. J. Multimed. Inf. Syst. **2**(2), 233–240 (2015)
6. Kwapisz, J.R., Weiss, G.M., Moore, S.A.: Activity recognition using cell phone accelerometers. ACM SIGKDD Explor. Newsl. **12**(2), 74 (2011)
7. Incel, O.D.: Analysis of movement, orientation and rotation-based sensing for phone placement recognition. Sensors **15**(10), 25474–25506 (2015)
8. Lo, B., Atallah, L., Aziz, O., El Elhew, M., Darzi, A., Yang, G.Z.: Real-time pervasive monitoring for postoperative care. IFMBE Proc. **13**, 122–127 (2007)
9. Atallah, L., Lo, B., King, R., Yang, G.Z.: Sensor positioning for activity recognition using wearable accelerometers. IEEE Trans. Biomed. Circuits Syst. **5**(4), 320–329 (2011)

10. Lee, S.H., Park, H.D., Hong, S.Y., Lee, K.J., Kim, Y.H.: A study on the activity classification using a triaxial accelerometer. In: Annual International Conference of the IEEE Engineering in Medicine and Biology Society - Proceedings, August, vol. 3, pp. 2941–2943 (2003)
11. Bouten, C.V.C., Koekkoek, K.T.M., Verduin, M., Kodde, R., Janssen, J.D.: A triaxial accelerometer and portable data processing unit for the assessment of daily physical activity. IEEE Trans. Biomed. Eng. **44**(3), 136–147 (1997)
12. Bonomi, A.G., Goris, A.H.C., Yin, B., Westerterp, K.R.: Detection of type, duration, and intensity of physical activity using an accelerometer. Med. Sci. Sports Exerc. **41**(9), 1770–1777 (2009)
13. Li, J.H., Tian, L., Wang, H., An, Y., Wang, K., Yu, L.: Segmentation and recognition of basic and transitional activities for continuous physical human activity. IEEE Access **7**, 42565–42576 (2019)
14. Lara, Ó.D., Labrador, M.A.: A survey on human activity recognition using wearable sensors. IEEE Commun. Surv. Tutorials **15**(3), 1192–1209 (2013)
15. Wang, Y., et al.: Smartphone-based human activity recognition. In: Device-to-Device based Proximity Service, pp. 199–239 (2017)
16. Ronao, C.A., Cho, S.B.: Human activity recognition with smartphone sensors using deep learning neural networks. Expert Syst. Appl. **59**, 235–244 (2016)
17. Subetha, T., Chitrakala, S.: A survey on human activity recognition from videos. In: 2016 International Conference on Information Communication Embeded System, ICICES 2016, ICICES (2016)
18. Gani, M.O.: A novel approach to complex human activity recognition. ProQuest dissertation theses, p. 155 (2017)
19. Nweke, H.F., Teh, Y.W., Al-garadi, M.A., Alo, U.R.: Deep learning algorithms for human activity recognition using mobile and wearable sensor networks: state of the art and research challenges. Expert Syst. Appl. **105**, 233–261 (2018)
20. Weiss, G.M.: WISDM Smartphone and Smartwatch Activity and Biometrics Dataset
21. Anguita, D., Ghio, A., Oneto, L., Parra, X., Reyes-Ortiz, J.L.: A public domain dataset for human activity recognition using smartphones. In: ESANN 2013 Proceedings, 21st European Symposium on Artificial Neural Networks, Computational Intelligence and Machine Learning, April, pp. 437–442 (2013)
22. Mannini, A., Intille, S.S., Rosenberger, M., Sabatini, A.M., Haskell, W.: Activity recognition using a single accelerometer placed at the wrist or ankle. Med. Sci. Sports Exerc. **45**(11), 2193–2203 (2013)
23. Ranasinghe, D.C., Torres, R.L.S., Wickramasinghe, A.: Automated activity recognition and monitoring of elderly using wireless sensors: research challenges. In: Proceedings of the 2013 5th IEEE International Working Advances Sensors Interfaces, IWASI 2013, pp. 224–227 (2013)
24. Cleland, I., et al.: Optimal placement of accelerometers for the detection of everyday activities. Sensors (Basel) **13**(7), 9183–9200 (2013)
25. Voicu, R.A., Dobre, C., Bajenaru, L., Ciobanu, R.I.: Human physical activity recognition using smartphone sensorsH. Sensors (Switzerland) **19**(3), 1–18 (2019)
26. Voicu, R.A., Dobre, C., Bajenaru, L., Ciobanu, R.I.: Human physical activity recognition using smartphone sensors. Sensors (Switzerland) **19**(3), 458 (2019)
27. Chernbumroong, S., Cang, S., Atkins, A., Yu, H.: Elderly activities recognition and classification for applications in assisted living. Expert Syst. Appl. **40**(5), 1662–1674 (2013)
28. Reyes-Ortiz, J.L., Oneto, L., Samà, A., Parra, X., Anguita, D.: Transition-aware human activity recognition using smartphones. Neurocomputing **171**, 754–767 (2016)

A Holistic Approach for Enhancing Critical Infrastructure Protection: Research Agenda

Livinus Obiora Nweke[1](✉) and Stephen D. Wolthusen[1,2]

[1] Information Security and Communication Technology, Norwegian University of Science and Technology (NTNU), Gjøvik, Norway
livinus.nweke@ntnu.no

[2] School of Mathematics and Information Security, Royal Holloway, University of London, Egham, UK

Abstract. Critical infrastructure is an asset or a system that is essential for the maintenance of vital societal functions. The protection of such an infrastructure requires more than a technical understanding of the underlying issues; it also needs an understanding of the organizational aspects. Although there are several standards and guidelines for the protection of critical infrastructure, they are usually vague and do not offer practical solutions. In this paper, we describe a 'work in progress' holistic approach for enhancing critical infrastructure protection. First, we introduce the theoretical background of this study. Then, based on this theoretical foundation, we propose a holistic approach which takes into account both organisational and technical measures. In addition, we provide a synopsis of our research outcomes so far and our ongoing work towards enhancing critical infrastructure protection.

Keywords: Critical infrastructure protection · NIS directive · Holistic approach

1 Introduction

There has been an increasing capability and intent to attack critical infrastructure. This is apparent from the growing number of reported attacks against critical infrastructure in the last few years. For example, it was recently reported that a ransomware attack was targeted at a critical infrastructure belonging to a United States (US) based natural-gas compression facility [13]. The situation is further exacerbated by state actors that have enormous resources for staging similar or more complex attacks. All these calls for rethinking about how we protect critical infrastructure considering the devastating effect a successful attack would have on the society.

Many efforts have been made around the world to address critical infrastructure protection (CIP). According to the European Commission, CIP can be defined as "the ability to prepare for, protect against, mitigate, respond to, and recover from critical infrastructure disruptions or destruction" [4]. To support efforts in that direction, the Commission launched the European Programme for Critical Infrastructure Protection (EPCIP) [5]. Similar efforts have been made in North America with the North American

J. H. Abawajy et al. (Eds.): EATI 2020, LNNS 254, pp. 220–228, 2021.
https://doi.org/10.1007/978-3-030-80216-5_16

Electric Reliability Corporation Critical Infrastructure Protection (NERC-CIP) program [16]. Even though the NERC-CIP mainly focuses on electric power, it demonstrates the importance of having a framework for the protection of critical infrastructure. Further, there is a consensus among all these efforts about the need for a holistic approach to enhancing CIP.

Several works in the literature have proposed a holistic approach for enhancing CIP [11, 12, 27]. The authors in [27] evaluate the risk assessment phase of critical infrastructure and propose an enhancement to the traditional risk management methods. They employ the cybernetic construct of the viable system model towards a holistic view of the risks against critical infrastructure. The holistic approach in [12] is concerned with critical infrastructure dependencies and draws on the results of a survey of critical infrastructure experts from several countries. Also, a holistic framework for building critical infrastructure resilience is presented in [11]. It consists of three parts: a set of resilience policies; an influence table that assesses the impact of policies on prevention, absorption and recovery stages; and an implementation methodology that defines the temporal order in which the policies should be implemented.

Unlike the works in the preceding paragraph, this study proposes a holistic approach for enhancing CIP which examines both organisational and technical measures. First, we introduce the theoretical background of this study. Then, based on this theoretical foundation, a holistic approach which consists of three parts is proposed. The first part considers policies, processes and procedures relating to CIP; the second part investigates the vulnerabilities, threats and attacks that the critical infrastructure may be exposed to; while the third part deals with prevention, detection and mitigation. Lastly, we provide a synopsis of our research outcomes so far and our ongoing work towards enhancing CIP.

2 Background

Critical infrastructure is an asset or a system that is essential for the maintenance of vital societal functions. The protection of such an infrastructure has become a major concern for countries around the world. This is because any successful attack against a critical infrastructure would have a devastating effect on the society. For example, the successful attack against a power transmission station, north of the city of Kiev, Ukraine, blacked out a portion of the Ukrainian capital equivalent to a fifth of its total power capacity [9]. Consequently, to protect the individual Member States and the Union as a whole against similar occurrence, the European Union (EU) enacted the Network and Information Systems (NIS) Directive [7]. Even though we use the NIS Directive in this study as the theoretical foundation, a similar approach has been introduced in the US with the creation of NIST (National Institute of Standards and Technology) framework for improving critical infrastructure cybersecurity [15].

The NIS Directive is the first EU-wide legislation on cybersecurity that came into effect November 2018. It was designed to facilitate a high common level of protection for network and information systems across the EU's critical infrastructure. The NIS Directive aims to achieve this by focusing on three top-level objectives: enhanced cybersecurity capabilities across the EU Member States, increased level of cooperation between the EU Member States and supervision of critical sectors. Also, there are two types of organizations that the NIS Directive is applicable to. They are the digital service providers and the operators of essential services (operators of critical infrastructure). However, this study focuses on the requirements of NIS Directive that applies to the operators of critical infrastructure.

Critical infrastructure operators are obliged by the NIS Directive to ensure the protection of their network and information systems. The NIS Directive clearly states that the "responsibilities in ensuring the security of network and information systems lie, to a great extent, with operators of essential services". It went further to encourage a culture of risk management, which involves risk assessment and the implementation of appropriate security measures [7]. These measures are to include measures to identify any risks of incidents through the understanding of vulnerabilities, threats and attacks, and countermeasures to prevent, detect and mitigate their impact.

Furthermore, the NIS Directive observes that the "technical and organisational measures imposed on the operators of essential services and digital service providers should not require a particular commercial information and communications technology product to be designed, developed or manufactured in a particular manner". This implies that critical infrastructure operators have the liberty to adopt any type of technical and organisational measures towards enhancing the protection of critical infrastructure. Although hardware manufacturers and software developers are not classified as critical infrastructure by the NIS Directive, it notes that their products enhance the protection of network and information systems within the critical infrastructure. Thus, they have an essential role in supporting critical infrastructure operators in order to protect their network and information systems.

Based on the above theoretical background, we propose a holistic approach for enhancing CIP in Sect. 3. This holistic approach draws from our experience in our ongoing research towards enhancing CIP. Similar to the suggestions made in the NIS Directive, we note that a holistic approach to enhancing CIP should not only include technical measures, but also, should consider organisational measures which cover policies, processes and procedures. In addition, we provide a synopsis of our research outcomes so far and our ongoing work towards enhancing CIP in Sect. 4.

3 The Holistic Approach

In this study, we propose a holistic approach for enhancing CIP Which consists of three parts. The first part considers policies, processes and procedures relating to CIP. In the second part, the concern is to understand the vulnerabilities, threats and attacks that the critical infrastructure may be exposed to. And the third part of the holistic approach deals with prevention, detection and mitigation of the vulnerabilities, threats and attacks that have been identified in the second part of the holistic approach. Whilst the first part deals with the organisational measures of CIP, the second and the third parts are mainly concerned with the technical measures related to CIP. This holistic approach can be viewed as a triangular framework as shown in Fig. 1.

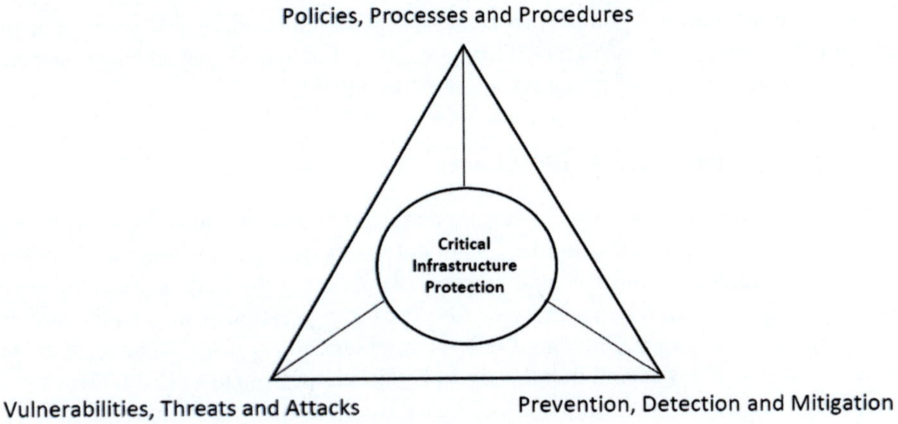

Fig. 1. The holistic approach

3.1 Policies, Processes and Procedures

Policies refer to rules, laws, regulations, or set of guidelines. They provide guidelines for the implementation of processes and procedures. With regards to enhancing CIP, we can define policies as rules, laws, regulations, or set of guidelines that are related to CIP, which are established to guide the activities of critical infrastructure operators. Considering the importance of critical infrastructure, these policies are usually enacted at the national and international levels. For example, in the EU, either directive or regulation could be used. When a directive is issued, the individual Member States are required to enact laws using the directive as the minimum standard and in the case of a regulation, it is usually binding for all Member States [6].

Processes can be defined as steps that must be taken to ensure compliance with a policy. An understanding of processes is an essential prerequisite for complying with the defined policies. Processes usually include specifying who is responsible for executing a task, what task should be performed, and when that task should be performed. For instance, given a critical infrastructure operator like the healthcare providers, processes are high set of things defined by the providers in order to ensure compliance with policies defined at the national and international levels.

Procedures are specific, detailed series of actions that must be performed and are geared towards implementing a process and complying with a policy. In the context of CIP, procedures are related to individuals involved in the actual operations within the critical infrastructure. They are the practical application of the processes and policies defined at the national or international levels, and management level of the critical infrastructure operations respectively. For example, in the healthcare sector, an understanding of procedures for enhancing CIP may require the modelling and analysis of healthcare professionals' security practices.

In general, policies, processes and procedures are related to the activities of humans involved in the protection of critical infrastructure. There is a consensus among security practitioners that humans are usually the weakest link in the security chain. Hence, technical measures alone are not enough for enhancing CIP. It is also important that there is a clear understanding of policies, processes and procedures among all those responsible for protecting critical infrastructure. This will ensure that with strong technical measures in place, the organizational measures are also strengthened.

3.2 Vulnerabilities, Threats and Attacks

Vulnerabilities are weaknesses in a system that can be exploited to wreak havoc on the system. These vulnerabilities could exist not only in the technology being used, but also in the configurations and the implemented security policies [28]. In the critical infrastructure sector, several technologies are used to support its operations, and an understanding of the vulnerabilities in such technologies is an essential step for enhancing their protection. For example, vulnerability discovery models have been used in the literature for predicting future software vulnerabilities based on their historical behaviour [2, 10]. It is possible to use a similar approach to understand and to predict future vulnerabilities of systems deployed for CIP.

Threats can be viewed as events that are likely to cause harm to the confidentiality, integrity or availability (CIA model [17]) of systems, through unauthorized disclosure, misuse, alteration or destruction [14]. They can be malicious or accidental. Malicious threats are those actions that could exploit the vulnerabilities in a system to cause harm to the system while accidental threats are those unintended insider actions that could cause harm to the system. The process of understanding the likelihood of those threats being realized is called threat modelling. Threat modelling approaches have been employed for enhancing CIP [1, 3, 25]. However, we have observed that the regular threat modelling approaches are not able to adequately capture the threats to cyber-physical system (CPS) which is a major part of critical infrastructure, due to timing, uncertainty and dependencies that exist between its entities [20]. Therefore, there is a need for an appropriate threat modelling approach that takes into account the unique features of CPS.

Attacks are actions that leverage one or more vulnerabilities to realize a threat. They are usually intentional acts aimed at causing damage to a system. Like other information systems, attacks against critical infrastructure can be classified according to the CIA model. The most severe type of attack for critical infrastructure is the attack that affects availability. A critical infrastructure is required to be available all the time, even in the face of a natural disaster. This makes attacks against the availability of a critical infrastructure like the denial of service (DoS) attack as one of the most extreme types of attack that needs to be considered for enhancing CIP.

Undeniably, enhancing CIP requires a good understanding of vulnerabilities, threats and attacks. As one of the technical measures in our proposed holistic approach, this understanding will provide practitioners with a tool for predicting future vulnerabilities of systems deployed for CIP, better threat modelling approach to adequately capture the threats to critical infrastructure and a good insight into the possible attacks against critical infrastructure. All these will provide the necessary knowledge needed for the third part of our proposed holistic approach which involves prevention, detection and mitigation.

3.3 Prevention, Detection and Mitigation

Prevention refers to actions that could stop attacks against a system from happening. It requires an understanding of all possible attacks that could be targeted against a system and then implementing appropriate measures to prevent them from occurring in the first place. For example, in CIP, a possible approach for preventing attacks against availability may involve the use of redundant systems. Here, two or more systems may be used to provide the same service such that if one of the systems fails, another system continues to provide the same service without any interruptions.

Detection is the process of uncovering attacks against a system. The overall goal of detection mechanisms is to be able to identify attacks and to implement appropriate countermeasures for minimising the impact of those attacks. With regards to CIP, there is an increasing complexity in the types of reported attacks. Thus, the use of traditional detection mechanisms like the intrusion detection systems no longer suffices for the protection critical infrastructure. There is a need for detection mechanisms that combine more than one approach for enhancing CIP.

Mitigation is concerned with minimising the severity of attacks against a system. It requires an understanding of when the system is under attack. This implies that the mitigation process is activated after the detection process has occurred. The mitigation strategies that should be implemented for a system depend on the functionality and the services provided by the system. For critical infrastructure, service availability is usually a priority. It follows that the mitigation strategies for critical infrastructure should ensure the availability of service even amid an attack or a restoration of service as quickly as possible.

In summary, a good grasp of the prevention, detection and mitigation techniques is essential for enhancing CIP. The prevention process allows practitioners to implement the required technical measures for ensuring that attacks against critical infrastructure never occur. However, the attack landscape is constantly evolving, and this makes it impossible for practitioners to implement all the necessary measures for preventing future attacks. Hence, detection mechanism is required to uncover when those attacks occur. With the detection of the attacks, countermeasures are needed to minimise the impact of those attacks. These countermeasures are implemented using the mitigation process. Therefore, a holistic approach to enhancing CIP is one that employs appropriate technical measures in addition to robust organisational measures which cover policies, processes and procedures implemented towards enhancing the protection of critical infrastructure.

4 Discussion

Already, we have observed that organisational measures are very essential for enhancing CIP. To this end, we studied the legal issues related to cyber threat information (CTI) sharing among private entities for CIP in [19]. This is because CTI sharing has been proposed as an efficient and effective method for improving CIP. The work provides guidance and incentives for private entities willing to participate in CTI sharing, especially for CIP. Also, it has been observed in [8] that practising law involves anticipating how and what a judge might decide when presented with an issue, but ethics appears as a superior and stable reference to which laws can refer to. Thus, we presented in

[21] the ethical implications of security vulnerability research for CIP. The result of this study shows that a security researcher could rely on the three different normative ethical frameworks to reason about the best course of action during security vulnerability research for CIP.

Furthermore, other works until now that have considered organisational measures related to CIP are presented in [24, 29]. There has been an increasing deployment of attribute-based access control (ABAC) in the healthcare sector, which is one of the sectors classified as critical infrastructure. This motivates our paper in [24] where we examined the existing literature on the application of ABAC in the healthcare sector. The work can serve as a basis for selecting and further advancing the use of ABAC in e-health systems. We also conducted a systematic literature review of artificial intelligence strategies and their hybrid aspects in [29]. The study identified appropriate artificial intelligence strategies and their hybrid aspects which can be employed to efficiently detect anomaly and malicious events in healthcare staff's security practices using the access logs.

Technical measures are the core aspects of the holistic approach and as such, they were divided into two parts. Considering the part which examines vulnerabilities, threats and attacks, we proposed a vulnerability discovery modelling with vulnerability severity in [26]. As we have already noted from the NIS Directive, hardware and software products enhance the protection of critical infrastructure. Hence, the ability to predict the future vulnerabilities of software deployed in critical infrastructure will ensure that appropriate resources are allocated for their protection. It is possible to use our proposed approach for predicting the future vulnerabilities of software deployed in critical infrastructure. We also conducted a review of asset-centric threat modelling approaches in [20]. The study observed that the existing threat modelling approaches are not able to capture all the threats to a CPS, which is one of the main components of critical infrastructure. This is because of the uncertainty, timing and dependencies that exist between the entities of a CPS. The main objective of the review is to serve as a foundation for determining the appropriate asset-centric threat modelling approaches that could be employed for a given scenario.

Moreover, our research outcomes which can be categorized into the second part of the holistic approach are published in [18, 22, 23]. We remarked in [23] that software-defined networks (SDN) are widely being adopted and are likely to be deployed in critical infrastructure. However, there is a need to evaluate the security and performance guarantees that can be given for the data plane of such critical systems. To this end, we studied the existing literature on the analysis of SDN using queueing networks and proposed ways in which models need to be extended in order to study attacks. We then employed one of the proposed models to study the effect of adversarial ow in software-defined industrial control networks using a queueing network model in [22]. This work provides useful insights for benchmarking and facilitates the identification of factors that could cause the network to breach the stringent QoS requirements in software-defined industrial control networks. In addition, we noted in [18] that real-time communication protocols are among the most commonly used communication protocols in critical infrastructure and they are used to monitor, and control industrial automation processes deployed in critical infrastructure. Thus, we proposed an adversary model for attacks against these communication protocols taking into account their unique properties. This is to facilitate the understanding of how adversarial actions may influence the communication protocols deployed for CIP.

One of our ongoing work is to develop a legal compliance framework that will assist private entities within the critical infrastructure sector to share CTI among themselves for the purpose of improving their overall cyber intelligence and defence. Other ongoing work is the asset-centric threat modelling of CPS using a formal technique to address the limitations of the existing approaches identified in [20]. We are also investigating the prevention, detection and mitigation strategies (third part of the holistic approach) for the vulnerabilities, threats and attacks we have identified in the second part of the holistic approach.

5 Conclusion

Enhancing CIP requires a combination of both organisational and technical measures in order to be effective. In this paper, we have proposed a holistic approach for enhancing CIP which consists of three parts. The first part considers policies, processes and procedures relating to CIP; the second part investigates the vulnerabilities, threats and attacks that the critical infrastructure may be exposed to; while the third part deals with prevention, detection and mitigation. The holistic approach draws from the NIS Directive and our experience in the ongoing study towards enhancing CIP. Lastly, we have presented a synopsis of our research outcomes till date and our ongoing work which were grouped according to the three parts of the holistic approach.

References

1. Abomhara, M., Gerdes, M., Køien, G.M.: A stride-based threat model for telehealth systems. Norsk informasjonssikkerhetskonferanse (NISK) **8**(1), 82–96 (2015)
2. Alhazmi, O., Malaiya, Y.: Prediction capabilities of vulnerability discovery models, pp. 86–91. IEEE (2006). https://doi.org/10.1109/RAMS.2006.1677355
3. Atif, Y., et al.: Cyber-threat analysis for cyber-physical systems (2018)
4. European Commission: Green paper on a European programme for critical infrastructure protection. Technical report, European Commission (2005)
5. European Commission: Communication from the commission on a European programme for critical infrastructure protection. Technical report, European Commission (2006)
6. European Union: Regulations, directives, and other acts. https://europa.eu/european-union/eu-law/legal-acts
7. European Union: Directive (EU) 2016/1148 of the European parliament and of the council of 6 july 2016. Offic. J. Eur. Union (2016)
8. Fuster, G.G., Gutwirth, S.: Ethics, law and privacy: disentangling law from ethics in privacy discourse. In: Proceedings of the Technology and Engineering 2014 IEEE International Symposium on Ethics in Science, pp. 1–6, May 2014
9. Greenberg, A.: 'crash override': The malware that took down a power grid (2017). https://www.wired.com/story/crash-override-malware/
10. Joh, H., Kim, J., Malaiya, Y.K.: Vulnerability discovery modeling using weibull distribution, pp. 299–300. IEEE (2008). https://doi.org/10.1109/ISSRE.2008.32

11. Labaka, L., Hernantes, J., Sarriegi, J.M.: A holistic framework for building critical infrastructure resilience **103**, 21–33 (2016)
12. Laugé, A., Hernantes, J., Sarriegi, J.M.: Critical infrastructure dependencies: a holistic, dynamic and quantitative approach. Int. J. Crit. Infrastructure Prot. **8**, 16–23 (2015)
13. Mühlberg, B.: U.S. critical infrastructure victim of ransomware attack, March 2020. https://www.cpomagazine.com/cyber-security/u-s-critical-infrastructurevictim-of-ransomware-attack/
14. National Institute of Standards and Technology: Information security: Guide for conducting risk assessments, September 2012
15. National Institute of Standards and Technology: Framework for improving critical infrastructure cybersecurity. Technical report. National Institute of Standards and Technology (2014)
16. NERC: Critical infrastructure protection (CIP) standards. Technical report, North American Electric Reliability Corporation (2020)
17. Nweke, L.O.: Using the CIA and AAA models to explain cybersecurity activities. PM World J. **6** (2017)
18. Nweke, L.O., Weldehawaryat, G.K., Wolthusen, S.D.: Adversary model for attacks against IEC 61850 real-time communication protocols. In: 16th International Conference on the Design of Reliable Communication Networks DRCN, pp. 1–8. IEEE (2020)
19. Nweke, L.O., Wolthusen, S.: Legal issues related to cyber threat information sharing among private entities for critical infrastructure protection. In: NATO CCDCOE 12th International Conference on Cyber Conflict (2020)
20. Nweke, L.O., Wolthusen, S.: A review of asset-centric threat modelling approaches. Int. J. Adv. Comput. Sci. Appl. **11**(2), 1–6 (2020)
21. Nweke, L.O., Wolthusen, S.D.: Ethical implications of security vulnerability research for critical infrastructure protection. In: 15th International Conference on Wirtschaftsinformatik (2020)
22. Nweke, L.O., Wolthusen, S.D.: Modelling adversarial flow in software-defined industrial control networks using a queueing network model. In: IEEE Conference on Communications and Network Security (2020)
23. Nweke, L.O., Wolthusen, S.D.: Resilience analysis of software-defined networks using queueing networks. In: 2020 International Conference on Computing, Networking and Communications (ICNC), pp. 536–542. IEEE (2020)
24. Nweke, L.O., Yeng, P., Wolthusen, S.D., Yang, B.: Understanding attribute-based access control for modelling and analysing healthcare professionals' security practices. Int. J. Adv. Comput. Sci. Appl. **11**(2), 683–690 (2020). https://doi.org/10.14569/ijacsa.2020.0110286
25. Rekik, M., Gransart, C., Berbineau, M.: Cyber-physical threats and vulnerabilities analysis for train control and monitoring systems. In: Proceedings of Computers and Communications (ISNCC) 2018 International Symposium Networks, pp. 1–6, June 2018
26. Shukla, A., Katt, B., Nweke, L.O.: Vulnerability discovery modelling with vulnerability severity. In: 2019 IEEE Conference on Information and Communication Technology, pp. 1–6. IEEE (2019). https://doi.org/10.1109/CICT48419.2019.9066187
27. Spyridopoulos, T., Topa, I.-A., Tryfonas, T., Karyda, M.: A holistic approach for cyber assurance of critical infrastructure with the viable system model. In: Cuppens-Boulahia, N., Cuppens, F., Jajodia, S., Abou, A., Sans, T. (eds.) SEC 2014. IAICT, vol. 428, pp. 438–445. Springer, Heidelberg (2014). https://doi.org/10.1007/978-3-642-55415-5_37
28. Whitaker, A., Newman, D.P.: Penetration Testing and Network Defense. Cisco Press, Indianapolis (2006)
29. Yeng, P.K., Nweke, L.O., Woldaregay, A.Z., Yang, B., Snekkenes, E.A.: Data-driven and artificial intelligence (AI) approach for modelling and analyzing healthcare security practice: a systematic review. In: Intelligent Systems Conference (IntelliSys) 2020 (2020)

Cursory View of IoT-Forensic Readiness Framework Based on ISO/IEC 27043 Recommendations

Phathutshedzo P. Mudau[1], H. S. Venter[1], Victor R. Kebande[2], Richard A. Ikuesan[3(✉)], and Nickson M. Karie[4]

[1] University of Pretoria, Pretoria, South Africa
[2] Malmö Universitet, Malmö, Sweden
[3] Qatar Community College, Doha, Qatar
richard.ikuesan@ccq.edu.qa
[4] Edith Cowan University, Joondalup, Australia

Abstract. The Internet of things (IoT) technology has been adopted by many organisations as a way of improving business operations and discovering new business opportunities. However, the threat-related landscape and security ramifications that comes with this adoption has become a subject of focus. As a result, there is a need to significantly explore IoT readiness across organisations. The authors in this paper propose an IoT Forensic Readiness (IoT-FR) Framework for organisations that leverages on ISO/IEC 27043 International Standard. The IoT-FR framework benefits organisations that intend to adopt and implement IoT.

Keywords: IoT · Digital forensics · Readiness · ISO/IEC 27043

1 Introduction

The Internet of Things (IoT) wave has shifted the era of computing from traditional desktop to smart connected, ubiquitous and pervasive way of delivering tasks. Additionally, IoT paradigm has basically enabled Information and Communication Technology (ICT) to be embraced as a driver of computing technologies, which has made a number of organisations to adopt and prioritize technologies that are more focused in addressing data security and privacy [1–3]. It is imperative to say that ICT has also changed how businesses are being conducted globally and IoT has mainly been the game changer. IoT has been projected to be a significant driver that has shaped based on its elements of digital revolution. As a result, many organisations have adopted the IoT technology with the promise of improving their operations and creating diverse business models [4]. International Data Corporation (IDC) and Gartner reports have predicted that by the year 2020, more than 20 billion IoT devices stands to be connected and the estimated revenue is estimated to be trillions of dollars.

Even though the adoption of IoT presents organisations with many digital business possibilities, it also presents new avenues, challenges and complexities as a result of

J. H. Abawajy et al. (Eds.): EATI 2020, LNNS 254, pp. 229–239, 2021.
https://doi.org/10.1007/978-3-030-80216-5_17

connectivity. These complexities also serve as an entry point for cyber-attackers and other emerging security threats [5, 6]. As a way to filtering and mitigating the cyber-attacks and managing the and IoT challenges, it is important to employ proactive approaches, i.e. Digital Forensic Readiness (DFR) as a strategic action. DFR mainly [7, 8] ensures that an organisation has incidental planning and preparation in order to further a digital investigation.

The remainder of this paper is organised as follows: Sect. 2 discuss the Background and Related literature, while Sect. 3 discuss the IoT-FR framework. This is followed by discussions in Sect. 4, then a conclusion and a mention of future work in Sect. 5.

2 Background and Related Literature

2.1 Internet of Things

Internet of Things (IoT) is a term used to depict the inter-connectedness of devices and in some cases, people, usually over the Internet [5]. Architecturally, the IoT system comprises of sensors mounted on a wireless-enabled medium that transmit sensor signal to a networked location/platform. This could also be in the form of an ad-hoc scenario where series of interconnected devices form part of a larger broadband of network. As asserted in [9], such inter-connectivity presents a means for computers to understand the world around them and become humanity's nervous system. Further, [9] defined IoT as "a collection of sub-systems which collect, store, process, interpret, visualise data and respond based on those interpretations". Similarly, study in [10] opined that these "subsystems" have introduced an added layer of dynamism which has revolutionized the Internet into a network of connected objects capable of autonomous remote sensing and interaction with the physical environment. The sensing capabilities of these sub-systems have permeated the human day-to-day living. This is reflected in the deployment of IoT in virtually every aspect of the human daily life. As asserted in [5], the IoT-assisted living has been established within a diverse sphere of the human life ranging from health, education, industries to smart homes and inter-connected service delivery process (supply chain management). Whilst several research works and international standardisation are currently being proposed, and or developed, there is, yet, no universally accepted standard of the architecture of these sub-systems. For this paper, various literature sources which addresses several diverse components of the IoT architecture were consulted.

2.2 IoT Reference Model

A reference IoT architecture is defined as an architectural composition that can reliably provide a baseline for the development and evaluation of an IoT solution. The integration of a reference architectural model in evaluating the effectiveness of any given IoT solution is essential for most organisation. However, the process of selecting the appropriate architectural model often present a challenge. Nonetheless, considerations often involve factors such as the degree of comprehensiveness, adaptability, and availability for integration. One possible, widely considered architecture, is the IoT Europe Research Cluster (ERC) [11]. The ERC was developed as a reference model which comprises, primarily, 4-layers. The layers include application, support, network, device. Security and management capabilities are also integrated across all layers. Furthermore, this model integrates the possibility of an application-specific requirements, which can be leveraged for adaptability, and enhanced comprehensiveness. Therefore, the ERC can satisfy some rudimentary acceptability in IoT solution deployment. This is because it highlights the distinction between application-specific and generic architectural capabilities across all its layers. Generic capabilities refers to common IoT application requirements while application-specific capabilities refers to capabilities attributable to a given application capabilities. A further exposition of the layer of the ERC is discussed thus.

The topmost layer of the ERC model is the application layer. This layer relates primarily with applications on which IoT infrastructures. IoT applications, such as agriculture and mobile money, are typical of the application layer. The second layer - support layer - is responsible for service and other application support services. Overall, the support layer addresses challenges that relates with support services such as data storage and data processing, and other application specific support capabilities. The third layer - network layer - comprises functionalities and capabilities that relates with networking and transport services. These services provide relevant network and connectivity function requirements for an IoT infrastructure. The fourth layer device management layer - consists of dual capabilities, which are device and gateway capabilities. The gateway capabilities involve ways and mechanisms that allow for the interfacing of IoT devices within the different types of technologies, networks and protocols. The device capabilities on the other hand, contains mechanisms that can be used to enable devices to interact within the IoT infrastructure communication process. Finally, the management capabilities enables the administration of IoT devices while the security capabilities aligns the security requirements within each layer of the ERC model. Although the ERC provided a generic security capability across the different layers of the architecture, they are not detailed enough to support the development of the IoT-FR framework. Furthermore, as stated earlier, the ERC was developed as a baseline on which further framework, and context specific IoT solutions can be developed. As a way to further explore how to build-on the ERC, the next section explores the notion of digital forensic readiness, which is further integrated into the proposed framework in this study.

2.3 Digital Forensic Readiness

Recent studies [7, 8, 12] have defined digital forensics as the application of forensic science to the investigation and interpretation of digital data collected from electronic devices in its latent form, in a manner that is tenable in litigation. However, findings in [7, 8] asserted that, digital forensics does not only consist of computer systems but all digital devices, including networking infrastructure and future technologies. Therefore, the growth of digital forensics can be closely aligned with the growth of technology. Furthermore, technologies advances tend to introduce newer threat factors. In this regard, therefore, new modalities for conducting investigations are continually required to ensure the integrity of digital forensic data. This is readily applicable to pre-incident, during incident and post-incident investigation process, and to demonstrate compliance to standardized laws and regulation [14].

Although organisations are aware of the relevance of digital forensic during investigations, there is, however, a need to further recognise the significance of "being ready" for forensic operations remains largely ignored. Digital forensic readiness (DFR) is a planning process developed to address the unavailability of a potential digital evidence, specifically in a forensically sound manner. Failure to implement digital forensic readiness processes and procedures may result in operations being disrupted and evidence being lost, altered or not being handled in an acceptable manner [7, 8]. Digital forensic readiness has proven to inhibit the occurrence of these problems by ensuring that the necessary pre-incident processes are developed through the implementations of forensically sound processes prior to incident occurrence [14, 15]. Thus, the implementation of DFR processes can ensure that digital forensic evidence is available, secure and credible for litigation prior to incident occurrence [7, 8, 13]. The notion of DFR has been further elucidated in a recent standardized forensic process -the ISO/IEC 27043 international standard. This standard [16], as asserted in existing studies, was developed to provide appropriate guidance for conventional digital forensic investigation processes with respect to diverse investigation scenarios. The standard is applicable in any digital investigation process and emphasizes the investigation processes from the pre-incident preparation phase to subsequent processes. A section of the readiness process group extracted from the ISO/IEC 27043is presented in Fig. 1.

The ISO/IEC 27043 comprises of five core investigation process classes which include the readiness-phase, initialisation-phase, acquisitive-phase, investigation phase, and concurrent processes. The readiness processes class covers those processes that can be used to help an organisation put in place the ability and capacity to maximise the potential use of digital forensic evidence before an incidence occurs. The initialisation processes class has process groups that are meant to initialise the investigation process. The acquisition process class on the other hand contains guidelines and steps that deal with the acquisition and gathering of potential digital evidence. The investigation process class, however, has process groups that take care of incidents investigations, analysis, interpretation and presentation of the final forensic evidence at any court of law or civil hearing. It also comprises of the investigation closure process. The concurrent process classes have all the process groups that can be done alongside any investigation process. These processes can be undertaken in no particular order, unlike the others which are sequential.

3 IoT Forensic Readiness Framework

The proposed IoT Forensic Readiness (IoT-FR) Framework is discussed in this section. The first section presents a high-level view of the framework and a detailed framework is given in the later section.

3.1 High-Level View of the Framework

The IoT-FR framework consists of five components namely; the organisational level, readiness, IoT security, reactive and concurrent processes, as highlighted in Fig. 2.

An organisation based on the (organisation level processes) may handle different commitments and actions in order for it to satisfy the legal compliance and regulatory frameworks. Also important to this are the policies and the necessary structures that are

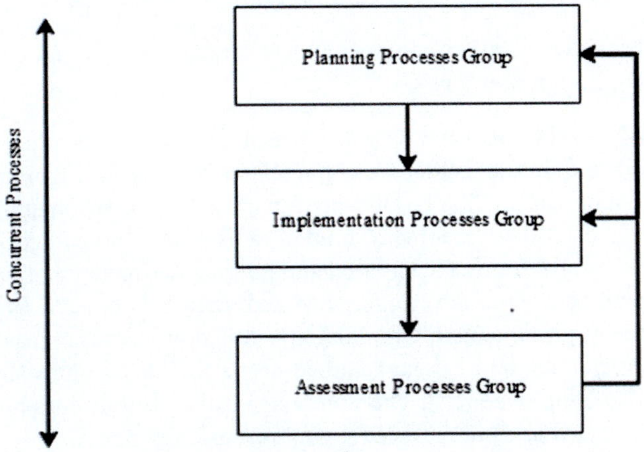

Fig. 1. Readiness process groups [16].

needed to ensure that the implementation and adoption of DFR procedures are successful. The task of the readiness processes are not only limited to incidental preparation in this context but also exploring relevant digital forensic data in order to proactively identify, collect, process, present and forensically store these evidence in line with the ISO/IEC 27043 guidelines. Next; the IoT security processes to a good degree of acceptability guarantees that the security of collected IoT data and forensic data is maintained throughout the forensic investigation cycle. It is important to note that the reactive (investigative) and concurrent processes (happen in tandem) are taken verbatim from ISO/IEC 27043.

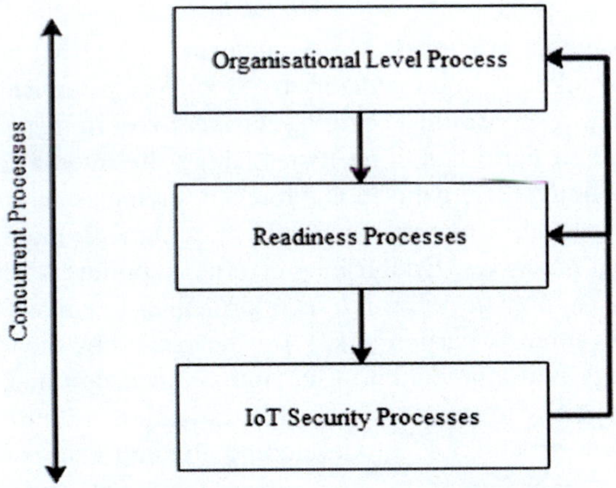

Fig. 2. High-level view of IoT-FR framework.

3.2 Detailed View of the IoT-FR Framework

This section provides detailed content of each component of the proposed IoT-FR framework, as highlighted in Fig. 2.

Organisational Level Processes: Organisational level processes address digital forensic investigations requirements that are applicable across the entire organisation. These processes set a tone and are a sign of commitment from the top management, that is, the need for digital forensic readiness in an organisation. Two categories of organisational processes will be discussed, external and internal organisational processes. External organisational processes are triggered by external factors such as regulatory and legal compliance requirements. Internal organisational processes consist of the internal analysis and documentation of organisational needs for forensic investigation, such as investigation of an employee [17]. The following section will discuss both external and internal organisational level process for digital forensic readiness.

- External Organisational Processes: External organisational processes are triggered by factors external to the organisation which highlight specific requirements that affect digital investigation processes. These external factors are the regulatory and legal investigation compliance requirements [13, 17]. Compliance with various regulatory requirements may force organisations to adopt DFR. [18] mentioned that some IoT devices might contain data that is regulated by various privacy and data protection laws. Therefore, it is important that organisations identify the jurisdiction they operate in and consider the ramifications of where they store and process data to ensure that they remain in compliance with the relevant laws and regulations. Also, digital evidence needs to satisfy legal requirements before it can be presented in a court of law or any other arbitrary cases. The burden of proof in on the organisation to demonstrate that the evidence is authentic and that the right tools and processes were followed when gathering digital evidence [13].
- Internal Organisational Processes: The components of internal organisation process include: Establishing legal and regulatory requirements, establishing forensic policy, defining roles and responsibility, defining business requirements, defining training requirements and scenario definition. Establishing the internal organisational processes include defining the organisational forensic strategies and policies by translating applicable legal and regulatory compliance requirements highlighted in the external organisational processes. Establishing a forensic policy is key to implementing DFR [13, 19, 20]. A forensic policy is a set of principles to guide day by day activities and individual for forensic purposes [13]. It is developed by the top management, to translate a strategy into a guideline for operational decision making [21]. Defining roles and responsibility to support a digital investigation is critical in designing and implementing DFR processes [20]. Developing a team to support the digital investigation process may include teams from other departments and sometimes external parties depending on the type and magnitude of the investigation [22]. The non-technical stakeholders may include senior management, legal team, human resources. Some non-technical stakeholders may be external to the organisation such as the Information regulator, auditors and law enforcement. The objective of training in DFR is

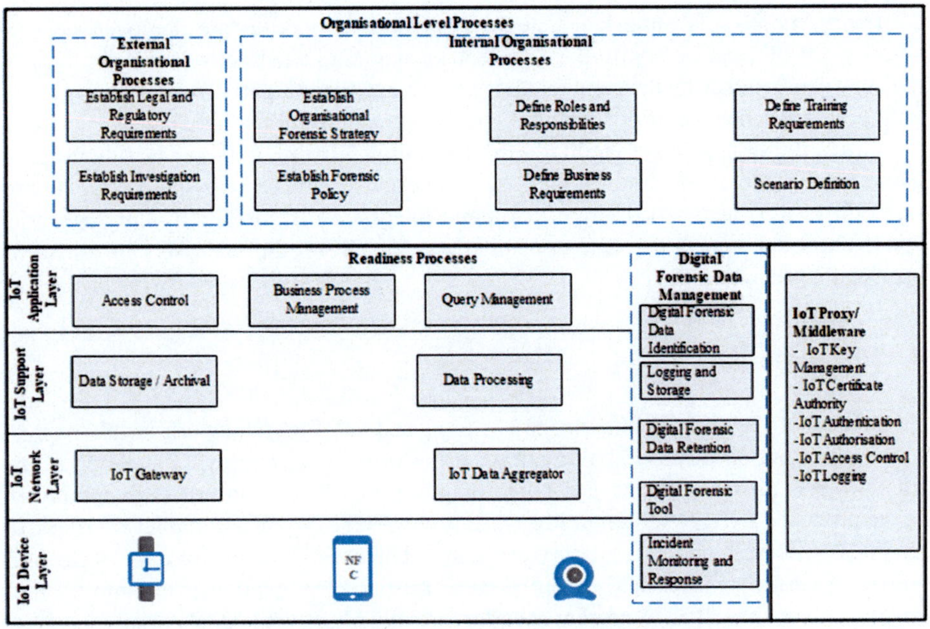

Fig. 3. Detailed IoT-FR framework.

to ensure that the human resources of an organisation and the various stakeholders can contribute effectively towards the detection and prevention of security incidents. Although the benefits of digital forensic readiness have been mostly about fulfilling the legal requirements and maintaining a chain of custody, it is also important that business requirements for forensic investigations are also identified. Scenario definition is a risk-based approach used by organisations to identify potential business processes or activities where potential digital forensic may be required [19, 20].

Readiness Processes: The readiness processes basically discuss the application of the readiness processes group highlighted in the ISO/IEC 27043 in order to achieve DFR in an organisation. The assessment and concurrent processes group are basically adopted as is in ISO/IEC 27043. Basically, identification of potential evidence sources is critical in building a DFR program. The identification of potential digital forensic data sources is guided by the organisational-level processes.

The IoT data management cycle has the following sub modules: Production, collection or delivery, storage or archival, processing or integration, delivery and consumption [23]. The IoT device layer is responsible for data production within which the connected devices, sensors are able to observe data from the environment which is then forwarded to the enterprise network. Additionally, it is important that there exist unique identities among the connected devices in order for them to be dynamically identified within the enterprise network for purposes of querying and retrieval.

Furthermore, it is important that the aggregation modules are deployed near the sensors, for purposes of optimizing the sensor data and to effectively aggregating the data that originates from the connected devices. Next, the IoT gateway or proxy, is tasked with providing interfacing IoT devices within the organisational network.

Lastly, the support layer facilitates the collection/delivery of the sensor where aggregators are put into use. Moreover, the data storage and archival are also held at the support layer. After this, data processing, fusion and integration is handled by the application layer. The final step is the data consumption that allows the end-users to utilise the provided information [23].

4 Discussions

To the best of the authors knowledge the proposed IoT-FR framework in this paper is a new contribution in the IoT domain as at the time of its writing. The scope of this framework is presented using a detailed diagram with the content of each component of the proposed IoT-FR shown in Fig. 3. This covers the organisational level processes, readiness processes and IoT security processes. The specific details of each of the individual processes as identified in Fig. 3 have further been explained to simplify their understanding as well as to present specific finer details of the Framework.

However, note that the different processes as identified in Fig. 3 are meant to facilitate this study and primarily focus on the proposed IoT-FR framework. Moreover, the proposed processes in this paper are by no means the final guaranteed processes to a perfect IoT-Forensic Readiness (IoT-FR) framework in an organisation. This is because, with the advancement in technology more processes can be added into the framework to perfect it as the IoT domain keeps evolving.

From Fig. 3, the Organisational Level Processes addresses the digital forensic investigations requirements that are applicable across the entire organisation. Two categories of organisational processes, the external and internal organisational processes were discussed. In this case the external organisational processes are triggered by external factors such as regulatory and legal compliance requirements while the internal organisational processes consist of the internal analysis and documentation of organisational needs for forensic investigations.

The Readiness processes are part of the readiness processes group that are defined in the ISO/IEC 27043 and pre-incident strategies/processes for computer forensics [24, 25]. These processes work across the different layers of the IoT architecture, as shown in Fig. 3, and concentrates on the digital forensic data management process which include digital forensic data identification, logging and storage, digital forensic data retention, digital forensic tools as well as incident monitoring and response. This is was primarily done to achieve digital forensic readiness in any IoT-enabled organisation. Finally, the IoT Security Processes are meant to cover key management, certificate authority, authentication, authorisation, access control and logging. This proposed framework, therefore, builds on the layers of the ERC while complying with the readiness composition of the respective ISO standardisation. By integrating these two components, this study has provide an additional structure on which the baseline of IoT security and forensic processes can be achieved within any organisation, size notwithstanding.

The proposed IoT-FR framework can be used in the IoT domain, for example, to help organisation maximize their potential to record activities and data in such a manner that the records are sufficient in their extent for subsequent forensic purposes, and the records are acceptable in terms of their perceived authenticity as evidence in subsequent forensic investigations. Besides, the proposed IoT-FR framework can also be helpful to organisation, for example, in reasoning about the application of specific new processes as the domain evolve. The proposed framework can also be used for academic purposes, especially on how to create and manage new processes. Developers can also use the proposed framework to develop new Forensic Readiness tools or improve on existing ones. This proposition will be further implemented in small and medium size organisation to further ascertain the feasibility of its implementation. Furthermore, our future studies will consider the development of a policy framework which can provide a guideline towards the implementation of this proposition. Studies have expounded on the implication of technology advances in IoT deployment within organisations. Our future works will also consider the impact of technology advances on IoT deployment particularly, the growing trend of 5G and 6G network.

5 Conclusion

The need for an IoT digital forensic readiness framework in organisations has been a topic of interest to many researchers and organisations alike. For this reason, the authors in this paper have proposed an IoT-FR framework for organisations based on the recommendations stipulated in the ISO/IEC 27043 international standard, and the model developed by the IoT Europe Research Cluster. This was motivated by the fact that; several organisations, both small and medium sizes, as well as large size organisations, are gravitating towards adopting IoT technology. This has been widely adopted towards the improvement of business operations as well as exploring the possibility of new business potentials. To the best of the authors' knowledge, this is a new contribution that can provide a significant direction towards advancing the security and forensic readiness of the IoT domain.

References

1. Kebande, V.R., Karie, N.M., Venter, H.S.: Adding digital forensic readiness as a security component to the IoT domain (2018)
2. Kebande, V.R., Ray, I.: A generic digital forensic investigation framework for Internet of Things (IoT). In: 2016 IEEE 4th International Conference on Future Internet of Things and Cloud (FiCloud), pp. 356–362. IEEE, August 2016
3. Kebande, V.R., Malapane, S., Karie, N.M., Venter, H.S., Wario, R.D.: Towards an integrated digital forensic investigation framework for an IoT-based ecosystem. In: 2018 IEEE International Conference on Smart Internet of Things (SmartIoT), pp. 93–98. IEEE, August 2018
4. Alaba, F.A., Othman, M., Hashem, I.A.T., Alotaibi, F.: Internet of Things security: a survey. J. Netw. Comput. Appl. **88**, 10–28 (2017)

5. Khorashadizadeh, S., Ikuesan, A.R., Kebande, V.R.: Generic 5G infrastructure for IoT ecosystem. In: Saeed, F., Mohammed, F., Gazem, N. (eds.) Emerging Trends in Intelligent Computing and Informatics: Data Science, Intelligent Information Systems and Smart Computing, pp. 451–462. Springer, Cham (2020). https://doi.org/10.1007/978-3-030-33582-3_43
6. Kebande, V.R., Venter, H.S.: Novel digital forensic readiness technique in the cloud environment. Aust. J. Forensic Sci. **50**(5), 552–591 (2018)
7. Lagrasse, M., Venter, H.: Digital forensic readiness framework for software-defined networks using a trigger-based collection mechanism. In: ICCWS 2020: 15th International Conference on Cyber Warfare and Security, p. 296. Academic Conferences and Publishing Limited, March 2020
8. Philomin, S., Singh, A., Ikuesan, A., Venter, H.: Digital forensic readiness framework for smart homes. In: International Conference on Cyber Warfare and Security, pp. 627–XVIII. Academic Conferences International Limited (2020)
9. Zdravković, M., Zdravković, J., Aubry, A., Moalla, N., Guédria, W., Sarraipa, J.: Domain framework for implementation of open IoT ecosystems. Int. J. Prod. Res. **56**(7), 2552–2569 (2018)
10. Gubbi, J., Buyya, R., Marusic, S., Palaniswami, M.: Internet of Things (IoT): a vision, architectural elements, and future directions. Future Gener. Comput. Syst. **29**(7), 1645–1660 (2013)
11. ITU: Erc: 2012 overview of the Internet of Things. Telecommunication standardization sector of ITU, vol. 1, no. 1, pp. 1–12 (2012)
12. Dominik Ernsberger, R., Adeyemi Ikuesan, S., Venter, H., Zugenmaier, A.: A web-based mouse dynamics visualization tool for user attribution in digital forensic readiness. In: Matoušek, P., Schmiedecker, M. (eds.) Digital Forensics and Cyber Crime, pp. 64–79. Springer International Publishing, Cham (2018). https://doi.org/10.1007/978-3-319-73697-6_5
13. Elyas, M., Maynard, S.B., Ahmad, A., Lonie, A.: Towards a systemic framework for digital forensic readiness. J. Comput. Inf. Syst. **54**(3), 97–105 (2014)
14. Mohlala, M., Ikuesan, A.R., Venter, H.S.: User attribution based on keystroke dynamics in digital forensic readiness process. In: 2017 IEEE Conference on Application, Information and Network Security (AINS), pp. 124–129. IEEE, November 2017
15. Singh, A., Ikuesan, A.R., Venter, H.S.: Digital forensic readiness framework for ransomware investigation. In: Breitinger, F., Baggili, I. (eds.) Digital Forensics and Cyber Crime: 10th International EAI Conference, ICDF2C 2018, New Orleans, LA, USA, September 10–12, 2018, Proceedings, pp. 91–105. Springer, Cham (2019). https://doi.org/10.1007/978-3-030-05487-8_5
16. ISO/IEC: 27043: 2015 international standard, information technology—security techniques—incident investigation principles and processes, vol. 1, no. 1, pp. 1–30. ISO.org (2015)
17. Trček, D., Abie, H., Skomedal, Å., Starc, I.: Advanced framework for digital forensic technologies and procedures. J. Forensic Sci. **55**(6), 1471–1480 (2010)
18. Zawoad, S., Hasan, R.: FAIoT: towards building a forensics aware eco system for the Internet of Things. In: 2015 IEEE International Conference on Services Computing, pp. 279–284. IEEE, June 2015
19. Kebande, V., Ntsamo, H.S., Venter, H.S.: Towards a prototype for achieving digital forensic readiness in the cloud using a distributed NMB solution. In: European Conference on Cyber Warfare and Security, p. 369. Academic Conferences International Limited, July 2016
20. Munkhondya, H., Ikuesan, A., Venter, H.: Digital forensic readiness approach for potential evidence preservation in software-defined networks. In: 14th International Conference on Cyber Warfare and Security: ICCWS 2019, vol. 268. Academic Conferences and Publishing Limited, February 2019

21. Park, S., Kim, Y., Park, G., Na, O., Chang, H.: Research on digital forensic readiness design in a cloud computing-based smart work environment. Sustainability **10**(4), 1203 (2018)
22. Pooe, A., Labuschagne, L.: A conceptual model for digital forensic readiness. In: 2012 Information Security for South Africa, pp. 1–8. IEEE, August 2012
23. Abu-Elkheir, M., Hayajneh, M., Ali, N.A.: Data management for the Internet of Things: design primitives and solution. Sensors **13**(11), 15582–15612 (2013)
24. Kebande, V.R., Venter, H.S.: On digital forensic readiness in the cloud using a distributed agent-based solution: issues and challenges. Aust. J. Forensic Sci. **50**(2), 209–238 (2018)
25. Karie, N.M., Kebande, V.R., Venter, H.S.: Diverging deep learning cognitive computing techniques into cyber forensics. Forensic Sci. Int. Synergy **1**, 61–67 (2019)

Regulation and Standardization of Blockchain Technology for Improved Benefit Realization

Udochukwu C. Enwerem[1(✉)] and Gloria A. Chukwudebe[2]

[1] Department of Computer Science, Federal University of Technology, Owerri, Nigeria
[2] Department of Electrical and Electronics Engineering, Federal University of Technology, Owerri, Nigeria
gachukwudebe@futo.edu.ng

Abstract. Regulation and standardization are critical success factors for benefit realization and wide adoption of any new technology. Although, Bitcoin, the first Blockchain implementation met with varied resistance, recent non-financial applications of Blockchain technology will play critical roles in the upcoming fourth Industrial Revolution (Industry 4.0) due to its characteristic features of trust, immutability, transparency and decentralization. In this paper, we presented Blockchain technology and highlighted some of its innovative applications in business, government, health, and oil and gas sectors that promise to be disruptive with great benefits. The efforts towards regulation of Blockchain technology by various national governments were examined and it was found that only a handful of countries have adopted and published regulations. Furthermore, the emerging new Blockchain Service Providers offering open source development environments is making it imperative for regulatory policies and standards to be in place for optimum benefit realization. Meanwhile, except for a few published standards on Blockchain technology by the IEEE, the status of standards development by other international organizations; ISO and ITU are still at the preliminary stages. We recommend that for adequate benefit realization of this technology and for the success of Industry 4.0, that the international standardization bodies should accelerate work on standards publication to enable various countries to make regulatory policies for safe and secure use of Blockchain technologies. Additionally, due to the numerous potentials of Blockchain in various areas of endeavor, we recommend an interdisciplinary approach to development of regulatory policies and standards for the technology.

Keywords: Cryptocurrency · Blockchain · Smart contracts · Regulation · Standardization · Bitcoin · Ethereum · Hyperledger · Distributed Ledger Technology

1 Introduction

A Blockchain is a distributed ledger system where records are stored in several independent computers called nodes with no form of centralized management. Besides management; trust is also decentralized and forms the basis for why a Blockchain can be

J. H. Abawajy et al. (Eds.): EATI 2020, LNNS 254, pp. 240–253, 2021.
https://doi.org/10.1007/978-3-030-80216-5_18

trusted. The fundamental concepts underlying Blockchain was successfully brought into the limelight by its application as a cryptocurrency called Bitcoin in 2008. Subsequently, Bitcoin was successfully put into circulation without the need of any intermediary, such as banks. At the time of the study, though not widely accepted, Bitcoin had gotten a total market capitalization of $178 billion with the price of a single bitcoin tagged at about $9,800 (Sam 2020).

One of the interesting features of Bitcoin is that it makes the centralized banking system unnecessary because every transaction has a unique digital record and a signature that can be validated and shared without any intermediaries. This feature slowed down many governments' adoption of Bitcoin application because of the regulatory requirement of all financial transactions. However, with the increasing research and development on distributed ledger technologies, new robust Blockchain platforms have emerged since 2015; Ethereum and Hyperledger. The Ethereum blockchain introduced the concept of smart contracts and is receiving support from an alliance of companies who have agreed to support innovative applications using the technology. Similarly, the Hyperledger is an open source platform for blockchain business applications supported by the Linux foundation.

These open source Blockchain platforms have opened up opportunities for development of innovative new applications in manufacturing, agriculture, real estate, politics, supply chains and logistics that will be useful for the fourth industrial revolution (Industry 4.0). Additionally, the technologies that will power Industry 4.0 such as Internet of Things (IoT), machine learning, cloud computing, will need to be established on trusted networks with the characteristic features of Blockchain.

More importantly, the fact that Blockchain is based on the Internet means that its applications will go beyond national and regional boundaries towards becoming a global phenomenon; hence it is an obvious disruptive technology that can no longer be overlooked. Presently, some countries have taken bold steps to understudy and adopt Bitcoin and some other Blockchain applications for some selected transactions. In view of the possible disruptive impact the applications can cause the world, international standards organizations; ISO, IEEE and ITU-T have set up working groups to develop standards that will aid regulation and ensure benefit realizations of the Blockchain applications.

In this paper, the key features of Blockchain technology are briefly described in Sect. 2. Section 3 presents the study of regulatory and standardization efforts of the technology. Section 4 highlights some of the emerging Blockchain applications that will impact benefit realizations of Industry 4.0. The conclusion and recommendations for further work are presented in Sect. 5.

2 The Blockchain Technology

2.1 Blockchain Architecture

The Blockchain technology uses the peer-to-peer computer network architecture instead of the traditional client-server architecture where resources are stored and managed in a central server and accessed from numerous clients that may be remotely located (Fig. 1). This traditional architecture (Fig. 2) has been the basis for creating network and Internet applications to manage various sectors of the economy including public and private corporations.

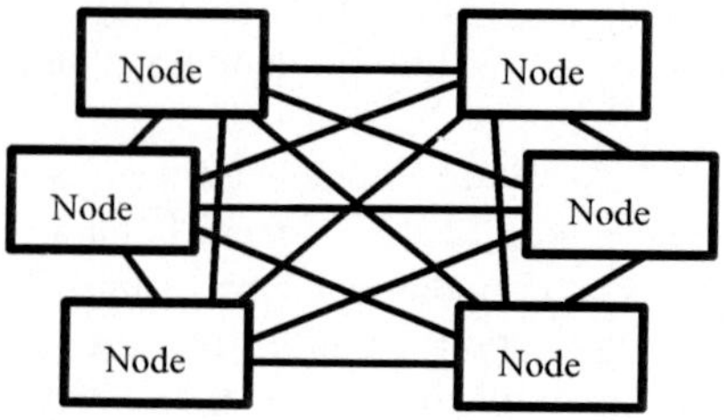

Fig. 1. Blockchain peer-to-peer architecture.

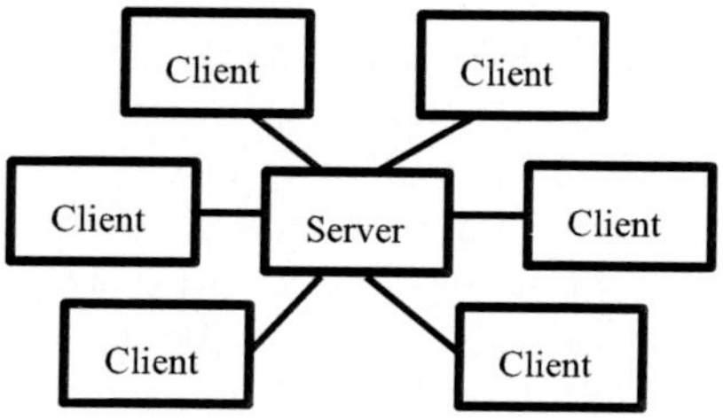

Fig. 2. Client-server architecture.

Until now, the default consensus in the banking and financial industry has been the centralized IT architecture whereby, each banking corporation has a central server or cluster of servers that host customer data and financial records. All actual execution of transactions that cause change to the customer record must occur at the central server. This architecture has the advantage of being easy to maintain because it has a single point of failure. However, this has the disadvantage of making it easy for data to be compromised by an intruder.

In contrast, blockchain applications use a peer-to-peer network architecture as shown in Fig. 1. Firstly, customer records (termed ledgers) are not stored in a central server but are stored and managed by all of several participating computers known as nodes. Each of these nodes has a copy of all records in the system. Secondly, though transactions can be authorized from a single node, all actual execution of transactions that cause change to customer records must occur on every node in the system. Furthermore, all nodes cooperate in verifying each transaction; hence, no record can be changed without consensus.

2.2 Blockchain Data Structure and Digital Signature

The major components of a blockchain are the blocks which hold transaction data and the digital signature which constitutes public and private key pairs that are used to sign and verify every new transaction. These components and their uses are discussed in this section.

Block. A blockchain is essentially a linked-list like data structure that consists of a sequence of multiple blocks. Each block points to an immediate previous block (known as the *parent block*), using a hash value of the address of the parent block. The first block on the blockchain with no parent is known as the *genesis block*. These blocks are distributed among the nodes in a blockchain peer-to-peer network. Each block has a header that contains the **Parent block hash, Timestamp** and **Nonce**, while the Transaction counter and Transactions make up the body of the block.

- Parent block hash is a 256-bit hash value that holds the address of the previous block.
- Timestamp is the time the block was created.
- Nonce is a 16-bit field, which usually starts with 0 and increases for every hash calculation. It is used for generating new blocks in the chain.
- Transactions field has a record of all the transactions in the system (including specific versions of transactions).
- Transaction counter is used to keep track of new transactions.

Digital Signature. The digital signature used in a blockchain comprises of two set of keys: a private key that is only visible to the owner and a public key that is visible to every user of the blockchain. The private key is used by the initiator of a transaction (say Alice) to digitally sign the transaction. The signed transaction and the user's public key are then bundled together and distributed to all blocks on the chain. When the signed transaction reaches the target user (say Bob), Bob uses Alice's public key to decrypt the signature on the transaction. The resulting value is compared to a value derived by running the same hash function as Alice on the transaction as a form of verification.

3 Regulation and Standardization of Blockchain Technology

Although, the speculative impact of Blockchain technology has been a global one, regional regulatory policies have been affecting its adoption in various countries, although some countries are not permitting the technology at all. In this section, some regulatory efforts by various national governments and the international standards organizations investigated will be presented.

3.1 China and East Asia

China. With China being the birthplace of the Bitcoin cryptocurrency, its government has been evolving regulatory measures to control the impact of blockchain technology. Most of the early stringent regulations in China were against private corporate users and operators of cryptocurrencies. The Chinese government has, on the other hand, been encouraging the development of a state-owned official cryptocurrency called Digital Currency for Electronic Payment ("DCEP") by the People's Bank of China. Also, the government had authorized the People's Bank of China to establish a cryptocurrency research institute to study cryptocurrency and its underlying technology – blockchain, with an aim to harness the potentials of blockchain technology to serve the Chinese financial sector to the fullest extent (Lefan and Luping 2018).

Japan. From the study, Japan is one of the largest cryptocurrency markets in the world and is the first country to enact a law defining Virtual Currency as a legal term. The law requires an entity to register as a Virtual Currency Exchange Service provider in order to provide such services to residents in Japan (Taro 2018). The Payment Services Act updated since 2017 stipulates that Cryptocurrency exchange businesses must be registered, keep records and take measures to protect customers, in addition to being subject to money laundering regulations.

Hong Kong. In Hong Kong, there is a high degree of autonomy as regards the introduction of new financial instruments such as cryptocurrencies due to the free market principles enshrined in its laws. This has resulted in many private and public enterprises being formed in Hong Kong to take advantage of the opportunities the new technology can offer, and to leverage Hong Kong's unique position as an international financial center (Yu 2018).

3.2 USA and Canada

USA. In the United States, due to the government structure, both federal and the states have varied regulations and policies regarding cryptocurrencies and blockchain technology. At the time of the survey, Wyoming, Colorado, Arizona and Georgia are some states who were making efforts in promoting the technology in USA whereas California, New Mexico and New York do not fully support cryptocurrency and the underlying blockchain technology (Josias 2018).

On the average, most of the policies are in the positive direction towards promoting the adoption and development of the technology in USA. The regulators are for legislations that would not deprive the country of investment opportunities in the technology. The Regulators involved in the US include the Securities and Exchange Commission (SEC), the Commodities and Futures Trading Commission (CFTC), the Federal Trade Commission (FTC) and the Department of the Treasury, and the Financial Crimes Enforcement Network (Josias 2018).

More recently, at the close of 2019, the United States Congress is processing the 'Crypto-Currency Act of 2020' primarily to clarify the federal agencies to regulate digital assets. Some suspect that this was prompted by the cryptocurrency project- Libra proposed by Facebook owner, Mark Zuckerberg.

Canada. In Canada, the use of cryptocurrencies and blockchain technology is allowed. However, they are not considered as legal tender within the country. There is also purposeful regulation regarding the use of cryptocurrencies. For instance, Canada's regulation governing tax laws and rules including the "Income Tax Act", also apply to cryptocurrency transactions (Tariq 2018).

3.3 Nigeria and Africa

Nigeria. Nigeria has not started actively regulating cryptocurrencies and blockchain technology. However, the financial regulatory bodies; the Central Bank of Nigeria (CBN) and Nigerian Deposit Insurance Corporation (NDIC) have issued series of public statements warning against the use of cryptocurrencies within the country. Although, it is reported by Jen et al. (2018), that Nigeria has the world's third largest bitcoin holdings as a percentage of gross domestic products. Most of the fears expressed by the Nigerian government regarding the adoption of cryptocurrency and blockchain technology evidently stem from the concerns that funding to terrorist groups using cryptocurrencies may be difficult to monitor.

Unlike Nigeria, several other countries in the African continent that have set out active and purposeful policy guidelines and regulation regarding cryptocurrencies and blockchain technology include South Africa, Tunisia, Sierra-Leone and Senegal.

South Africa. In South Africa, the two major financial regulators are the South African Reserve Bank (SARB) and the South African Revenue Service (SARS). The two bodies have explicitly declared all forms of virtual currencies including cryptocurrencies as not being legal tender in South Africa but do not prohibit their use as a form of exchange. Furthermore, the SARB has been very active in revising its policy on the issue and has recently collaborated with several other federal financial agencies and a private firm named ConsenSys (a New York based blockchain technology company), to lunch a project codenamed "Project Khokha" for use in understudying Distributed Ledger Technologies (DLT). According to Jen et al. (2018), there have been two successful public Initial Coin Offerings (ICO's) for two cryptocurrencies in South Africa. These include the "Rhino Coin" and the "SafCoin".

Tunisia. Tunisia seems to be in the lead as regards the utilization and regulation of cryptocurrency and blockchain technology in Africa. Foremost among its achievements is the pilot launch of its own national digital currency, the eDinar that is powered by Blockchain technology using smart contracts (Adrian 2019). This Tunisian government project was in partnership with a Russian startup (Universa Blockchain) (Danny and Anna 2019).

Another effort by the government to feed about 400,000 under-privileged Tunisian school children in 6,000 schools is in partnership with another Blockchain startup - Devery.io (Jen et al. 2018). The project involved building a blockchain-based supply chain management system to track the delivery of lunch to school children.

3.4 Standardization of Blockchain Technology

There have been efforts by various International standards bodies such as the International Organization of Standardization (ISO), Institute of Electrical and Electronics Engineers (IEEE) and the International Telecommunication Union (ITU) to make standards for regulation of blockchain technology. However, most of the standards have not been published yet with many of them still at their infant stages of development. This is partly due to the fact that many existing implementations of blockchain technology vary widely in their conceptual methodology. In this section, we will examine the efforts by International standards bodies.

International Organization of Standardization (ISO). The ISO is currently working on ISO/TC 307 family of standards for Blockchain and Distributed Ledger Technologies since 2016 (Heung and Skylar 2019). At the time of writing, ISO/TR 23455:2019 is one of the family of ISO/TC 307 standards already published. About 10 other standards are still under development. The published standard is mainly on technical specifications for smart contracts and suggestions on some application areas (ISO 2019). The other ISO/TC 307 standards under development relate to Security risks and vulnerabilities, reference architecture, security management of digital asset and guidelines for governance.

International Telecommunication Union (ITU). The ITU is also working towards the development of new standards for blockchain technology. Among the mandates given to the Standardization group by the ITU, is the development of a standardization roadmap for interoperable Distributed Ledger Technology (DLT) based services. This effort has led to a liaison between the ITU-T and ISO/TC 307 group (Anjum et al. 2017). The ITU-T focus group completed its mission on the 1st of August, 2019 and submitted deliverables covering the following; blockchain technology and DLT overview, concepts, ecosystem; DLT standardization landscape; reference architecture; Assessment criteria for DLT platforms; DLT regulatory framework and Outlook on DLTs (ITU 2019).

Institute of Electrical and Electronics Engineers (IEEE). IEEE recognizes the vital role standards will play in the development and adoption of blockchain technologies; hence it set up the IEEE blockchain group under the IEEE standardization Association (IEEE-SA). At the time of writing, besides the IEEE P2140 family of standards developed for cryptocurrency and anti-money laundry, the IEEE blockchain group has active standards projects under 26 different thematic areas. For instance, the IEEE P2412.1 is on Recommended Practice for E-Invoice using Blockchain, while IEEE P2418, is a family of standards for the frameworks for Blockchain use in Government, Digital Asset Management, IoT, Agriculture, Autonomous Vehicles, Healthcare, and Social Sciences.

4 Emerging Blockchain Applications and Benefits

Blockchain technology benefits are due to its characteristic features of trust, immutability, transparency and decentralization. In this section, we present some of the application areas and the benefits.

4.1 Medicine and Healthcare

Blockchain technology can be used to improve healthcare data-sharing and diagnosis applications. Some concrete examples include the Healthcare Data Gateway (HDG) method developed using a private blockchain for monitoring and storing clinical data with functionality to ensure patient privacy (Yue et al. 2016). Also, Griggs et al. (2018) built a private blockchain, using the Ethereum protocol, which was able to manage the secure use of medical sensors while eliminating the security risks associated with remote patient monitoring.

Dr Abdullah Albeyatti and his team took it to another level by developing MedicalChain, a Smart mobile and desktop App for health records using blockchain technologies. This App puts patient's medical records at the fingertips of the patient, the doctor, and any other relevant stakeholders. This team is already collaborating with various stakeholders, UK medical doctors, the Blockchain Research Institute, and the Linux foundation. MedicalChain's capabilities is still evolving, it has telemedicine features and API for innovators to create more Apps (Albeyatti 2017).

4.2 Politics and Election

In the area of politics, blockchain technology is already proving useful in electoral processes and has been applied in developing e-voting applications. An example of this is the Kaspersky Lab's Polys, an Ethereum-based e-voting platform that uses smart contracts to verify ballots and tally votes (Oliver 2019). Also, there was the recent case of Sierra-Leone employing blockchain technology developed by Agora, a Swiss foundation offering digital voting solutions, using permissioned blockchain (Yomi 2018). Though it was claimed by the report to be the world's first blockchain-supported election, several other sources claimed that Agora's involvement in the election was limited and that they only had independent observers use blockchain to collate the result independently from polling stations (Neer 2018). Also, Kristin (2018) claimed that the use of blockchain in the Sierra-Leone's election was only a proof-of-concept experiment.

4.3 Supply Chain Management

Another area where Blockchain technology is promising to be disruptive is the supply chain industry. Blockchain is already proving to be invaluable in ameliorating the challenges faced by the Supply Chain industry by providing overall cost reduction in tracking and moving materials in a supply chain and establishing trusts between the numerous participants (employees, managers, customers, and transporters) in the supply chain.

Blockchain enhanced supply chain use-cases already exist in the Drugs and Pharmaceuticals industry where it is being used for tracking drug chemical supplies and to curb the counterfeiting of drugs and proliferation of unlicensed pharmaceuticals. Another use-case can be found in the Food industry where it is currently being used by Walmart to track food materials from farms and food supplies to customers in other to ensure food safety (Forkast. Insights 2020).

Blockchain applications in Oil and Gas offer more transparency, efficiency, effectiveness and cost reduction. Typically, the applications have functionalities for crude oil import trading (digital bill of laden and smart contracts), tracking products in the Oil and Gas supply chain, audit trails of equipment used throughout the lifecycle, etc. According to Lu, since 2017, a handful of Oil and Gas industry Blockchain projects worldwide have gone live. Energy giants in the industry, BP and Shell were the pioneers in Blockchain application technology (Lu et al. 2019).

5 Proposed Framework for Blockchain Regulation in Nigeria

As mentioned earlier, Nigeria is yet to harness the benefits of blockchain and its related technologies and have no regulatory framework currently in place. Hence, in this section, we propose a framework that will aid the government of Nigeria regulate the new technology in order to maximize its benefit and control abuse of the technology.

5.1 Factors Influencing the Adoption of Blockchain in Nigeria

Before delving into a framework proper, certain factors that are precipitated by the unique state of the Nigerian economy and the country's status as a developing country need to be highlighted. This is because these factors will greatly influence the adoption of the new technology by both public and private institutions. Though, change may be difficult at first due to the disruptive nature of the technology, Allen et al. (2020) was able to prove that innovations and introduction of institutional technologies such as blockchain technology in governance was highly beneficial though rare.

Currently, Nigeria is still majorly a cash driven economy. Though, several efforts have been made by the government for the country to transit into a cashless economy, majority of the population are still below the poverty line and need to transact business using cash. This is in conflict with the idea that blockchain technology is an electronic and cashless technology.

Secondly, the country's current electricity power generation and supply capacity is inadequate. This force major businesses in the country to depend mainly on other private sources of electricity thereby increasing their operational cost. This fact eliminates the possibility of setting up cryptocurrency mining nodes in the country as these nodes need to be up and running almost all the time.

Thirdly, the above-mentioned points have further hampered the development and adoption of Information Technology (IT) in the country. Though the IT infrastructures in Nigeria (such as mobile and telecommunication facilities) have grown over the years, most advance services provided by these facilities are under-utilized. These services include the use and transaction of electronic assets; electronic and mobile assess to the Nigerian Stock Exchange, amongst others. Furthermore, bandwidth availability across the country is evidently insufficient for the number of subscribers to telecommunication services in the country.

Lastly, the slow judicial process and long periods between policy formulation and implementation may hamper the full adoption of blockchain technology. This is further exacerbated by the problem of corruption in the form of money laundry in Nigeria.

5.2 Proposed Blockchain Regulatory Framework

Taking the above factors into cognizance, our proposed framework will be built around 4 major aspects that will not only help in early adoption of the technology but will also mitigate some of associated problems that may be unique to Nigeria. The aspects include: Legislature, Finance, Technology and Energy. Under a broader consideration, the Finance and Technology aspects can be classified as core aspects, while the Legislature and Energy aspects are termed as supportive aspects. This is depicted in Fig. 3.

Legislature. The legislative process is the first step in the development of our blockchain regulatory framework. Without this being first in place, abuse of the technology can become imminent. Besides providing an overall supportive role, it is continuous throughout the lifecycle of the framework development process. Elements of the aspect of legislature that lend support to our proposed regulatory framework include:

- Ongoing approval for major disruptive projects associated with the other aspects of the framework.
- Approval and disbursement of budgetary resources for blockchain related projects.

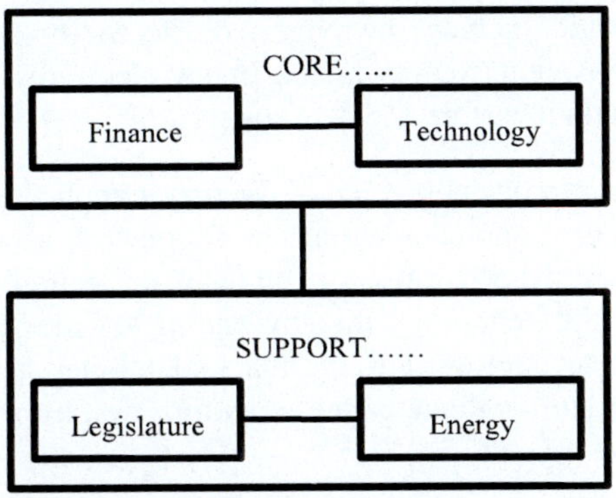

Fig. 3. Proposed blockchain regulatory framework.

- Approval of policy guidelines to regulate the other aspects of the framework to meet minimal standards for the takeoff of blockchain technology.
- Approval of policy guidelines to regulate the disruptive effect of blockchain technology.
- Formulation and enforcement of laws to curb the abuse of blockchain technology.
- Setting up of special tribunals for the speedy resolution of electronic and internet related crimes

Major stakeholders here will include the Presidency, members of the houses of Senate and Representatives, members of the Judicial Council and law enforcement agencies.

Energy. There are several characteristics of blockchain technology that require undisrupted electricity supply. Hence, boosting energy supply to locations that will be housing blockchain related facilities will aid the early adoption of the technology in the country. These locations include universities and research institutes where development of the technology may be ongoing. However, there have been recent calls for the reduction of emissions by disruptive technologies, globally. Also, Nigeria is a signee to the Paris Agreement (List of Parties to the Paris Agreement, n.d). According to Truby (2018), there should be the need to compel industries – both public and private, to bear the cost of negative environmental impacts caused by disruptive technologies like blockchain, in order to motivate them to switch to less energy intensive substitutes. In order to achieve this, the framework proposes the following measures:

- Provision of solar powered electricity sources for major cryptocurrency mining nodes in universities and research institutes in the country especially at off-peak hours.
- Provision of stable power supply to blockchain research labs and mining node especially at peak hours during the day when utilization is high.
- Placement of major blockchain mining nodes and research labs under the national energy grid distribution system rather than that of the local distribution system.
- Development of strategic policies for monitoring and measuring energy consumption by public and private blockchain businesses.

Stakeholders here would include the Nigerian Electricity Regulatory Commission (NERC), public and private university management, electricity distribution companies, private sector blockchain investors and startups.

Finance. With cryptocurrency being one of the major applications of blockchain technology, the role of financial policies within the framework becomes eminent. Without these policies in place prospective users of blockchain technology will be deterred to transact business using the technology, thereby leading to widespread public mistrust and under-utilization of the technology. Several financial policies and measures that are proposed by our framework to facilitate benefit maximization from blockchain technology include:

- Creation of a state-owned cryptocurrency.
- Inclusion of cryptocurrencies and cryptoassets for trading in the Nigerian Stock Exchange (NSE).
- Provision of insurance cover for legally owned cryptocurrencies.
- Ensuring that all cryptocurrencies and cryptoassets are declared and that provision of tax inclusion is done for same.
- Ensuring that commercial banks and other financial institutions involved in the exchange of cryptocurrencies are certified and registered.

The major stakeholder here is the Central Bank of Nigeria (CBN) which has an oversight role over every financial institution in the country. Others include the Federal Inland Revenue Service (FIRS), commercial banks operating within the country, the Nigerian Stock Exchange (NSE) and other registered financial operators.

Technology. Technology remains the driving force of blockchain technology, hence it features as a core aspect of our proposed regulatory framework. In other for blockchain technology to remain viable and relevant, Nigeria must keep up with the pace of development in several core technological aspects that drive the blockchain technology around the globe. These aspects mainly are telecommunications and cryptographic technologies. In the aspect of telecommunications, there are already propositions for the adoption of 5G technology and smart contracts in order to ameliorate the problems posed by the resource intensive nature of blockchain technology (Unal et al. 2020). Also, Nguyen et al. (2020) were able to identify major aspects where 5G can support blockchain technology that include security, system performance, and resource management. However, 5G technology is not fully deployed in Nigeria at the moment but promises useful application in blockchain technology when deployed.

Major stakeholders here will include academics and researchers in telecommunications and cryptographic technologies.

6 Recommendations

Firstly, for adequate benefit realization of this technology, the international standardization bodies need to accelerate work on publication of standards on blockchain and

DLT to ensure safe and secure use of the technologies. Secondly, the availability of the international standards will facilitate regulatory policy implementation and enforcement in various countries and worldwide. Furthermore, the fact that blockchain technology is still rapidly evolving, we recommend that countries that would like to maximize its benefits and at the same time mitigate the risks that may be involved, must put in enough resources to fund research and development.

Presently, open source tools for developing blockchain applications are available and service providers are growing in numbers; there may not be interoperability of independently developed blockchain systems, hence we recommend that various national governments need to be vigilant and take prompt legislative action when necessary to cover all possible use cases and prevent the unethical use of the technology.

A final consideration is the fact that blockchain technology application areas cuts across disciplines relating to privacy, law, governance, business, medicine, agriculture and engineering. These disciplines will directly impact the framework and fundamentals of blockchain technology as it evolves. Hence, we recommend that an interdisciplinary approach to development of regulatory policies and standards for blockchain technology be adopted.

7 Conclusion

Blockchain technology is both emerging and disruptive and will play a vital role in the forthcoming Industry 4.0 because the technology has the potential to provide solution to challenges of trust and security in IoT and Cloud applications. Furthermore, besides Bitcoin, non-financial blockchain applications built using Ethereum and Hyperledger have gone live that have solutions for transparency, immutability and integrity that were challenges for existing traditional applications.

China, Japan, and Hong Kong are ahead in development of regulatory policies and laws for use of blockchain technology, although some countries are not permitting the technology at all, their citizens are trading in bitcoin. Hence, it has become obvious that since the technology is Internet-based it cannot easily be stopped. Furthermore, there are already Blockchain service provider companies who are unleashing non-financial applications that will boost productivity and efficiency in the health, manufacturing, oil and gas sectors. Thus, it has become important for Standards Organizations to step up activities in publishing standards on blockchain and DLT that will enable governments to make policies and regulations for the adoption of blockchain technology. Governments that do not put legislation in place for blockchain technology in the coming years may find their economy at a disadvantaged position relative to the global economy.

References

Albeyatti, A.: MedicalChain. MedicalChain, 1–42 (2017). https://medicalchain.com/en/whitepaper/

Adrian, Z.: Tunisia to Launch E-Dinar National Currency Using Blockchain. Cointelegraph, 9 November 2019. https://cointelegraph.com/news/tunisia-to-launch-e-dinar-national-currency-using-blockchain. Accessed 10 Feb 2020

Allen, D.W.E., Berg, C., Markey-Towler, B., Novak, M., Potts, J.: Blockchain and the evolution of institutional technologies: implications for innovation policy. Res. Policy **49**(1), 103865 (2020)

Anjum, A., Sporny, M., Sill, A.: Blockchain standards for compliance and trust. IEEE J. Cloud Comput. **4**(4), 84–90 (2017)

Danny, N., Anna, B.: Tunisia's Central Bank Denies Reports Claiming It Issued an E-Dinar. Coindesk, 12 November 2019. https://www.coindesk.com/tunisias-central-bank-denies-reports-claiming-it-issued-an-e-dinar. Accessed 10 Feb 2020

Truby, J.: Decarbonizing Bitcoin: Law and policy choices for reducing the energy consumption of Blockchain technologies and digital currencies. Energy Res. Soc. Sci. **44**, 399–410 (2018)

Forkast. Insights: China Blockchain Report 2019–2020: Blockchain is not China's Future – It's the present (2020)

Griggs, K.N., Ossipova, O., Kohlios, C.P., Baccarini, A.N., Howson, E.A., Hayajneh, T.: Healthcare Blockchain system using smart contracts for secure. J. Med. Syst. **42**(7), 130 (2018)

Heung, Y.Y., Skylar, H.: Distributed Ledger Technology Standardization Landscape. Technical Report FG DLT D1.3. Telecommunication Standardization Sector of ITU (ITU-T), p. 6, 1 August 2019

IEEE: IEEE Blockchain Standards (2020). https://blockchain.ieee.org/standards. Accessed 7 March 2020

ISO: ISO/TR 23455:2019, Blockchain and distributed ledger technologies – overview of and interactions between smart contracts in Blockchain and distributed ledger technology systems (2019). https://www.iso.org/standard/75624.html. Accessed 7 Mar 2020

ITU: Focus group on application of distributed ledger technology, 1 August 2019. https://www.itu.int/en/ITU-T/focusgroups/dlt/Pages/default.aspx. Accessed 7 Mar 2020

Jen, S.P., Ashlin, P., Emma, S.: Blockchain and Cryptocurrency in Africa: A Comparative Summary of the Reception and Regulation of Blockchain and Cryptocurrency in Africa. Baker McKenzie, Johannesburg (2018)

Josias, D.: Global Legal Insights – Blockchain & Cryptocurrency Regulation 2019, 1st edn., pp. 279–287. Global Legal Group Ltd. (2018)

Kristin, H.: Hold up: what actually happened in Sierra Leone's "Blockchain" Election? 2 April 2018. https://futurism.com/sierra-leone-election-blockchain-agora. Accessed 10 Feb 2020

Lefan, G., Luping, Y.: Global Legal Insights – Blockchain & Cryptocurrency Regulation 2019, 1st edn., pp. 262–267. Global Legal Group Ltd. (2018)

List of parties to the Paris Agreement. (n.d.). Wikipedia. https://en.wikipedia.org/wiki/List_of_parties_to_the_Paris_Agreement. Accessed 22 June 2020

Lu, H., Huang, K., Azimi, M., Guo, L.: Blockchain technology in the oil and gas industry: a review of applications, opportunities, challenges, and risks. IEEE Access **7**, 41426–41444 (2019). https://doi.org/10.1109/ACCESS.2019.2907695

Neer, V.: Sierra Leone didn't really use blockchain in their election, 20 March 20. https://thenextweb.com/hardfork/2018/03/20/sierra-leone-blockchain-elections/. Accessed 10 Feb 2020

Nguyen, D.C., Pathirana, P.N., Ding, M., Seneviratne, A.: Blockchain for 5G and beyond networks: a state of the art survey. J. Netw. Comput. Appl. **166**(2020). 102693 (2020)

Oliver, P.: Blockchain technology is transforming global politics. Etorox, 7 April 2019. https://www.etorox.com/news/blockchain-technology-is-transforming-global-politics-by-making-elections-transparent/. Accessed 10 Feb 2020

Sam, B.: $10,000 Bitcoin Looks Imminent Based on this One Crucial Chart, 7 February 2020. Cryptocurrency News. https://www.ccn.com/10000-bitcoin-looks-imminent-based-on-this-one-crucial-chart/. Accessed 10 Feb 2020

Tariq, A.: Regulation of cryptocurrency in selected jurisdictions Canada. In: The Law Library of Congress, Global Legal Research Center, pp. 25–29 (2018)

Taro, A.: Global Legal Insights – Blockchain & Cryptocurrency Regulation 2019, 1st edn., pp. 349–358. Global Legal Group Ltd. (2018)

Unal, D., Hammoudeh, M., Kirazc, M.S.: Policy specification and verification for blockchain and smart contracts in 5G networks. ICT Express **6**(2020), 43–47 (2020)

Yomi, K.: The world's first blockchain-supported elections just happened in Sierra Leone. Quartz Africa, 13 March 2018. https://qz.com/africa/1227050/sierra-leone-elections-powered-by-blockchain/. Accessed 10 Feb 2020

Yu, P.H.: Global Legal Insights – Blockchain & Cryptocurrency Regulation 2019, 1st edn., pp. 325–334. Global Legal Group Ltd. (2018)

Yue, X., Wang, H., Jin, D., Li, M., Jiang, W.: Healthcare data gateways: found healthcare intelligence on Blockchain with novel privacy risk control. J. Med. Syst. **40**(10), 218 (2016)

Deep Learning Solutions for Protein: Recent Development and Future Directions

Haruna Chiroma[1], Ali Muhammad Usman[2,3](✉), Fatsuma Jauro[4], Lubna A. Gabralla[5], Kayode S. Adewole[6], Emmanuel Gbenga Dada[7], Fatima Shittu[8], Aishatu Yahaya Umar[9], Julius O. Okesola[10], and Awodele Oludele[11]

[1] Computer Science Department, Kaduna State University, Kaduna, Nigeria
[2] Computer Science Department, Federal College of Education (Technical), Gombe, Nigeria
[3] School of Computer Sciences, Universiti Sains Malaysia, 11800 Pulau Pinang, Malaysia
[4] Ahmadu Bello University, Zaria, Nigeria
[5] Department of Computer Science and IT, Princess Nourah Bint Abdulrahman University, Riyadh, Saudi Arabia
[6] Department of Computer Science, University of Ilorin, Ilorin, Nigeria
[7] Department of Computer Engineering, University of Maiduguri, Maiduguri, Nigeria
[8] Department of Computer Science, Federal Polytechnic, Damaturu, Damaturu, Nigeria
[9] Department of Mathematics, Gombe State University, Gombe, Nigeria
[10] Department of Compter Science, First Technical University, Ibadan, Nigeria
[11] Computer science Department, School of computing and Engineering Sciences, Babcock University Ilisan-Remo, Ikenne, Ogun State, Nigeria

Abstract. The prediction of protein structure has a major role in drugs design and network pharmacology. However, complexity of the protein structure, time consumption and expensive cost incurred by the state-of-the-art methods of predicting protein structure motivated researchers to propose deep learning solutions. Despite the growing applications of deep learning in protein structure, no dedicated comprehensive survey on the exploration of deep learning in protein structure. In this paper, a comprehensive survey on the exploration of deep learning in protein structure is presented. This survey starts by addressing a quantitative and synthesis analysis to show an insight on deep learning in protein structure prediction. Based on the synthesis analysis provided, the survey created a taxonomy of the literature related to exploring deep learning in protein structure. Research challenges and new perspective for future research direction for developing deep learning solutions for protein structure prediction are discussed in details. Specifically, the survey identified that the exploration of nature inspired algorithms in deep learning for protein structure remain untapped. In addition, hybrid/ensemble deep learning architecture is receiving unprecedented attention and growing in number at an exponential rate. In another twist, no single contributor from Africa contributed to the development of this research area. We believed that our survey will served as a starting point for novice researchers and

J. H. Abawajy et al. (Eds.): EATI 2020, LNNS 254, pp. 254–271, 2021.
https://doi.org/10.1007/978-3-030-80216-5_19

it can help expert researchers to quickly identify research direction for developing deep learning solutions for exploring protein structure.

Keywords: Deep learning · Convolutional neural network · Stacked auto-encoder · Deep belief network · Generative adversarial network · Protein structure · Genome · Protein interaction

1 Introduction

The raise of large scale genome sequence data motivated the emergence of bioinformatics. The bioinformatics is a synergy between biology and computer science. The bioinformatics is a research area in health informatics that is gaining tremendous attention from both the industry and academia. Exploring protein structure is one of the hot topics in bioinformatics gaining attention because of it is significance in the health community. In each major biological process, protein [2] and interaction of the protein plays a critical role [34,51]. Structure based drugs design and network pharmacology [43] require the acquisition of knowledge for residual-residual contact information for 2 protein interactions [1,5].

Proteins houses primary structure that contained amino acids chain. The folding of the chain yields 3D structure of the protein. The 3D structure of the protein plays a noteworthy role in it is biological function. Gaining insight on the interaction of proteins structure with proteins, molecules and DNA/RNA required the knowledge of proteins structure. The prediction of proteins structure is needed to unveil the uncovered structure in the proteins sequence. As such, the prediction of the proteins structure becomes a significant task in bioinformatics as well as computational biology. Prediction of the proteins structure is a difficult optimization problem [7]. There are many conventional methods for identification of protein structure but the three main conventional methods used for identifying proteins structure are X-ray diffraction of protein crystals, nuclear magnetic resonance and electron crystallography [48]. However, those conventional methods are very expensive, it takes long time [7], the process is difficult and even impossible to detect structure for some of the proteins [48].

Therefore, researchers in bioinformatics resort to deep learning solutions for the prediction of protein structure to mitigate the challenges of the conventional methods. The deep learning is currently a hot theme in artificial intelligence and it has proven to perform excellently in many other domains. The deep learning has the capability to uncover complex structure from large scale data. The application of the deep learning in protein structure prediction is gaining an extraordinary attention from the bioinformatics research community because it has able to avoid the limitations of the conventional methods and performs better or comparable to the conventional methods.

Despite the growing applications of deep learning architectures in protein structure prediction, to be best of the authors knowledge, no dedicated comprehensive survey on the exploration of deep learning in the prediction of protein structure. This paper, propose to comprehensively survey the exploration of deep learning architectures in the prediction of protein structure.

2 Deep Learning: The Study Background

This section discussed the basic background of deep learning architectures. Many deep learning architectures are found to be applied in the prediction of protein structure. However, we discussed only the major deep learning architectures that received the highest attention in the literature as regard to their application in protein structure prediction. The deep learning architectures are discussed as follows: First, convolutional neural network (ConvNet) is a deep learning architecture of artificial neural network (ANN). The ConveNet is first proposed by [23]. The ConvNet is subsequently modified [22] to improve it is effectiveness. The ConvNet comprises of many layers. The architecture of the ConvNet comprised of convolutional layer, pooling layer, sub-sampling, normalization and fully connected layer all within the hidden layer. In addition, the ConvNet composed of input and output layer, it is training is based on backpropagation algorithm to capture complex pattern in a large scale data set. Secondly, stacked Auto-Encoder (SAE): Basically, the operation of the SAE involved two phases, namely, encoding and decoding. The SAE architecture is constructed by stacking many Auto-Encoders. Typically, the Auto-Encoders are feed forward ANNs. Input vectors are transformed to the hidden layer at the encoder phase through the encoder function. In the decoding phase, the hidden representations are converted back to the original input vector. Both the encoder and decoder functions are non-linear mapping functions [12,39]. Thirdly, hybrid/ensemble (H/E) deep learning architecture: The H/E deep learning models are emerging form of deep learning architectures that combined the power of two or more deep learning architectures to make the H/E deep learning model to be more powerful than the constituent architectures. The hybrid integrates two or more deep learning architectures to produce a powerful result while in ensemble, two or more deep learning architecture are combined but each operate to produce it is own result before the result of individual deep learning models are combined to produce a single result. The other deep learning architecture includes: deep belief network (DBN), generative adversarial network (GAN), deep recurrent neural network (DRNN) and deep neural network (DNN).

3 Exploring Deep Learning in Proteins Structure Prediction

The deep learning architectures were explored for the prediction of proteins structure. Figure 1 is the taxonomy created based on the survey of the papers that published deep learning solutions to protein structure prediction. The taxonomy shows the application of deep learning for protein structure, the variant of the deep learning architecture and the nature of the protein predicted.

The discussion of the deep learning in protein structure prediction is presented as follows:

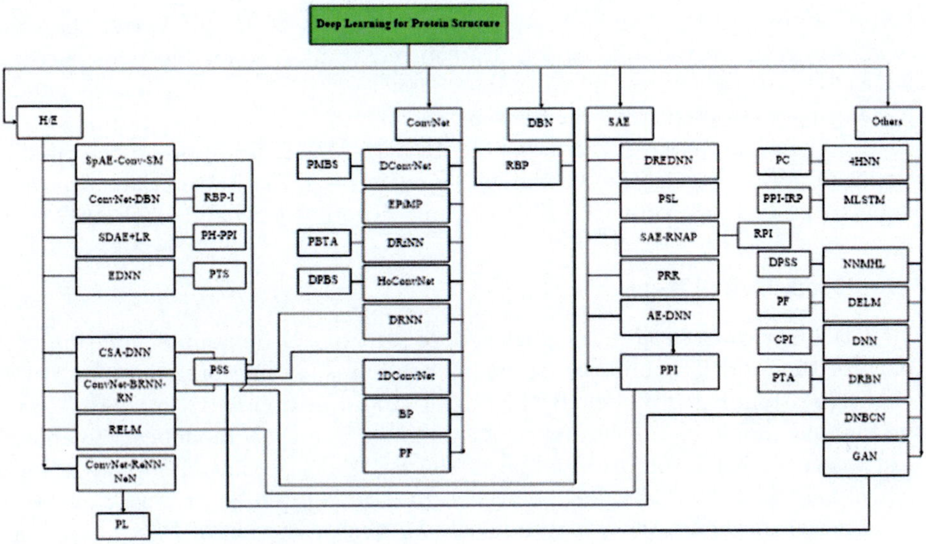

Fig. 1. Taxonomy of the deep learning proteins prediction

3.1 Stacked Auto-Encoder

The SAE and it is variants are found to be applied in protein structure prediction. For example, [49] explore SAE for the prediction of RNA-Protein interaction (RPI) to reduce computational cost and fasten convergence. The high level features of the RNA and proteins sequence were mined by the SAE. The deep learning model built based on the SAE for the RNA-Protein (SAE-RNAP) is applied to predict RPI. The performance is compared with classical prediction methods. The SAE-RNAP is found to be better than the classical tools: RPI-Pred, RPISeq-RF, IPMiner and IncPro. Similarly, [8] proposes SAE to predict protein residue-residue (PRR) contact matrix of interfacial region to mitigate the challenges of its prediction. The Fisher score of the protein sequence was extracted by Hidden Markov Models. Subsequently, the SAE is used to extract and learn the protein hidden abstraction features. The SAE model is evaluated by comparing it with the performance of support vector machine (SVM) based models. The result suggested that the SAE model significantly perform better than the SVM based models. In addition, [46] used SAE to predict protein subcellular locations (PSL) to deviate from manual feature extraction commonly practice by researchers in bioinformatics. The SAE learn the high level abstraction of the protein features representation. The SAE is found to be better than the classical SVM and RF.

[19] propose the hybrid of autoencoder like DNN (AE-DNN) and regularized Laplacian kernel (RLK) (SAE-RLK) to predict protein-protein interaction (PPI). The difference between the AE-DNN and the SAE is that the whole node at the input layer have zero input. The AE-DNN is applied to predict PPI and the performance is compared with the RLK. The result shows that the accuracy of the AE-DNN is better than the RLK in predicting PPI. [40] used SAE for

the prediction of PPI. The SAE achieved an accuracy of 97.19% better than the SVM. [45] propose the application of deep recurrent encoder- decoder network (DREDN) to predict protein secondary structure (PSS). The DREDN model the features extracted from the protein, secondary structure and interaction of the proteins. The experiment indicated that the DREDN improve the prediction accuracy of the PSS (Q3 and Q8) and it performs better than the deep generative stochastic network (GSN), SSpro, RaptorX-SS8 and PSIPRED.

3.2 Deep Belief Network

The DBN is used to explore the structure of proteins. For example, [53] proposes DBN for the modelling and prediction of structural binding preference and sites of the RNA-binding proteins (RBP). General and flexible DBN model is developed for the modelling of the structural binding. The DBN model can be able to model and predict structure binding of the RBP motifs and RBP binding sites respectively. The DBN model achieved comparable or superior performance to the classical methods. [37] proposes DBN for the prediction of PSS to enhance accuracy and convergence speed. The DBN is trained using the protein sequence dataset. The DBN is applied to predict the PSS and it was found to improve the accuracy of the state-of-the-art methods: SSpro, PSIPRED and RaptorX.

3.3 Convolutional Neural Network

The ConvNet is the deep learning architecture that received the highest attention from the researchers in exploring protein structure. For example, [13] applied deep learning architecture to predict binding potential (BP). The deep learning architecture is referred to as the ConvNet. The ConvNet generate a window size fingerprint of the protein as well as the tiny molecules. Operation is performed separately by the atom convolution and softmax to each of the tiny molecules. Subsequently, the fingerprints were transformed in a non-linear fashion. The products were computed and applied for the prediction of the BP. The propose ConvNet is found to be better than the commonly used AutoDock Vina and Smina in the prediction of the BP. [9] used deep neighbor residual network (DNRN) to predict protein secondary structure (PSS). The DNRN comprised on multiple convolution layers, concatenation of features, filters and the neighbor residual unit. The DNRN is model on multiple protein datasets. The DNRN predicted the PSS in either one of eight states (Q8) or one of three states (Q3). The DNRN performance is found to be better than the SPINE-X, PSIPRED, JPRED, RaptorX-SS8, SSPro and Psipred. [52] proposes high-order (Ho) ConvNet (HoConvNet) to predict DNA-protein binding sites (DPBS). The encoding technique is used for the building of the Ho dependencies amongst the nucleotides. The Conv layer of the ConvNet captured the different lengths of the motif features. The HoConvNet is applied to predict the DPBS. Results suggested that the HoConvNet performs better than the DeepBind.

[10] propose deep residual inception neural networks (DRiNN) to predict protein backbone torsion angles (PBTA). The DRiNN model is based on the

combination of inception networks and residual networks. The DRiNN model is applied to predict the PBTA and it is performance is found to be better than the SPIDER3. [17] proposes the use of ConvNet for prediction of protein ubiquitylation site (PUS). The ConvNet is built based on protein data set and subsequently used to predict PUS. The ConvNet is found to be better than the ESA-Ubisite and UbiProber. [41] put forward ConvNet to predict efflux proteins in membrane and transport proteins (EPiMP). The ConvNet is built based on membrane, transport and efflux proteins datasets. The ConvNet model is applied to predict EPiMP and the result indicated that the performance is superior over the QuickRBF. [42] proposes ConvNet for the recognition of protein function (PF). The ConvNet model is developed based on PTS. The ConvNet is able to recognise the PF with high accuracy (Maximum of 88% and average of 81%) compared to already published results. [28] propose 2D ConvNet (2DConvNet) for the prediction of PSS. The 2DConvNet has 6 and 5 Conv and pooling layers respectively. The 2DConvNet is modelled based on protein sequence. The 2DConvNet predict the PSS with better accuracy compared to most of the classical methods e.g. SSpro, SPINE-X, PSIPRED, JPRED among others.

[25] applied ConvNet for the detection of PSS in protein density maps from cryo-electron microscopy. The ConvNet classifier is built based on the secondary structure of the protein. The ConvNet is subjected to extract global features and detect the PSS. The performance of the ConvNet is compared to SVM. The results suggested that the ConvNet performs better than the SVM. [45] proposes Deep fast ConvNet (DF-ConvNet) for the prediction of membrane protein (MP) contact. The DF-ConvNet comprised of 2 residual network, 2 ReLu activation layer and 2 convolution layers. The DF-ConvNet is trained on none MP for the prediction and it is performance is evaluated. The DF-ConvNet performance is better than the other prediction methods such as EVfold, PSICOV, CCMpred and MetaPSICO. [15] proposes deep ConvNet (DConvNet) for the prediction of protein metal binding sites (PMBS). The DConvNet is trained on the features extracted from protein sequence. Subsequently, the DConvNet is applied to predict the PMBS. Experimental result indicated that the DConvNet performs better than the SVM.

[21] proposed a ConvNet based model for identifying clathrin proteins. The model uses 2D-convnet and PSSM profiles for the clathrin protein identification. The convnet accepts parameters from the PSSM profile and predicts if the sequence is a clathrin protein. Comparison with KNN, SVM, MLP, RandomForest and 1D-convnet proved that the 2D-convnet is the best. However, there is need to improve performance by feeding all the PSSM information into the convnet.

3.4 Hybrid/Ensemble Deep Learning Architecture

The H/E architecture of the deep learning is gaining acceptability in proteins structure prediction because of it is performance. For example, [33] propose hybrid of ConvNet and DBN (ConveNet-DBN) to predict the RBP interaction (RBP-I) sites and motifs on the RNAs to ease the interpretation of the results. The ConvNet has better advantage in handling sequence data whereas DBN works well with numerical data. As such, the ConveNet and DBN were hybridized to create ConveNet-DBN for handling multiple heterogeneous data sources. The ConveNet-DBN is applied to predict RBP-I and it is performance is compared with baseline methods. Result suggested that the ConveNet-DBN perform better than the baseline methods. [52] used ensemble deep learning architectures to predict PSS. The ensemble architecture comprised of ConvNet, Bidirectional recurrent neural network (BRNN) and residual network (RN) (ConvNet-BRNN-RN). In the architecture, the ConvNet captured local sequence of features, the BRNN captured the context features and the RN enhance the flow of information between hidden layers and cascaded RNN. The ConvNet-BRNN-RN is trained on benchmark dataset to predict PSS and it was found to achieve better performance compared to the state-of-the-art methods. [48] hybridized colonal selection algorithm (CSA) and DNN with multiple hidden layers (CSA-DNN) for the prediction of PSS. The quality of the protein dataset was enhanced by the CSA and the prediction is performed by the DNN. The CSA-DNN is applied to predict PSS and it is performance is not compared with other methods. The CSA-DNN achieved a success rate of 96.61%. [50] proposes ensemble DNN (EDNN) for the prediction of protein-protein interaction (PPI). The EDNN is built based on amino acid sequences. Different DNN were model with different positive and negative set PPI and output of each is ensemble to create the EDNN to predict PPI. The performance of the EDNN is evaluated. Results demonstrated that the EDNN performs better than the variants of SVM and kNN. [2] propose the hybrid of ConvNet and RBM (ConvNet-RBM) for the prediction of protein tertiary structure (PTS). In the ConvNet-RBM architecture, the RBM extract features that can enhance the prediction of PTS. The ConvNet learn high level features from the protein sequences. The ConvNet-RBM experiment performed on benchmark protein dataset prove it is effectiveness. It is found that the ConvNet-RBM performs better than the SP-Score, CoMOGPHOG-Score and TM-Score.

[4] used hybrid of stacked denoising autoencoder and linear regression (LR) (SDAE+LR) to predict pathogen-host PPI (PH-PPI). The SDAE+LR is modelled based on PH-PPI dataset. The SDAE+LR is subsequently applied to predict the PH-PPI and the result indicated better performance compared to LR. [32] propose combination of deep learning method for the modelling of protein loop (PL). The deep learning method comprises of ConvNet, residual neural network (ReNN) and nearest neighbour (NeN) (ConvNet-ReNN-NeN), the constituent algorithms operates in different capacity before it produced the complete PL model. The ConvNet-ReNN-NeN performance is compared with Nextgeneration KIC (NGK), GalaxyLoop-PS1 (Galaxy PS1) and GalaxyLoop-PS2 (Galaxy PS2). The results show that the performance achieved by ConvNet-ReNN-NeN is better than the state-of-the-art methods or comparable in other case.

[24] hybridize multimodal deep polynomial network (MDPN) and regularized ELM (RELM) for the prediction of PPI. The hybrid MDPN- RELM is model based on the protein sequence. The MDPN fuse modalities to improve PPI prediction accuracy. Subsequently, the RELM predict the PPI. It is found that the MDPN- RELM performs better than the classical methods: ELM and SVM. [6] propose hybrid of DNN for the prediction of PSS. The hybrid DNN architecture

comprised of sparse auto-encoder (SpAE), ConvNet and the softmax classifier (SM). The ConvNet is responsible for extracting features, 13 polypeptides are learned by the SpAE and the polypeptides extracted by the Conv layer are transmitted to the SM. The SpAE-Conv-SM is created for the prediction of the PSS. The results show that the SpAE-Conv-SM performs better than the SVM+ position-specific scoring matrix (PSSM).

[30] proposed an ensemble deep learning for the prediction of bacterial protein function. The ensemble deep learning was developed by stacking five most outstanding DNN models trained using subsets of some dataset and combination of four groups of features; sequence, motif count based features, anomaly, and annotation. The DNNs accepts as input the four categories of features and makes multi-label classification on the created dataset. The ensemble deep learning outperformed DNN, Random forest classifier, and Gaussian Naïve Bayes. However, the model does not consider other protein organisms.

[16] proposed an ensemble deep learning for the prediction of protein intrinsic disorder. The ensemble deep learning architecture was formed by a sequential combination IncReSeNet, LSTM and FC. The ensemble model uses evolutionary information to predict protein disorder. Evaluation results proved the model outperformed an ensemble of protein disorder predictors ESpritz-D (prof), Sequence-based Prediction Online Tools for disorder (SPOT-Disorder), Multilayered Fusion-based Disorder predictor (MFDp), protein disorder regions-various short long PONDR-VSL, DISOPRED, MetaDisorder-md, MobiDB-lite, MetaDisorder-md2, NetSurfp-2.0, meta-server for prediction of intrinsic protein disorder (MetaDisorder), ESpritz-N (prof), MetaDisorder-3D, MFDp2, SPOT-Disorder, ESpritz-X (prof), and Sequence based Prediction with Integrated NEural network for Disordered residues (SPINE-D).

[31] proposed an ensemble deep learning model for classification of protein. The ensemble model was formed by combining different ConvNet models (AlexNet, GoogleNet, and ResNet) trained on a set of 2D multi-view images extracted from 3D protein structures. Evaluation results showed that the ensemble model outperformed autocovariance (AC), quasi residue couple (QRC), TXT, Flexible structure AlignmenT by Chaining Aligned Fragment Pairs with Twists (FATCAT), combinatorial extension (CE), template modeling (TM), and Grassmann Discriminant Analysis GDA.

[47] proposed LSTM autoencoder model for predicting plant IncRNA-protein interaction. The model uses a fusion of multiple features including structural information, gapped k-mer, binary profile, tri-nucleotide, and recursive complement for prediction. LSTM autoencoder with attention layer is adopted for extraction of high level generative features. The model showed better performance compared to Interaction Pattern Miner (IpMiner), RNA-protein interaction with stacked autoencoder (RPI-SAN), and Extreme Gradient Boosting classifier based RNA-protein interaction (XRPI).

[14] proposed a ConvNet-LSTM model for prediction of protein structure property. Specifically, Bagging MSA method was used to compute the PSSM features for training of the model. The ConvNet was used for the extraction of local features and the LSTM (bidirectional) was used to capture long term

dependencies. Experiment showed that the proposed model produces more useful PSSM features for the prediction.

3.5 Other Deep Learning Architecture

This section presents the deep learning models that are unique to the once discussed in the previous sections, as such, classified under others. For example, [43] proposes the application of DNN to predict compound protein interaction (CPI) so as to improve the prediction performance of the CPI. The chemical fingerprints of the compound and the features of the protein were extracted from the datasets. The DNN is applied to learn the CPI representation effectively and efficiently. The DNN CPI prediction performance is compared with the state-of-the-art methods: SVM and linear regression models (LR) on imbalance and balance datasets. The DNN model performs better than the SVM and LR. [20] used deep extreme learning machine (DELM) for the prediction of protein fold (PF). Four features extraction techniques were used to extract features from amino acid sequence for modelling DELM to improve it is performance. The DELM is applied to predict the PF. The result performs better than the already published SVM based methods.

[11] applied the neural network with multiple hidden layers (NNMHL) to predict docked protein structure similarity (DPSS). The unbound complexes are used for the modelling of the NNMHL. The NNMHL model is applied to predict DPSS. The performance of the NNMHL is compared with the DBN and two-layer neural network (2LNN). Results indicated that the NNMHL performs better. [54] proposes multi-layered long-short term memory networks (MLSTM) to predict PPI interface residue pair (PPI-IRP). The MLSTM model is built based on dimers in the unbound state. Physicochemical features for the local residue pairs is used in place of single residue pairs. The MLSTM model is applied to predict the PPI-IRP. The performance indicated improvement over PPiPP and PAIRPred methods.

[18] propose the use of deep neural block cascade networks (DNBCN) for translating sequence (TS), PSS as case study. Each layer in DNBCN has one or more feature extractor. Experimental study show that the DNBCN has the ability to model internal relationship between amino acid and PSS as well as modelling of complex sequence and dependency. The DNBCN achieved better performance compared to SSPro, RaptorX, MSC, and stacks of ConvNet and conditional random fields, and the Multi-scale ConvNet (DeepCNF). [27] propose the application of generative adversarial network (GAN) to predict protein loop (PL). The generative component of the GAN captured the loop region and the missing region is predicted. The adversarial component of GAN makes the PL prediction to appear as the actual. The GAN performs better than the state-of-the-art methods: NGK, Galaxy PS1, and Galaxy PS2.

[25] compared the performance of four different deep learning architectures, SPIDER2 and ANGLOR state of the art methods for the prediction of protein torsion angle (PTA). The deep learning architectures involved: DNN, deep restricted Boltzmann machine (DRBN), DRNN and deep recurrent restricted

Boltzmann machine (DRERBM). All the methods predict the PTA. The performance of the deep learning architectures perform better or comparable to the SPIDER2 and ANGLOR. [38] proposes 4-multiple hidden layer network (4HNN) for the prediction of protein contact (PC). The propose method combine evolutionary, physicochemical, and sequence-based information. The deep (D4HNN) is used to predict long-range contacts. The performance of the D4HNN is found to be better than the CASP11.

[44] proposed a DNN approach for protein quality assessment. The proposed approach follows a two-step process (training data preparation and DNN training). Important sequence and feature information are first extracted from adequate amount of proteins whose structures have been known to be similar to the aimed protein sequence, they are then used to train the DNN to predict global distance test-total score (GDT-TS scores). The proposed algorithm outperformed MUfoldQA-S, Modfold, Proq3, and MUFold2. However, there is need to minimize the training time.

[26] proposed a DNN for prediction of species specific S-glutathionylation. The model uses binary encoding profiles (BE), amino acid composition (AAC), position-specific scoring matrix (PSSM), and composition of k-spaced amino acid pairs (CKSAAP) to extract feature sequences. The DNN uses particle swarm optimization (PSO) for optimization of hyperparameters to predict S-glutathionylation sites. The model outperformed Sun's method, predictions of S-glutathionylated sites (PGlus), and species-specific protein S-glutathionylation (SSGlu). More features such as protein-protein interaction, evolutionary information, and others need to be included to improve accuracy of the model.

4 Performance Metrics

Evaluating the performance of deep learning application in protein structure prediction is based on measurement metrics. The deep learning solutions to protein structure prediction heavily depends on several performance metrics as depicted in Fig. 2. The performance metrics depicted in Fig. 2 were extracted from the literature as used by researchers to measure the performance of their deep learning models in predicting structure of protein.

5 General Overview of the Research Area

It has been shown from the survey that the deep learning solutions to protein structure prediction has received a tremendous attention from the researchers. The deep learning solutions are able to enhance the accuracy and speed up convergence. The use of laborious state-of-art methods of protein structure prediction is on the declining state while the deep learning solution is on the increase. The application of deep learning in protein structure exploration has shown good performance. It is able to perform better or comparable to the state-of-the-art methods. The deep learning has gain prominence likely because of the ability

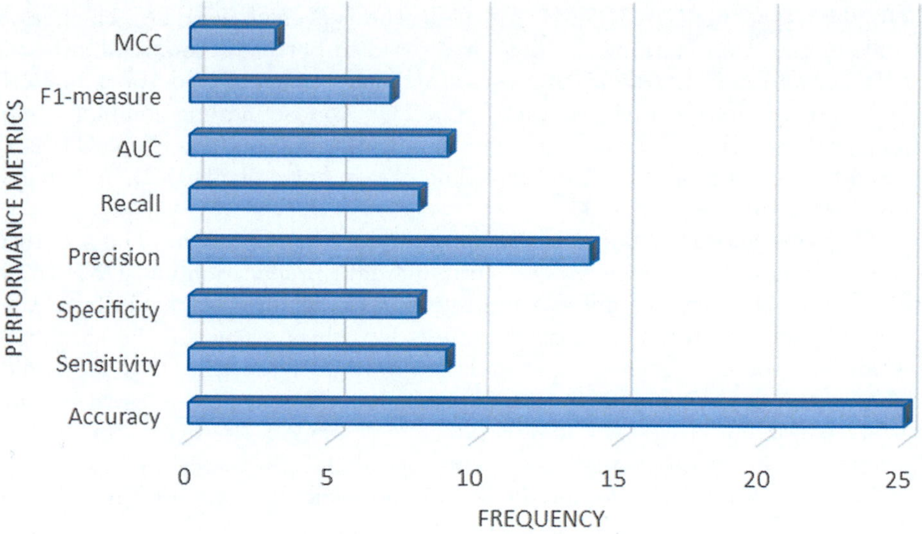

Fig. 2. Performance metrics extracted from the survey

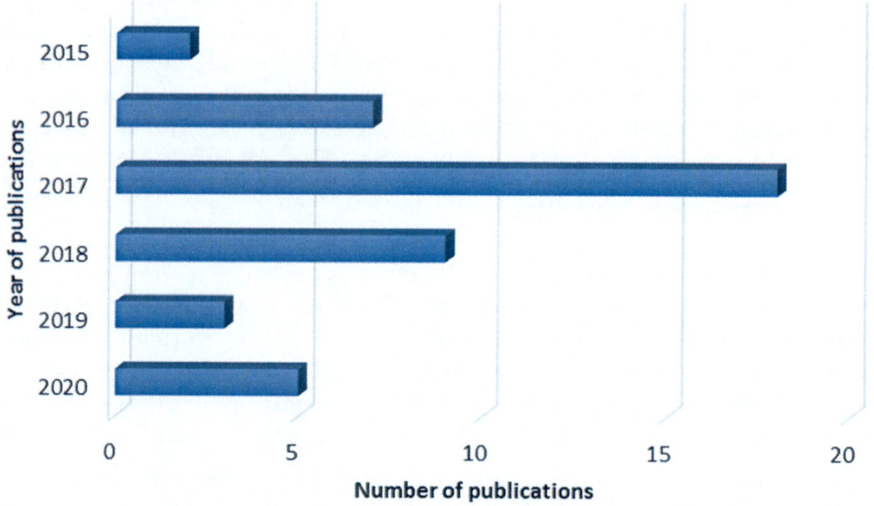

Fig. 3. Trend of publications

of the models to deal with large scale dataset, model complex protein structure, reduces time consumption, reduce cost and intensive labour required by the state-of-the-art methods. The deep learning has the ability to capture long range of protein sequence interaction as agued in [28]. The trend of publications in this research area is increasing as depicted in Fig. 3. It shows that the deep learning has started gaining attention in the protein structure prediction from

2015. From 2015, the trend of the publications is increasing up to current times. The highest number of publications is witnessed in 2017, already 2018 publications is on the increase but not up to 2017. Few publications were obtained in 2019. There are a number of publications in 2020 and being the active year, more publications should be expected. The trend of the publications is expected to grow into the future because of the attention been witness from researchers in this research area. Figure 4 indicated the deep learning architectures found to be applied for the prediction of protein structure. The proposing of H/E got the highest number of applications. This shows that the application of H/E deep learning architecture in protein structure prediction supersedes single form deep learning architecture. Evidence from the literature suggested that the H/E deep learning architecture performance better than the constituent architectures.

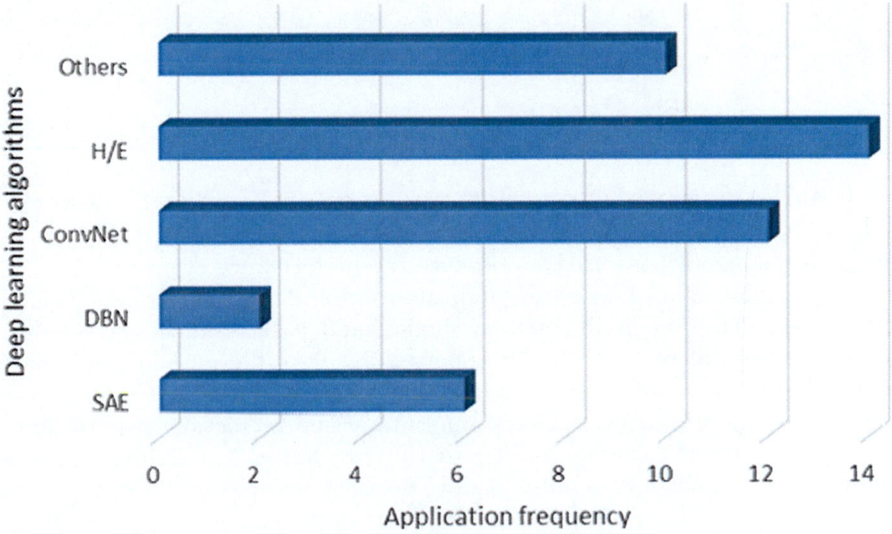

Fig. 4. Deep learning architectures

The ConvNet received the highest number of attention (in constituent algorithms) from the researchers likely because of it is ability to capture high level abstraction in protein sequence data and its excellent performance in dealing with classification problems.

Figure 5 depicted the countries and number of authors that contributed to the development of this research. As shown on Fig. 5, the highest number of contributors are from China as indicated by the longest bar. Followed by the USA, the competition is more of between China and USA as regard to the development of this research area. In Africa, only one contributor from Kenya contributed to the development of the research area. This signifies that research in this area is very low or absent in Africa despite the significance of the research area in drugs design.

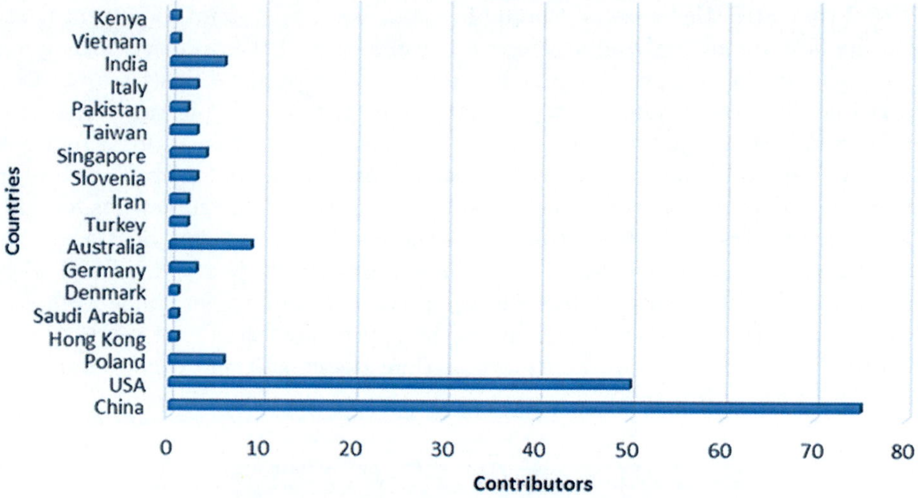

Fig. 5. Contributing countries and authors

6 Challenges and New Perspective for Future Research

The performance metrics used to measure the performance of protein structure prediction is not uniform as clearly indicated in Fig. 2. The performance metrics in Fig. 2 are extracted from different studies and it has indicated that different studies use different metrics. It is very difficult to find any justification for chosen the performance metrics used in the different studies. This is a serious issue because performance of the protein interaction prediction or modelling of the protein structure can mean different thing to different individual. As such, it can result to confusion or miss leading decision because of different decision that might arise as a result of this inconsistent performance metrics. Having a uniform performance metrics can help to provide universally acceptable performance metrics that can be relied upon for researchers or decision makers to take uniform decision. We recommend the research community to devote high level synergy to start proposing a systematic framework for uniform performance metrics to measure the success or failure of the prediction of protein structure.

Despite the significant efforts made by the research community in exploring PPI, yet, [50] argued that copious un-expose PPI still exist. We recommend the research community to further apply deep learning models such as GAN, DBN and DRNN that received little attention to be explored to uncover additional PPI structure.

It has been unveiled that the application of nature inspired meta-heuristic algorithm e.g. but not limited to cuckoo search algorithm, firefly algorithm, Bat algorithm, and flower pollination algorithms for optimising the hyperparameters e.g. weights, of the deep learning architecture to predict the structure of protein is not been explored in the literature. Presently, the research community heavily relied on manual/trials setting of the deep learning hyperparameters. The manual setting is tedious, time consuming and prompt to error. The application of the nature inspired algorithm to optimize the hyperparameters of the deep learning architecture can automatically determine the optimal hyperparameters of the deep learning architecture without human intervention. Thereby making the deep learning architecture to be heading towards autonomous deep learning model. The optimization of the deep learning model through nature inspired algorithms has improved accuracy and reduce computational time as evident in [29]. The optimization of the deep learning hyperparameters has been proven to enhance the accuracy of the conventional deep learning model [3,35]. In the future, to improve the performance of the deep learning in modelling the structure of protein interaction and prediction, we suggest researchers to apply nature inspired algorithms in the optimization of the deep learning architecture as a new perspective in this research area.

The dataset mostly used for conducting research in this area are benchmark datasets. Evaluating deep learning model on only benchmark dataset might not show the actual performance of the deep learning model on protein structure. It is possible for an algorithm to perform well on benchmark dataset but perform differently when applied on real world dataset. Therefore, it is worth to evaluate a propose deep learning model on both real world and benchmark datasets. To effectively evaluate deep learning architecture for the prediction of protein structure and it is modelling, we believed that both benchmark and real world datasets should be used. This will really unveil the true performance of the deep learning model. Large scale data set is required to discover more structure and sequence of proteins [28]. Big data platforms that support deep learning like apache spark can be used by researchers to handle large scale data set to model very large sequence and structure.

Despite the strength of the deep learning models in predicting the structure of protein, those deep learning models have their weaknesses. The common issue among the deep learning model is that many hyperparameters requires settings and large scale dataset is required for a typical deep learning model to perform excellently. In addition, deep learning required high performance computing system equipped with GPUs for training the models. As regard to individual deep learning models each has it is own unique challenges. For example, in ConvNet, too much layers are needed to find the whole hierarchy of the features. The initial process and selection in DBN causes the architecture to incur high computational cost. The Auto-Encoder require pre-training phase before full engagement of the Auto-Encoder learning process. In the future, we recommend researchers to devote resources towards mitigating these limitations. This will make the deep learning application in bioinformatics – structure of protein interaction modelling and prediction to be more flexible, robust and realistic in real world applications especially in the discovery of drugs.

The training of the deep learning models through reinforcement learning in this domain received little attention from research despite the fact that it is feasible and has it is own merits as shown in [36]. We suggest the exploration of applying reinforcement learning in deep learning for predicting protein structure. More deep learning models should be explored in the prediction of interface residue pairs from double monomer proteins because it remained an open challenge as pointed in [54].

7 Conclusions

The paper proposed a survey, find out challenges and suggest future research directions to researchers regarding deep learning exploration in protein structure. We surveyed researches conducted on the prediction of protein structure based on deep learning. For each of the paper surveyed, the deep learning architecture applied, protein problem solved, solutions, performance metrics and evaluation technique used were pointed out. A new taxonomy is created based on the deep learning architecture and the protein structure predicted. The surveyed unveiled that deep learning is receiving tremendous attention from researchers especially ConvNet, SAE and H/E architectures. The trend of publication shows that the exploration of deep learning in protein structure is rapidly growing and we belief it will continue to grow at an exponential rate into the future. Lastly, the challenges and future research directions are outlined for easy identification by researchers intending to contribute to the development of protein structure based on deep learning.

References

1. Azzarito, V., et al.: Inhibition of α-helix-mediated protein-protein interactions using designed molecules. Nat. Chem. **5**(3), 161–173 (2013)
2. Bai, L., Yang, L.: A unified deep learning model for protein structure prediction. In: 2017 3rd IEEE International Conference on Cybernetics (CYBCONF), pp. 1–6. IEEE (2017)
3. Baldominos, A., Saez, Y., Isasi, P.: Evolutionary convolutional neural networks: an application to handwriting recognition. Neurocomputing **283**, 38–52 (2018)
4. Chen, H., et al.: Leveraging stacked denoising autoencoder in prediction of pathogen-host protein-protein interactions. In: 2017 IEEE International Congress on Big Data (BigData Congress), pp. 368–375. IEEE (2017)
5. Xue-wen Chen and Jong Cheol Jeong: Sequence-based prediction of protein interaction sites with an integrative method. Bioinformatics **25**(5), 585–591 (2009)
6. Chen, Y.: Long sequence feature extraction based on deep learning neural network for protein secondary structure prediction. In: 2017 IEEE 3rd Information Technology and Mechatronics Engineering Conference (ITOEC), pp. 843–847. IEEE (2017)
7. Cheng, J., et al.: SCRATCH: a protein structure and structural feature prediction server. Nucleic Acids Res. **33**(*suppl_2*), W72–W76 (2005)
8. Tianchuan, D., et al.: Prediction of residue-residue contact matrix for protein-protein interaction with Fisher score features and deep learning. Methods **110**, 97–105 (2016)

9. Fang, C., Shang, Y., Xu, D.: A new deep neighbor residual network for protein secondary structure prediction. In: 2017 IEEE 29th International Conference on Tools with Artificial Intelligence (ICTAI), pp. 66–71. IEEE (2017)
10. Fang, C., Shang, Y., Dong, X.: Prediction of protein backbone torsion angles using deep residual inception neural networks. IEEE/ACM Trans. Comput. Biol. Bioinf. **16**(3), 1020–1028 (2018)
11. Farhoodi, R., Akbal-Delibas, B., Haspel, N.: Accurate prediction of docked protein structure similarity using neural networks and restricted Boltzmann machines. In: 2015 IEEE International Conference on Bioinformatics and Biomedicine (BIBM), pp. 1296–1303. IEEE (2015)
12. Gehring, J., et al.: Extracting deep bottleneck features using stacked auto-encoders. In: 2013 IEEE International Conference on Acoustics, Speech and Signal Processing, pp. 3377–3381. IEEE (2013)
13. Gonczarek, A., et al.: Interaction prediction in structure-based virtual screening using deep learning. Comput. Biol. Med. **100**, 253–258 (2018)
14. Guo, Y., et al.: Bagging MSA learning: enhancing low-quality PSSM with deep learning for accurate protein structure property prediction. In: International Conference on Research in Computational Molecular Biology, pp. 88–103. Springer (2020)
15. Haberal, I., Ogul, H.: DeepMBS: prediction of protein metal binding-site using deep learning networks. In: 2017 Fourth International Conference on Mathematics and Computers in Sciences and in Industry (MCSI), pp. 21–25. IEEE (2017)
16. Hanson, J., et al.: SPOT-Disorder2: improved protein intrinsic disorder prediction by ensembled deep learning. Genomics Proteomics Bioinform. **17**(6), 645–656 (2020)
17. He, F., et al.: A multimodal deep architecture for large-scale protein ubiquitylation site prediction. In: 2017 IEEE International Conference on Bioinformatics and Biomedicine (BIBM), pp. 108–113. IEEE (2017)
18. Yu, H., et al.: Sequence translating model using deep neural block cascade network: taking protein secondary structure prediction as an example. In: 2018 IEEE International Conference on Big Data and Smart Computing (BigComp), pp. 58–65. IEEE (2018)
19. Huang, L., Liao, L., Wu, C.H.: Completing sparse and disconnected protein-protein network by deep learning. BMC Bioinform. **19**(1) (2018). Article number: 103
20. Ibrahim, W., Abadeh, M.S.: Protein fold recognition using Deep Kernelized Extreme Learning Machine and linear discriminant analysis. Neural Comput. Appl. **31**(8), 4201–4214 (2019)
21. Le, N.Q.K., et al.: Identification of clathrin proteins by incorporating hyperparameter optimization in deep learning and PSSM profiles. Comput. Methods Programs Biomed. **177**, 81–88 (2019)
22. LeCun, Y., et al.: Gradient-based learning applied to document recognition. Proc. IEEE **86**(11), 2278–2324 (1998)
23. LeCun, Y., et al.: Handwritten digit recognition with a back-propagation network. In: Advances in Neural Information Processing Systems, pp. 396–404 (1990)
24. Lei, H., et al.: Protein-protein interactions prediction via multimodal deep polynomial network and regularized extreme learning machine. IEEE J. Biomed. Health Inform. **23**(3), 1290–1303 (2018)
25. Li, H., et al.: Deep learning methods for protein torsion angle prediction. BMC Bioinform. **18**(1) (2017). Article number: 417

26. Li, S., et al.: Deep learning based prediction of species-specific protein S-glutathionylation sites. Biochimica et Biophysica Acta (BBA)-Proteins Proteomics **1868**(7), 140422 (2020)
27. Li, Z., et al.: Protein loop modeling using deep generative adversarial network. In: 2017 IEEE 29th International Conference on Tools with Artificial Intelligence (ICTAI), pp. 1085–1091. IEEE (2017)
28. Liu, Y., et al.: Protein secondary structure prediction based on two dimensional deep convolutional neural networks. In: 2017 3rd IEEE International Conference on Computer and Communications (ICCC), pp. 1995–1999. IEEE (2017)
29. Ma, M., Sun, C., Chen, X.: Discriminative deep belief networks with ant colony optimization for health status assessment of machine. IEEE Trans. Instrum. Measur. **66**(12), 3115–3125 (2017)
30. Mishra, S., et al.: A deep learning ensemble for function prediction of hypothetical proteins from pathogenic bacterial species. Comput. Biol. Chem. **83**, 107147 (2019)
31. Nanni, L., et al.: iProStruct2D: identifying protein structural classes by deep learning via2D representations. Expert Syst. Appl. **142**, 113019 (2020)
32. Nguyen, S.P., et al.: New deep learning methods for protein loop modeling. IEEE/ACM Trans. Comput. Biol. Bioinform. **16**(2), 596–606 (2017)
33. Pan, X., Shen, H.-B.: RNA-protein binding motifs mining with a new hybrid deep learning based cross-domain knowledge integration approach. BMC Bioinform. **18**(1) (2017). Article number: 136
34. Pitre, S., et al.: Computational methods for predicting protein-protein interactions. In: Protein-Protein Interaction, pp. 247–267. Springer (2008)
35. Rosa, G., et al.: Learning parameters in deep belief networks through firefly algorithm. In: IAPR Workshop on Artificial Neural Networks in Pattern Recognition, pp. 138–149. Springer (2016)
36. Šilc, J., Papa, G., et al.: Construction of heuristic for protein structure optimization using deep reinforcement learning. In: International Conference on Bioinspired Methods and Their Applications, pp. 151–162. Springer (2018)
37. Spencer, M., Eickholt, J., Cheng, J.: A deep learning network approach to ab initio protein secondary structure prediction. IEEE/ACM Trans. Comput. Biol. Bioinf. **12**(1), 103–112 (2014)
38. Stahl, K., Schneider, M., Brock, O.: EPSILON-CP: using deep learning to combine information from multiple sources for protein contact prediction. BMC Bioinform. **18**(1) (2017). Article number: 303
39. Sun, M., Zhang, X., Zheng, T.F., et al.: Unseen noise estimation using separable deep auto encoder for speech enhancement. IEEE/ACM Trans. Audio Speech Lang. Process. **24**(1), 93–104 (2015)
40. Sun, T., et al.: Sequence-based prediction of protein protein interaction using a deep-learning algorithm. BMC Bioinform. **18**(1), 1–8 (2017)
41. Taju, S.W., Ou, Y.-Y., et al.: Using deep learning with position specific scoring matrices to identify efflux proteins in membrane and transport proteins. In: 2016 IEEE 16th International Conference on Bioinformatics and Bioengineering (BIBE), pp. 101–108. IEEE (2016)
42. Tavanaei, A., et al.: Towards recognition of protein function based on its structure using deep convolutional networks. In: 2016 IEEE International Conference on Bioinformatics and Biomedicine (BIBM), pp. 145–149. IEEE (2016)
43. Tian, K., et al.: Boosting compound-protein interaction prediction by deep learning. Methods **110**, 64–72 (2016)

44. Wang, J., Wang, W., Shang, Y.: A new approach of applying deep learning to protein model quality assessment. In: 2019 IEEE International Conference on Bioinformatics and Biomedicine (BIBM), pp. 2387–2392. IEEE (2019)
45. Wang, Y., Mao, H., Yi, Z.: Protein secondary structure prediction by using deep learning method. Knowl.-Based Syst. **118**, 115–123 (2017)
46. Wei, L., et al.: Prediction of human protein subcellular localization using deep learning. J. Parallel Distrib. Comput. **117**, 212–217 (2018)
47. Wekesa, J.S., Meng, J., Luan, Y.: Multi-feature fusion for deep learning to predict plant lncRNA-protein interaction. Genomics **112**(5), 2928–2936 (2020)
48. Yavuz, B.Ç., Yurtay, N., Ozkan, O.: Prediction of protein secondary structure with clonal selection algorithm and multilayer perceptron. IEEE Access **6**, 45256–45261 (2018)
49. Yi, H.-C., et al.: A deep learning framework for robust and accurate prediction of ncRNA-protein interactions using evolutionary information. Mol. Ther.-Nucleic Acids **11**, 337–344 (2018)
50. Zhang, L., et al.: Protein-protein interactions prediction based on ensemble deep neural networks. Neurocomputing **324**, 10–19 (2019)
51. Zhang, Q.C., et al.: Structure-based prediction of protein-protein interactions on a genome-wide scale. Nature **490**(7421), 556–560 (2012)
52. Zhang, Q., Zhu, L., Huang, D.-S.: High-order convolutional neural network architecture for predicting DNA-protein binding sites. IEEE/ACM Trans. Comput. Biol. Bioinform. **16**(4), 1184–1192 (2018)
53. Zhang, S., et al.: A deep learning framework for modeling structural features of RNA-binding protein targets. Nucleic Acids Res. **44**(4), e32–e32 (2016)
54. Zhao, Z., Gong, X.: Protein-protein interaction interface residue pair prediction based on deep learning architecture. IEEE/ACM Trans. Comput. Biol. Bioinform. **16**(5), 1753–1759 (2017)

Sentiment Analysis of Student Evaluations of Teaching Using Deep Learning Approach

Edi Sutoyo[1(✉)], Ahmad Almaarif[1], and Iwan Tri Riyadi Yanto[2]

[1] Department of Information Systems, Telkom University, Bandung, West Java, Indonesia
{edisutoyo,ahmadalmaarif}@telkomuniversity.ac.id

[2] Departement of Information System, Universitas Ahmad Dahlan, Yogyakarta, Indonesia
yanto.itr@is.uad.ac.id

Abstract. One part of improving and developing lecturer pedagogical competence and evaluating the substance of the courses taught in the lecture process is to use Lecturer Evaluation Effectiveness by Students (EDOM). The EDOM is a feedback questionnaire that can be used by educators to assess and evaluate pedagogical competencies that they have and aims to make continuous improvements to various matters related to the teaching process in the classroom. In addition to questions in the form of a Likert scale, EDOM also accommodates open-ended questions. However, this type of question tends not to be considered in the feedback assessment and is not explored more deeply to gain insight. Therefore, in this study text mining was carried out to find out the sentiments given by students, whether positive or negative sentiments using Convolutional Neural Network (CNN). The experimental results showed that CNN was able to achieve accuracy, precision, recall, and F1-Score of 87.95%, 87%, 78%, and 81%, respectively.

Keywords: Sentiment analysis · Opinion mining · Teaching evaluation · Convolutional Neural Network · Deep learning

1 Introduction

Evaluation is one of a series of activities in improving the quality, performance, or productivity of an institution in implementing its program. The focus of the evaluation is the individual, that is, the learning achievement achieved by the group or class. Through the evaluation will be obtained information about what has been achieved and what has not been achieved. Furthermore, this information is used to improve the program. The main function of evaluation, in this case, is to provide information that is useful for the decision-maker to determine the policy to be taken based on the evaluation that has been done [1].

In educational institution entities, one of the usual evaluations is giving feedback on evaluating lecturer performance. Feedback is a statement sent to the entity about its past behavior from which the entity can analyze future and current behavior to achieve the expected results. Feedback plays an important role in education and learning by helping to adopt new knowledge and prevent recurring mistakes. Feedback is a process

J. H. Abawajy et al. (Eds.): EATI 2020, LNNS 254, pp. 272–281, 2021.
https://doi.org/10.1007/978-3-030-80216-5_20

that helps organizations to monitor, evaluate, and manage the overall work environment. Good feedback practices provide useful information for organizations in improving teaching and learning experiences. At the end of the semester, in every university, it is common to conduct a lecturer evaluation process to determine the effect of lecturer teaching on students. Good teaching will certainly greatly help students to achieve good learning. The quality of teaching and academic standards certainly needs to be evaluated and improved because higher education is an important activity [2].

Quality of teaching, academic standards and student satisfaction are closely related to the match between expectations and the reality of the quality of educational services obtained from universities. If this educational institution treats students as customers by evaluating the gap between expectations and reality felt by students on the quality of educational services, the university can prepare an appropriate strategic plan to improve its quality. Evaluation of student satisfaction can be used to determine factors of quality of educational services that need to be improved, maintained, or even reduced will result in misallocation of resources such as funds, labor, and time. Incorrect allocation of resources causes ineffective efforts to improve quality and reduce student satisfaction.

Teaching and learning processes and the creation of a conducive learning atmosphere are carried out through innovative learning processes using electronic facilities/infrastructure and the latest methods in the teaching and learning process. To ensure a good teaching and learning process, regular monitoring of the implementation of learning activities is carried out, both regarding the frequency of attendance of lecturers/students and the suitability of the lecture substance discussed with the Semester Learning Plan. Evaluating and monitoring using Lecturer Evaluation Effectiveness by Students (EDOM) is conducted once per semester, in addition to evaluating lecturer performance, it is also used as a tool to evaluate how far learning targets have been met. The evaluation aims to improve the quality of learning including evaluating the way the material is delivered and the presence of lecturers, the material content of each course, student motivation and difficulties arising in the process of interaction between lecturers and students.

This feedback is distributed to students, while the result of the EDOM is intended to be information for stakeholders to improve the quality of learning. In the personal structure of lecturers, EDOM is expected to be a reflection as well as a means to improve the lecturers and develop potential of the lecturers. Whereas in the level of management/governance the results of EDOM are expected to be information that can be used as a reference in developing work programs related to improving the quality of the learning process and lecturer performance.

The feedback mechanism on EDOM, questions can be categorized in the form of scale questions (score based on Likert scale) and open-ended questions. The Likert Scale is a psychometric response scale mainly used in questionnaires to obtain participant preferences or the level of agreement with statements or sets of statements. Students are asked to indicate the level of the agreement through statements given by the ordinal scale. Likert scale-based score questions are given to students and asked to answer these questions using a ranking-based scale. These types of questions do not usually adequately represent student sentiment, while open-ended questions are sometimes not considered in feedback assessments and are not explored deeper to gain insight.

Sentiment Analysis (SA) has a significant role in various fields, including education because it can be used to assess the effectiveness of learning technology from student feedback [3]. Learning outcomes can be assessed in two ways, directly or indirectly. Direct assessment considers the results of student work such as midterm exam, final exam, assignments, quizzes, and project reports. Meanwhile, indirect assessment is based on feedback made by students about learning experiences and the quality of teaching. This indirect assessment can be done by conducting student feedback sentiment analysis. This sentiment analysis analyzes student feedback, whether it is a formal survey conducted at the end of the semester or informal comments on social media, which can be used to find out student opinions, evaluate lecturers' performance and to identify areas that can be improved through continuous improvement actions [3].

One of the most important processes in sentiment analysis is the determining of feature selection used to sentiment classifiers that classify sentiments. Algorithms in machine learning are commonly used as an approach to feature selection [4–6]. However, the approach of using machine learning in some cases still has not reached high accuracy. Accordingly, several researchers propose an approach using deep learning for improving the accuracy and computational time [7–10]. Therefore, in this study, one of the deep learning algorithms called Convolutional Neural Network (CNN) is used for sentiment analysis of student evaluations of teaching.

The remainder of this article is explained as follows: Sect. 2 explains the related work and also theoretical background. Section 3 describes the dataset used, elaborates on the results of experiments, and evaluating performance. And finally, this research is concluded in Sect. 4.

2 Theoretical Background

Deep learning is about learning several levels of representation and abstraction which help to understand data such as images, sounds, and text. Convolutional neural network (CNN) is one of the deep learning methods that can be applied to classify text documents. Along with the development of computing using the Graphical Processing Unit (GPU) to make the process of training the model on the CNN algorithm also becomes faster [11]. In this study, we apply the convolutional neural network method to classify texts based on positive and negative polarity which includes sentiment analysis.

2.1 Convolutional Neural Network (CNN)

The first layer stores words in a low dimensional vector. The next layer runs convolutions using multiple filter sizes. For example, shifting 3, 4, or 5 words at a time. Next, max-pool the results of the convolutional layer into a long feature vector, add a dropout regularization, and classify the results using the SoftMax layer [12]. In this experiment, the architectural pattern of the model is shown as in Fig. 1.

CNN is a mathematical construct that is usually composed of 3 types of layers, namely convolution, pooling, and a fully connected layer. The first two layers, convolution, and pooling do feature extraction, while the third layer, places extracted features into the final output, such as classification [13].

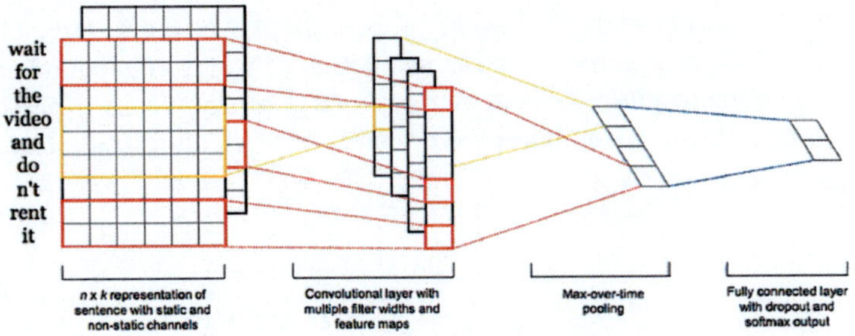

Fig. 1. The architecture of the multichannel CNN [7]

Convolution is a special type of linear operation used for feature extraction, where an array of numbers, called the kernel is implemented at the input, which is an array of numbers called tensors. Calculations are performed on each kernel matrix element and tensor input and summed to get an output called a feature map. This stage is repeated until a feature map is obtained that represents the characteristics of the tensor input. Therefore, the kernels of other sizes can have different extraction results. Convolution propagation calculations are shown in Eq. 1.

$$a_{ij}^{(k)} = \sum_{s=0}^{m-1} \sum_{t=0}^{n-1} W_{st}^{(k)} x_{(i+s)(j+t)} + b^{(k)} \tag{1}$$

where
x: is the input
a^k: after convolution
k: kernel index (weight filter)
W: kernel (weight filter)
b: bias

Pooling is also called down sampling which reduces the dimensionality of the field while maintaining important information. The pooling operation that is often used is max pooling, which is capable of extracting from the filter features input filter, then selecting the highest value in each filter and discarding other values. Pooling has several types, in this case, 1D global max pooling is used. 1D Global max pooling takes a vector and calculates the maximum value of all values for each input channel.

The Fully Connected (FC) Layer is a fully connected layer of neurons. All neurons in the FC layer have a relationship with all activations in the previous layer. This layer is placed before the results of the CNN classification. The Dropout layer is also applied to CNN to overcome overfitting. The dropout technique overcomes overfitting by temporarily removing the contribution to neuron activation during the forward pass, and any weight updates are not applied to the neurons when the backward pass.

The output of feature maps from the final convolution or pooling layer is usually flattened as it is transformed into a vector dimension and connected to one/more fully connected layers (FC) or also called dense layers, where each input connected to each output has a weight. After going through feature extraction at the convolution layer and reducing the sample at the pooling layer, then the output is mapped by the FC layer to the final output of the network, such as the probability of each class on the classification task. The final FC layer usually has the same number of nodes as the number of classes.

The output of the penultimate convolutional and pooling layers x is passed to a fully connected softmax layer. It computes the probability distribution using Eq. 2.

$$P(y = j|\mathrm{x}, \mathrm{s}, \mathrm{b}) = \mathrm{softmax}_j(x^T w + b) = \frac{e^{x^T} w_j + b_j}{\sum_{k=1}^{K} e^{x^T} w_k + b_k} \tag{2}$$

where w_k is the weight and b_k is the bias of the k-th class.

2.2 Sentiment Analysis

Sentiment Analysis is a method used to process various opinions given by consumers or experts through various media, regarding a product, service, or an agency. Sentiment Analysis is a method used to understand, extract opinion data, and process textual data automatically to get a sentiment contained in an opinion. In sentiment analysis, there are 3 types of opinions, namely positive, neutral, and negative opinions, so that the 3 types of opinions can be used by companies to determine consumer responses to a product or service [14]. But many researchers only use 2 types of sentiments, namely positive and neutral [16–18].

Sentiment analysis is one of the fields of Natural Language Processing (NLP) that builds systems for recognizing and extracting opinions in text form. Information in the form of text is currently widely available on the internet in the format of forums, blogs, social media, and sites containing reviews. The data can explain public opinion about products, brands, services, politics, or other topics. Companies, governments, and other fields then use these data to make marketing analyzes, product reviews, product feedback, and community services.

3 Result and Discussion

3.1 System Architecture

In general, the system architecture consists of three parts: the first is preprocessing for data collection, preparing data, and labeling, then the second is the formation of a data matrix as input, and the formation of a Convolutional Neural Network model as a classification feature and the third is the result of evaluation and validation. The system architecture is designed as shown in Fig. 2.

Data Collection. The dataset used in this study was sourced from student feedback comments on the EDOM application at Department of Information Systems, Telkom University. Preprocessing is crucial in the process of determining sentiment with CNN and the classification of tweets becomes more accurate. Preprocessing is also used to get clean data. The preprocessing stage consists of several processes namely Case Folding, Symbol Removal, Tokenization, Slang word Conversion, and Stopword Removal [19].

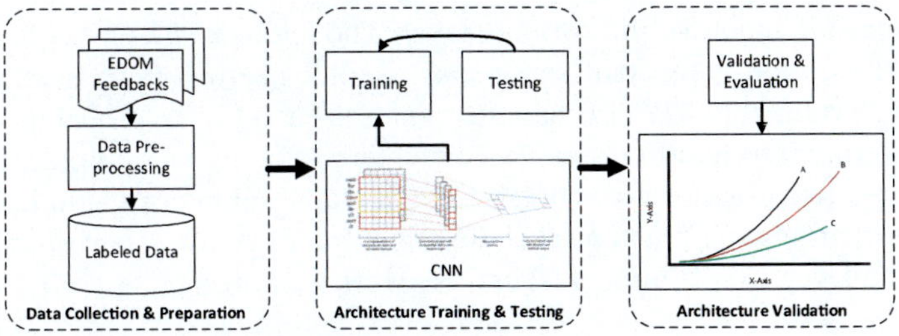

Fig. 2. System architecture

Due to existing limitations, the amount of feedback from EDOM that was successfully obtained was 1800 feedback. The feedback data is then labeled manually with positive and negative sentiments. Figure 3 illustrates the amount of feedback for each sentiment polarity, i.e. for positive sentiment is 1,300 feedback represented with 1 while negative sentiment is 420 labeled with 0.

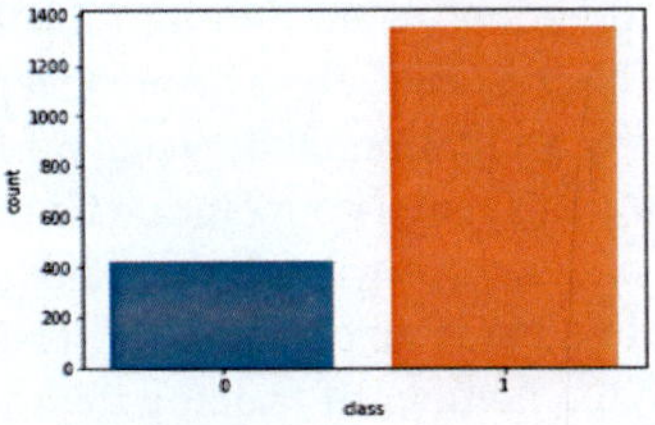

Fig. 3. Sentiment polarity distribution on the EDOM feedback dataset

Architecture Training & Testing. Convolutional Neural Network is a network model that accepts matrix input. The application of data in the form of images is done by using pixel numbers in the image as the contents of the matrix. This is different from text data that does not have numbers so it needs to be converted into numbers. Kim [7] conducted research by comparing the initialization of numbers used for text data on the Convolutional Neural Network method. The process of converting text into numbers is done by using pre-trained word vectors or Word2Vec.

To avoid overfitting models, a dropout layer is needed which functions to randomly choose neurons not to be used during the training process [20]. The probability of the dropout layer used in this study is 0.5. The training process continues with pooling on the feature map. The pooling method used is MaxPooling, which takes the maximum value $\hat{c} = \max(c)$ as a feature based on a filter. The purpose of the pooling process is to get the most important features that represent other features for each feature map. The features that have been obtained from the pooling results are used for the classification process at the fully connected layer and are trained with the Backpropagation algorithm [21]. Each feature will be connected to each neuron in the fully-connected layer with independent weights and a total of 50 neurons. Each neuron from the fully-connected layer becomes the input for the softmax layer. The number of neurons in the softmax layer is based on the number of sentiments used, 2 neurons for 2 sentiments and 3 neurons for 3 sentiments. The softmax function is used to get the probability values that have been normalized in all classes. In conducting training, an optimizer is needed to reduce the loss function or errors contained in neurons and filters. Optimizer is done by updating the neurons and filters on each iteration. In this study, Adam optimizer is used in conducting Deep Convolutional Neural Network training [22].

Architecture Validation. At this stage an evaluation of the architecture that has been formed. The architecture needs to be seen its performance in carrying out the desired task. There are several ways of evaluation that can be done in classification tasks on an architecture, including through accuracy, precision, recall, and F1-score. All evaluation metrics are from the confusion matrix.

3.2 Result

After all the designs have been completed and the model has been built, the next step is to use the model to produce dividing the dataset into 2 parts, namely training data and validation data of 80% and 20%, respectively. The next step is to test the accuracy of the model by testing data and using validation data so that performance can be determined.

In this study, we trained and evaluated convolution neural network (CNN) for several numbers of epochs and batch sizes, we trained the model using Adam optimizer [22] with a learning rate of 0.001 and batch sizes of 32, 64, 128, and 256. An early stopping technique is used to optimize the number of training iterations or epochs automatically. It has been supported by TensorFlow to determine when to stop training the model by monitoring validation errors. This method is used in order to avoid the overfitting of hyper-parameters by tuning the number of epochs. Dropout is also used to overcome overfitting on the model used. In the following Table 1 is the result of the training and validation process by considering the results of accuracy and loss with the number of epochs is 5.

Table 1. The result of the training and validation process with 10 epochs

No. of epoch	Loss	Accuracy	Validation loss	Validation accuracy
1	0.5692	0.7611	0.5342	0.7646
2	0.5302	0.7611	0.5041	0.7646
3	0.4723	0.7611	0.4441	0.7646
4	0.3812	0.8152	0.3583	0.8531
5	**0.2686**	**0.8975**	**0.2916**	**0.8795**

In the training and validation process, overfitting is observed for model experiments. Figure 4 shows the training accuracy and loss results for the CNN model used in this study. The results show that, if the error rates on the training dataset and dataset tests are low and the difference in error rates tends to be not far and stable, it means that it can be determined that the result is a good fit.

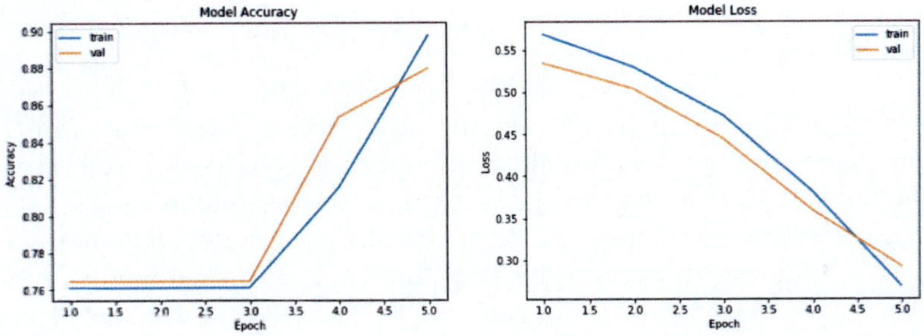

(a) (b)

Fig. 4. Training and validation accuracy (a) and training and validation loss (b) results

Based on the experimental results for the number of epochs = 5, the performance evaluation of the results obtained is 87.95% for accuracy (as shown in Table 1). Detailed performance evaluation from the point of view of the confusion matrix can be depicted in Fig. 5.

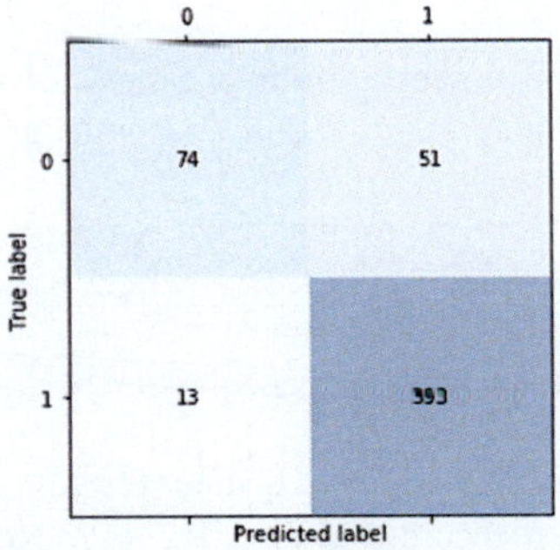

Fig. 5. Performance evaluation from the confusion matrix

From the results of the matrix confusion, other evaluation results such as precision, recall, and F1-score can be calculated. Table 2 shows the results for each positive and negative sentiment, namely class 1 and 0, respectively. While the results for macro average precision, macro average recall, and macro average F1-score are 87%, 78%, and 81%, respectively.

Table 2. Results detail of the performance evaluation

Class	Precision	Recall	F1-Score
0	83%	73%	77%
1	92%	95%	94%

4 Conclusion

The feedback mechanism on EDOM, questions can be categorized in the form of scale questions (score based on Likert scale) and open-ended questions. Likert scale-based score questions are given to students and asked to answer these questions using a ranking-based scale. These types of questions do not usually adequately represent student sentiment, while open-ended questions are sometimes not considered in feedback assessments and are not explored deeper to gain insight. Therefore, in this study text mining was carried out to find out the sentiments given by students' feedbacks, whether positive or negative sentiments using Convolutional Neural Network (CNN). To evaluate performance, accuracy, recall, precision, and F1-Score is calculated and the evaluation has shown positive results. Based on the results of experiments have shown that CNN is able to achieve accuracy, precision, recall, and F1-Score of 87.95%, 87%, 78%, and 81%, respectively. Suggestions for future research are to increase the amount of data for positive and negative sentiment and increase the accuracy of the algorithm used.

References

1. Hunston, S.: Evaluation and organization in a sample of written academic discourse. In: Advances in Written Text Analysis. pp. 205–232. Routledge (2002). https://doi.org/10.4324/9780203422656-16.
2. Chen, Y., Hoshower, L.B.: Student evaluation of teaching effectiveness: an assessment of student perception and motivation. Assess. Eval. High. Educ. **28**, 71–88 (2003). https://doi.org/10.1080/02602930301683
3. Rani, S., Kumar, P.: A sentiment analysis system to improve teaching and learning. Computer (Long. Beach. Calif). **50**, 36–43 (2017)
4. Sharma, A., Dey, S.: A comparative study of feature selection and machine learning techniques for sentiment analysis. In: Proceedings of the 2012 ACM Research in Applied Computation Symposium, pp. 1–7 (2012)
5. Neethu, M.S., Rajasree, R.: Sentiment analysis in twitter using machine learning techniques. In: 2013 4th International Conference on Computing, Communication and Networking Technologies (ICCCNT 2013) (2013). https://doi.org/10.1109/ICCCNT.2013.6726818
6. Anindtiya, A., Hasibuan, M.A., Sutoyo, E.: Text mining approach using TF-IDF and naive bayes for classification of exam questions based on cognitive level of bloom's taxonomy. In: 2019 IEEE International Conference on Internet of Things and Intelligence System (IoTaIS), pp. 112–117. IEEE (2019)
7. Kim, Y.: Convolutional neural networks for sentence classification. arXiv Preprint arXiv:1408.5882 (2014)

8. Severyn, A., Moschitti, A.: Twitter sentiment analysis with deep convolutional neural networks. In: Proceedings of the 38th International ACM SIGIR Conference on Research and Development in Information Retrieval, pp. 959–962. ACM (2015)
9. Vateekul, P., Koomsubha, T.: A study of sentiment analysis using deep learning techniques on Thai twitter data. In: 2016 13th International Joint Conference on Computer Science and Software Engineering (JCSSE), pp. 1–6 (2016)
10. Ay Karakuş, B., Talo, M., Hallaç, I.R., Aydin, G.: Evaluating deep learning models for sentiment classification. Concurr. Comput. **30**, 1–14 (2018). https://doi.org/10.1002/cpe.4783
11. Kalchbrenner, N., Grefenstette, E., Blunsom, P.: A convolutional neural network for modelling sentences. arXiv preprint arXiv:1404.2188 (2014). https://doi.org/10.3115/v1/p14-1062
12. Collobert, R., Weston, J., Bottou, L., Karlen, M., Kavukcuoglu, K., Kuksa, P.: Natural language processing (almost) from scratch. J. Mach. Learn. Res. **12**, 2493–2537 (2011)
13. Yamashita, R., Nishio, M., Do, R.K.G., Togashi, K.: Convolutional neural networks: an overview and application in radiology. Insights Imaging **9**(4), 611–629 (2018). https://doi.org/10.1007/s13244-018-0639-9
14. Srivastava, A., Singh, V., Drall, G.S.: Sentiment analysis of twitter data. Int. J. Healthc. Inf. Syst. Informatics. **14**, 1–16 (2019). https://doi.org/10.4018/ijhisi.2019040101
15. Pak, A., Paroubek, P.: Twitter as a corpus for sentiment analysis and opinion mining. In: Proceedings of the Seventh International Conference on Language Resources and Evaluation, vol. 5, pp. 1320–1326 (2010). https://doi.org/10.17148/ijarcce.2016.51274
16. Anstead, N., Loughlin, B.O.: Social media analysis and public opinion: the 2010 UK general election. J. Comput. Commun. **20**, 204–220 (2015). https://doi.org/10.1111/jcc4.12102
17. Li, Q., Li, S., Hu, J., Zhang, S., Hu, J.: Tourism review sentiment classification using a bidirectional recurrent neural network with an attention mechanism and topic-enriched word vectors. Sustain **10** (2018). https://doi.org/10.3390/su10093313
18. Sutoyo, E., Almaarif, A.: Twitter sentiment analysis of the relocation of Indonesia's Capital City. Bull. Electr. Eng. Inform. **9**, 1620–1630 (2020). https://doi.org/10.11591/eei.v9i4.2352
19. Liu, B., Liu, B.: Sentiment analysis and subjectivity. In: Handbook of Natural Language Processing, 2nd edn., Taylor & Francis Group, Boca Raton (2010)
20. Srivastava, N., et al.: Dropout: a simple way to prevent neural networks from overfitting. J. Mach. Learn. Res. **15**, 1929–1958 (2014)
21. Hecht-Nielsen, R.: Theory of the backpropagation neural network. In: Neural Networks for Perception, pp. 65–93. Elsevier (1992). https://doi.org/10.1016/b978-0-12-741252-8.50010-8
22. Kingma, D.P., Ba, J.L.: Adam: a method for stochastic optimization. In: 3rd International Conference on Learning Representations (ICLR 2015) - Conference Track Proceedings, pp. 1–15 (2014)

Analysis of Factors Affecting Successful Adoption and Acceptance of Electronic Health Records at Hospitals

Aniza Jamaluddin(✉) and Jemal H. Abawajy

School of Information Technology, Faculty of Science, Engineering and Built Environment, Deakin University, Geelong, VIC, Australia
ajamalud@deakin.edu.my, jemal.abawajy@deakin.edu.au

Abstract. There is ample evidence that electronic health records (EHRs) stand to benefit both the healthcare providers and the patients in many ways. As a result, healthcare providers in offices, clinics, and hospitals all over the world are increasingly looking into leveraging the potentials provided by HER systems. Although there are increasing body of knowledge on EHR initiatives, anecdotal evidence reveals that a significant number of EHR initiatives do not survive post-experimentation phase. This has led us to examine why, despite EHRs potential to provide a range of benefits, EHR initiatives do not outlast post-experimentation phase. In this paper, we examine factors that influence a successful adaption and acceptance of EHRs. Unlike the exiting research that largely focus on the technical factors, we will consider the non-technical factors that impede the successful adaption and acceptance of EHRs in hospitals and clinics. We undertake extensive analysis of the literature to investigate research landscape focusing on non-technical factors that affect the acceptance of EHRs by the healthcare providers. Our results show that social, culture and behavioural are as important as the technical factors in successful HER implementation.

Keywords: Electronic health records · Technical factors · Health professionals · Social factors · Cultural factors · Behavioural factors · Acceptance of EHRs · Adoption of EHRs

1 Introduction

The rapid growth of information technology in healthcare has led to the increasing use and sharing of electronic health records (EHRs). An EHR can be defined as "an application environment that is composed of clinical data repository, clinical decision support, controlled medical vocabulary, order entry, computerized provider order entry, pharmacy, and clinical document applications that support the patients' electronic medical record across inpatient and outpatient environments, and is used by healthcare practitioners to document, monitor and manage health care delivery within a care delivery organization" (Garets and Davis 2006). Furthermore, EHR contains new forms of patient-physician

J. H. Abawajy et al. (Eds.): EATI 2020, LNNS 254, pp. 282–293, 2021.
https://doi.org/10.1007/978-3-030-80216-5_21

interaction and poses challenges and threats related to ethical issues (AlHamad et al. 2014).

EHR is touted to provide effective and efficient management of medical records in the hospitals and other healthcare institutions (Herian et al. 2014). EHR integrate patientcare history into computer-process able files that improve in terms of accessibility, quantity, and quality of patientcare information that are available for the provider to plan and coordinate their patient care (Blumenthal and Tavenner 2010). Electronic records have many significant benefits with potential to reduce gaps in the quality of care of patient care (Jha et al. 2009). As a result, many countries initiated a project for the implementation of the EHRs.

Recently, anecdotal evidences suggest that non-technical factors such as social, culture and political factors do have impact on the implementation of EHRs as well (Morton and Wiedenbeck 2010; Finnell and Dixson 2016). Unfortunately, there is little or no research on the implications of non-technical factors on the successful adoption and diffusion of EHRs system. Simon et al. (2013) stated that the best way in managing perceptions may be vary, and likely depends on the underlying culture of the institution (or unit within the institution) itself. Therefore, the factors that relates to the social, culture and political is imperative to be investigated in ensuring the successful adoption and acceptance of EHRs.

2 Methods

2.1 Search Strategy

A wide-ranging search of the literature from 2004 to 2020 was performed using Journal of Medical Systems, MEDLINE, Journal of the American Medical Informatics Association (JAMIA), American Health Information Management Association (AHIMA), Science Direct, Health Care Management Review Journal Systems Research and Behavioural Science, African Journal for Physical Activity and Health Sciences (AJPHERD) and current databases on health. Search strategies specific with key words that reflected EHRs system and workflow. Furthermore, search strategy was involving the terms health information, information systems, electronic records, medical records systems, medical informatics, health informatics, computerized patient records, acceptance and adoption, clinicians, and user satisfaction. Thus, full-text papers that are published in journals and proceedings only were selected for review.

Based on the results of search strategy, the authors read and assessed the abstract of each publications according to research question and the papers' full text then were retrieved for further review. Reference lists then were inserted into Mendeley to help the authors to explore more related papers and helps in identifying other available relevant articles. Hence, this review consisting current and previous studies that reported on the relationship between the non-technological factors such as social and behavioural factors on the successful EHRs adoption and acceptance.

3 Electronic Health Records (EHR)

Several terms are available and used interchangeably in literature that are related to EHR such as computer-based patient record (CPR), electronic medical record (EMR), and electronic patient record (EPR) (Al-Aswad et al. 2013). National Alliance for Health Information Technology (NAHIT) defines EMR as "the electronic record of health-related information on an individual that is created, gathered, managed, and consulted by licensed clinicians and staff from a single organisation who are involved in the individual's health and care", while EHR is "the aggregate electronic record of health-related information on an individual that is created and gathered cumulatively across more than one health-care organisation and is managed and consulted by licensed clinicians and staff involved in the individual's health and care" (Amatayakul 2009).

The implementation of EHRs to replace the use of paper-based records which is more effective and efficient resulting to the efficiency of the management of the medical records in the hospitals and other healthcare institutions. However, there is lots of consideration need to be taken when implementing the EHRs especially when it comes to the repository parts. All the necessary clinical data need to be stored and kept in the electronic format appropriately. Some problems that will be highlighted sooner may occur in the process of storing and keeping those records in the database. The computerized applications and systems may bring some challenges that need to be faced by the practitioners and clinical staff. Kose et al. (2020) opined that the larger size of hospitals in the USA may have larger financial and human resources to dedicate to EHR adoption. They found that smaller country such as Turkey may take advantage of their size to actively control all state hospitals and able to change quickly as more financial and human resources are available to support the hospital's administration.

4 Factors Influence the Implementation of EHRs

Leaders in healthcare provider demonstrate that EHRs implementation is a high priority to support their care delivery. Thus, governments and healthcare providers in Europe and other parts of the world are currently engaged in health care system reforms designed to improve quality and safety of care which also involve costs (Silow-Carrol 2012). Mosadeghrad (2014) in his study discusses factors that influence the quality of healthcare services are patient related factors, provider related factors, and environmental factors. In many developing countries, the health status is still miserable and unacceptable. A large region in developing countries is underprivileged to access to simple health care (Al-washali et al. 2015). The implementation of EHR in the less-developed countries has risen up lots of issues and the factors of successful implementation of the system should be recognized which consist of technical and non-technical factors.

Most of previous studies were focusing on the technical factors of EHR's implementation. Thus, according to (Kaplan 2001; Berg 1999; Cresswell et al. 2012), sociotechnical approach is important to be studied as there are several issues related to people, organizational and social elements which includes human-computer interaction, socio-technical, cultural, and ethical concerns. According to Boonstra et al. (2014), studies focus on EHR implementation were conducted where EHR implementation been divided into three groups which are the context, the content, and the implementation process. The context consists of type of hospital, vendor selection, staff's ICT experience, organizational culture, care activities, and hospital bureaucracy. Thus, the content itself involves hardware and system reliability, safeguards of patient's privacy, fitting of technology and work process, and vendor adaptation. The final implementation process then involves strong involvement of management and clinicians, training, resistance of clinicians, finding champions, and having adequate time and resources. Hence, it is evident that it is important for the non-technical factors be determined to overcome the issues related to social, cultural and user attitudes towards a successful EHR system's adoption.

4.1 Social, Culture and Behavioural Factors

Social factors are things that affect lifestyle, such as politics, religion, education, age, race, sex and class. Social influence been defined as "the degree to which an individual perceives that important other believe he or she should use the new system" (Kijsanayotin et al. 2009; Yu et al. 2009; Wu et al. 2007). According to Lakbala and Dindarloo (2014), social issues have decelerated the pace of implementation or prevent the widespread plan to the EHRs system implementation. Bandyopadhyay and Bandyopadhyay (2010) in their study has found users' intention to use the prepayment system between United States and India are influenced by effort expectancy, performance expectancy, and social influence. Thus, it is evident that each country with different cultures possess significant effects to the system's behavioural intention. Oshlyansky and Thimbleby (2007) in their cross-culturally study (Czech Republic, Greece, India, Malaysia, New Zealand, Saudi Arabia, South Africa, United Kingdom, and the United States) on technology acceptance has found that social influence has significant affects towards website acceptance in Saudi Arabia more than in the other countries.

Social influence is one of direct determinant of user attitude towards system use and behavioural intentions been signified as a subjective norm and image. Social influence is "the degree to which an individual values the importance of others persistence that he or she should use the new system." (Jeng and Tzeng 2012). Interpersonal relationship is also one of the key issues when dealing with virtual community where small society shows circulation of behaviour and resource of an actual organization with shared value (vision and language), identification, and trust. Attitude is determined by two constructs: perceived usefulness (i.e., "the information technology system will enhance job performance") and perceived ease of use of the technology (i.e., "the information technology system will be free of effort") (Holden and Karsh 2010). Culture is defined as characteristics and knowledge of a particular group of people that consists of language, religion, social habits, behaviours and interactions, understanding. According to Ancker et al. (2013), successful of EHR implementation is correlated with positive attitudes about EHRs. This this then followed by a research conducted by Bishop et al. (2015) highlighted that once an EHRs is implemented, prior comfort with computers and attitudes about the EHR does not predict future use of that EHR.

According to Kralewski et al. (2008), the culture shown by the physicians contribute to the passive or active resistance whereby it is hard to change culture and might result in physician turnover and shape the innovation process in accommodating the culture. Morton and Wiedenbeck (2010) in their study highlighted that to have successful adoption of EHR, social and behavioural factors must be well addressed during the system planning.

4.2 Organizational and Governance in EHR Adoption

Political factor is very important as a significant factor that contribute to the successful implementation of EHR. Administration and the environmental itself play enormous role in making sure the successful adoption of EHRs system in the hospitals as well as bringing the national healthcare industry to another level suit with other sector's development. Bentley et al. (2016) belief that successful implementation of EHR is certainly an important milestone in the EHRs journey and the whole benefits of the system are not fully be seen until the platform and organization mature and begin to fully utilize the technology. This is then specified by Boonstra et al. (2014) where supportive collaboration and teamwork in organizational culture contribute to the successful implementation of EHR as the trust among employees are increased. However, Boyer et al. (2010) suggested that it is important to have a favourable culture that easy to adapt to EHR implementation even though it is not an easy task as it involves comprehensive approach that consists of incentives, allocation for resources, and responsive team.

According to Mohd and Syed Mohamaed (2005), the acceptance level of Total Hospital Information System (THIS) in IT-based hospital in Malaysia is still moderately accepted. Precisely, assigning a multidisciplinary team that specialised in dealing with issues related to EHR may avoid conflict and stimulate collaboration (Boyer et al. 2010). Korst et al. (2011) conducted hospital readiness study and found that quality improvement culture promoted by hospital leaders is the key to data sharing activities among users. This evident that as a policy and decision maker, it is important for this top management to foster a positive culture in improving the EHRs system throughout the hospitals. This is then highlighted by Boonstra et al. (2014) where management's support and involvement in EHR implementation posses' positive impact in the successful implementation and at the same time counterbalances the dominance of physicians. This is due to study conducted by Rivard et al. (2011) that found difficulties in collaboration and teamwork with existence of physicians' medical dominance and autonomy of other health professionals which finally complicate the implementation process.

Boonstra et al. (2014) highlighted that studies showed strong leadership may protect the project by performing project management techniques, standards establishment, and adequate staff's training to ensure such problem did not affect the system implementation. This is also agreed by Raglan et al. (2017) that stressed on the important of providing well-tailored and additional training may increase EHRs adoption as well as satisfaction among physicians. Hossain et al. (2018) in their research findings suggested the adoption of EHRs system should be increased by developing social strategies that may encourage physician to stimulate among them in system use, provide training and ensuring the technical sufficiency when using EHR. This is evident that leadership plays significance role in communicating on the implementation stage as well as monitoring the adoption of EHRs system throughout the hospital. Therefore, this study is important to be conducted which investigate the institutions, organizational, administration, environmental, and government factors as major factors of successful implementation of EHRs system.

5 User Attitudes, Acceptance and Perception Towards EHRs System

Practitioners' acceptance and satisfaction is crucial to be taken into consideration as they are the primary indicator to successful adoption of EHR system throughout the hospital. The clinicians and non-clinicians depending on the efficient and effective EHRs system that brings significance support in their daily task. The push toward implementation of EHRs has raised issues related to the acceptance of the technology (Carayon et al. 2011). Ancker et al. (2013) suggested that providers and staffs' positive attitude may bring long-term quality benefits to the system user. User's satisfaction (clinicians and non-clinicians) measurements are important to understand the adoption rates and usage patterns. Holanda et al. (2012) studied the physicians' satisfaction reported that slightly more than half of the respondents were satisfied with the system while almost half of the respondents were not. It also found that the structural issues are the major barriers.

Recent research shows that there is limited acceptance of EHR by physicians thus posing a risk to a successful EHR adoption (Boonstra and Broekhuis 2010). Meigs and Solomon (2016) reported an increasing physician's dissatisfaction of EHR in USA. Therefore, determining factors that largely influence the level of end user acceptance and their intention to use the EHR system in their daily work is an open problem. Many studies regarding acceptance of EHR system suggested that the physician is differ among the users in terms of their attitude. Therefore, Morton (2008) suggested that the physicians needs prior to the EHR implementation may help the healthcare facility in system identification as well as a proper implementation process.

Physicians acceptance is depending on how their work practices being transformed in the system (Aarts and Berg 2006). Therefore, it is important for each physician's needed to be taking care of to support their daily task and helps in better decision making. Ovretveit et al. (2007) proposed that physicians should be involved in the implementation process from the early stage, during the vendor requirement's formulation, as well as the beginning of use phase. This is also agreed by Boyer et al. (2010) that stressed on the important to form multidisciplinary work groups that identify the content of the EHR system and the rules in information sharing. Secginli et al. (2014) conducted a study where most of the respondents agreed that EHR gave them benefits and show their satisfaction towards EHR system. It is also indicated that the users acknowledge d the usefulness and importance of the EHR system and this result are consistence with other studies conducted by (Moody 2004; Heselmans 2012; Al-Harbi 2011). The survey was conducted during transitional phase and it was reported that the respondent's perception during the phase may reflected by the users who just learned on how to use the system.

Further study to access the practitioner's usefulness attitude in using EHRs may provide better detailed understanding of the end-user's attitudes and perception towards EHRs adoption. Ghani et al. (2008) state that in Malaysia, with various individuals involved in healthcare services such as specialist, nurses, pharmacist, medical assistants, administrator, and other health care staff, majority of them are having limited computer-skills and most of the users are get used to their existing work processes that has been part of their lives for decades. Thus, changes in their work routine are viewed with some resistance and hostility. Therefore, further study on those human factors that focusing on the user acceptance and satisfaction is important to be conducted especially in developing country such as Malaysia context where the current studies are more focusing on the technical aspects rather than non-technical aspects.

6 Conclusion

More research is needed to contextualize the fences of EHRs system in healthcare settings, especially in relation to EHRs' adoption and acceptance. Further research is needed to develop an evidence base where, social issues have slowed the pace of implementation or prevent the widespread plan to the EHRs system implementation (Lakbala and Dindarloo 2014). Varying ways of scholars and researchers seeing discussing on the factors that influencing the implementation of EHRs. Overview of studies reviewed on factors influence in EHRs system adoption/implementation is presented in Table 1. Most of the studies conducted are really focusing on the technical factors which are commonly found in developed countries. Furthermore, most studies conducted shown the failure of technology implementation always been related to technical problem while the people and the organizational issues are not fully been addressed (Eason 1988; Carayon et al. 2009). Hence, Karim et al. (2011) suggested the urge to establish better understanding on EHR implementation that look into implementation, technological and behavioural perspectives. Therefore, study on social, culture and behavioural, organizational and governance, user's attitudes, acceptance, and perception towards EHRs system is crucial to be conducted as EHR systems involves users with varying backgrounds, experiences, and values.

Table 1. Factors influencing EHR system adoption/implementation

Author/year	Hossain et al. (2018)	Sidek and Martins (2017)	Ismail et al. (2016)	Cucciniello et al. (2015)	Lakbala et al. (2014)	Secginli et al. (2014)	Or et al. (2014)
Country	Bangladesh	Brunei	Malaysia	Scotland	Iran	Turkey	Hong Kong
Title	"Investigating Factors Influencing the Physicians' Adoption of EHR In Healthcare System of Bangladesh: An Empirical Study"	"Perceived Critical Success Factors of Electronic Health Record System Implementation in A Dental Clinic Context: An Organisational Management Perspective"	"Hospital Information System (HIS) implementation in Malaysian public hospitals"	"Understanding Key Factors Affecting Electronic Medical Record Implementation: A Sociotechnical Approach"	"Physicians' Perception and Attitude Toward Electronic Medical Record"	"Attitudes of Health Professionals Towards Electronic Health Records in Primary Health Care Settings: A Questionnaire"	"Private Primary Care Physicians' Perspectives on Factors Affecting the Adoption of Electronic Medical Records: A Qualitative Pre-Implementation"
Objective	To identify the critical factors affecting the physicians' adoption of EHR in healthcare system of Bangladesh	To identify the perceived critical success factors of EHR system implementation in a dental clinic context	To identify factors affecting the HIS adoption across different categories of HIS's hospitals	Examined the interaction of sociological and technological factors in the implementation of an Electronic Medical Record (EMR) system by a major national hospital	To explore physicians, attitude and perceptions of importance EMRs function, anticipated utilization of EMR functions and issue impact EMRs	To assess the attitudes of health professionals towards electronic health records (EHRs) in primary health care settings in Turkey	To Identify the Factors Perceived as Relevant by Private Physicians When Considering EHR Adoption

(continued)

Table 1. *(continued)*

Author/year	Hossain et al. (2018)	Sidek and Martins (2017)	Ismail et al. (2016)	Cucciniello et al. (2015)	Lakbala et al. (2014)	Secginli et al. (2014)	Or et al. (2014)
Country	Bangladesh	Brunei	Malaysia	Scotland	Iran	Turkey	Hong Kong
Method	Partial least square (PLS) and statistical analysis technique based on the structural equation modeling (SEM)	Interpretive Grounded Theory approach followed by the stages of open, axial and selective coding	Cross-sectional survey	Actor Network Theory (ANT)	Cross-sectional study	Cross-sectional, descriptive design	Multilevel Work System Conceptual Framework for Technology Adoption
Factors	Social strategies Technical sufficiency Training	System usability User behaviour Requirements analysis Training Change management Project organisation	Human skills Experience Expert Satisfaction Information quality	Organizational Cultural Technological Financial	Policy interventions Individual-level responses Government role	Technology refinements Standardized terminologies EHR training	Technology Environment Organization Task Person

References

Aarts, J., Berg, M.: Same systems, different outcomes–comparing the implementation of computerized physician order entry in two Dutch hospitals. Methods Inf. Med. **45**(1), 53–61 (2006)

Abd Ghani, M.K., Wen, L.C.: The design of flexible Pervasive Electronic Health Record (PEHR). In: 2011 IEEE Colloquium on Humanities, Science and Engineering (CHUSER), pp. 249–254, 5–6 December 2011

Al-Aswad, A.M., Brownsell, S., Palmer, R., Nichol, J.P.: A review paper of the current status of electronic health records adoption worldwide: the gap between developed and developing countries. J. Health Inform. Dev. Countries **7**(2), 153–164 (2013)

AlHamad, A.Q., Al Omari, F., AlHamad, A.Q.: Recommendation for managing patients' privacy in an integrated health information network. J. Inf. Technol. Econ. Dev. **5**(1), 47 (2014)

Al-Harbi, A.: Healthcare providers' perceptions towards health information applications at King Abdul-Aziz medical City Saudi Arabia. IJACSA **2**, 14–22 (2011)

Amatayakul, M.K.: Electronic Health Records: A Practical Guide for Professionals and Organizations, 4th edn. AHIMA, Chicago (2009)

Ancker, J.S., et al.: Predictors of success for electronic health record implementation in small physician practices. Appl. Clin. Inf. **4**(1), 12–24 (2013). https://doi.org/10.4338/ACI-2012-09-RA-0033

Bandyopadhyay, K., Bandyopadhyay, S.: User acceptance of information technology across cultures. Int. J. Intercult. Inf. Manag. 2 (2010). https://doi.org/10.1504/IJIIM.2010.037862

Bentley, T., et al.: The journey from precontemplation to action: transitioning between electronic medical record systems. Health Care Manage. Rev. **41**(1), 22–31 (2016)

Bishop, T.F., Smith Ryan, M., McCullough, C.M., Shih, S.C., Casalino, L.P., Ryan, A.M.: Do provider attitudes about electronic health records predict future electronic health record use? Healthcare **3**(1), 5–11 (2015). https://doi.org/10.1016/j.hjdsi.2014.04.002

Blumenthal, D., Tavenner, M.: The "meaningful use" regulation for electronic health records. New Engl. J. Med. **363**(6), 501–504 (2010)

Boonstra, A., Broekhuis, M.: Barriers to the acceptance of electronic medical records by physicians from systematic review to taxonomy and interventions. BMC Health Serv. Res. **10**, 231 (2010). https://doi.org/10.1186/1472-6963-10-231

Boonstra, A., Versluis, A., Vos, J.F.: Implementing electronic health records in hospitals: a systematic literature review. BMC Health Serv. Res. **14**(1), 1 (2014)

Boyer, L., Samuelian, J.-C., Fieschi, M., Lancon, C.: Implementing electronic medical records in a psychiatric hospital: a qualitative study. Int. J. Psychiatry Clin. Pract. **14**(3), 223–227 (2010). https://doi.org/10.3109/13651501003717243

Carayon, P.: The balance theory and the work system model. Twenty years later. Int. J. Hum. Comput. Interact. **25**(5), 313–327 (n.d.). https://doi.org/10.1080/10447310902864928

Finnell, J.T., Dixon, B.E.: Clinical Informatics Study Guide: Text and Review. Springer, Cham (2016)

Carayon, P., et al.: ICU nurses' acceptance of electronic health records. J. Am. Med. Inform. Assoc. **18**(6), 812–819 (2011)

Cresswell, K.M., Worth, A., Sheikh, A.: Integration of a nationally procured electronic health record system into user work practices. BMC Med. Inform. Decis. Mak. **12**(1), 15 (2012). https://doi.org/10.1186/1472-6947-12-15

Cucciniello, M., Lapsley, I., Nasi, G., Pagliari, C.: Understanding key factors affecting electronic medical record implementation: a sociotechnical approach. BMC Health Serv. Res. **15**(1), 1 (2015)

Davis, M., Garets, D.: Electronic medical records vs. electronic health records: yes, there is a difference. Rep. HIMSS Anal. 26 January 2006. Web. 6 April 2016
Al-washali, H.A., et al.: Impression about healthcare services at medical hospitals in Dhamar city Yemen. Am. J. Health Res. **3**, 10–15 (2015)
Heeks, R.: Health information systems: failure, success and improvisation. Int. J. Med. Inform. **75**, 125–137 (2006)
Herian, M.N., Shank, N.C., Abdel-Monem, T.L.: Trust in government and support for governmental regulation: the case of electronic health records. Health Expect. **17**(6), 784–794 (2014). https://doi.org/10.1111/j.1369-7625.2012.00803.x
Holanda, A., CarmoeSá, H., Vieira, A., Catrib, A.: Use and satisfaction with electronic health record by primary care physicians in a health district in Brazil. J. Med. Syst. **36**(5), 3141–3149 (2012). https://doi.org/10.1007/s10916-011-9801-3
Holden, R.J., Karsh, B.-T.: The technology acceptance model: its past and its future in health care. J. Biomed. Inf. **43**(1), 159–172 (2010). https://doi.org/10.1016/j.jbi.2009.07.002
Ancker, J.S., et al.: Predictors of success for electronic health record implementation in small physician practices. Appl. Clin. Inform. **4**(1), 12–24 (2013). https://doi.org/10.4338/ACI-2012-09-RA-0033
Hossain, A., Quaresma, R., Rahman, H.: Investigating factors influencing the physicians' adoption of electronic health record (EHR) in healthcare system of Bangladesh: an empirical study. Int. J. Inf. Manag. **44**, 76–87 (2018). https://doi.org/10.1016/j.ijinfomgt.2018.09.016
Jeng, D.J.-F., Tzeng, G.-H.: Social influence on the use of clinical decision support systems: revisiting the unified theory of acceptance and use of technology by the fuzzy DEMATEL technique. Comput. Ind. Eng. **62**(3), 819–828 (2012). https://doi.org/10.1016/j.cie.2011.12.016
Jha, A.K., et al.: Use of electronic health records in US hospitals. N. Engl. J. Med. **360**(16), 1628–1638 (2009)
Kaiser Permanente's experience of implementing an electronic medical record: a qualitative study. Br. Med. J. (International Edition), **331**(7528), 1313–1316 (2005). https://doi.org/10.1136/bmj.38638.497477.68
Karim, N.S.A., Ahmad, M., Mohamed, N.: A framework for electronic health record (EHR) implementation impact on system service quality and individual performance among healthcare practitioners. In: Proceedings of the 10th WSEAS International Conference on E-activities. World Scientific and Engineering Academy and Society (WSEAS) (2011)
Kijsanayotin, B., Pannarunothai, S., Speedie, S.M.: Factors influencing health information technology adoption in Thailand's community health centers: applying the UTAUT model. Int. J. Med. Inform. **78**(6), 404–416 (2009). https://doi.org/10.1016/j.ijmedinf.2008.12.005
Korst, L.M., Aydin, C.E., Signer, J.M.K., Fink, A.: Hospital readiness for health information exchange: development of metrics associated with successful collaboration for quality improvement. Int. J. Med. Inform. **80**(8), e178–e188 (2011). https://doi.org/10.1016/j.ijmedinf.2011.01.010
Kose, I., Rayner, J., Birinci, S., Ulgu, M.M., Yilmaz, I., Guner, S.: Adoption rates of electronic health records in Turkish Hospitals and the relation with hospital sizes. BMC Health Serv. Res. **20**(1), 967 (2020). https://doi.org/10.1186/s12913-020-05767-5
Kralewski, J.E., Dowd, B.E., Cole-Adeniyi, T., Gans, D., Malakar, L., Elson, B.: Factors influencing physician use of clinical electronic information technologies after adoption by their medical group practices. Health Care Manage. Rev. **33**(4), 361–367 (2008)
Lakbala, P., Dindarloo, K.: Physicians' perception and attitude toward electronic medical record. Springerplus **3**(1), 1–8 (2014). https://doi.org/10.1186/2193-1801-3-63
Meigs, S.L., Solomon, M.: Electronic health record use a bitter pill for many physicians. Perspect. Health Inf. Manag. **13**(Winter), 1–17 (2016).
Morton, M.E.: Use and acceptance of an electronic health record: factors affecting physician attitudes. Drexel University (2008)

Morton, M.E., Wiedenbeck, S.: EHR acceptance factors in ambulatory care: a survey of physician perceptions. Perspect. Health Inf. Manag. **7**(1), 1–12 (2010)

Mosadeghrad, A.M.: Factors influencing healthcare service quality. Int. J. Health Policy Manag. **3**, 77–89 (2014)

Or, C., Wong, K., Tong, E., Sek, A.: Private primary care physicians' perspectives on factors affecting the adoption of electronic medical records: a qualitative pre-implementation study. Work **48**(4), 529–538 (2014). https://doi.org/10.3233/WOR-131808

Oshlyansky, L., Thimbleby, H.: Validating the unified theory of acceptance and use of technology (UTAUT) tool cross-culturally, pp. 83–86 (2007). https://doi.org/10.1145/1531407.1531429

Ovretveit, J., Scott, T., Rundall, T.G., Shortell, S.M., Brommels, M.: Improving quality through effective implementation of information technology in healthcare. Int. J. Qual. Health Care **19**(5), 259–266 (2007)

Ozkan, O.: Attitudes and opinions of people who use medical services about privacy and confidentiality of health information in electronic environment. Middle East Technical University, The Graduate School of Informatics (2011)

Raglan, G.B., Margolis, B., Paulus, R.A., Schulkin, J.: Electronic health record adoption among obstetrician/gynecologists in the United States: physician practices and satisfaction. J. Healthcare Qual. Promot. Excell. Healthcare **39**(3), 144–152 (2017). https://doi.org/10.1111/jhq.12072

Rivard, S., Lapointe, L., Kappos, A.: An organizational culture-based theory of clinical information systems implementation in hospitals. J. Assoc. Inf. Syst. **12**(2), 123–162 (2011)

Secginli, S., Erdogan, S., Monsen, K.A.: Attitudes of health professionals towards electronic health records in primary health care settings: a questionnaire survey. Inform. Health Soc. Care **39**(1), 15–32 (2014)

Sidek, Y.H., Martins, J.T.: Perceived critical success factors of electronic health record system implementation in a dental clinic context: an organisational management perspective. Int. J. Med. Inform. **107**, 88–100 (2017). https://doi.org/10.1016/j.ijmedinf.2017.08.007

Silow-Carroll, S., Edwards, J.N., Rodin, D.: Using electronic health records to improve quality and efficiency: the experiences of leading hospitals. Issue Brief (Commonw Fund) **17**, 1–40 (2012)

Simon, S.R., et al.: Lessons learned from implementation of computerized provider order entry in 5 community hospitals: a qualitative study. BMC Med. Inform. Decis. Mak. **13**(1), 67 (2013)

Author Index

J. H. Abawajy et al. (Eds.): EATI 2020, LNNS 254, pp. 295–296, 2021.
https://doi.org/10.1007/978-3-030-80216-5